The Physiology and Pathophysiology of Exercise Tolerance

The Physiology and Pathophysiology of Exercise Tolerance

Edited by

Jürgen M. Steinacker
University of Ulm
Ulm, Germany

and

Susan A. Ward
South Bank University
London, England

Plenum Press • New York and London

Library of Congress Cataloging-in-Publication Data

The physiology and pathophysiology of exercise tolerance / edited by
   Jürgen M. Steinacker and Susan A. Ward.
       p.   cm.
   "Proceedings of the International Symposium on the Physiology and
Pathophysiology of Exercise Tolerance held September 21-24, 1994 in
Ulm, Germany"--T.p. verso.
   Includes bibliographical references and index.
   ISBN 0-306-45492-0
   1. Exercise--Physiological aspects--Congresses. 2. Fatigue-
-Congresses. 3. Muscles--Physiology--Congresses. 4. Muscles-
-Pathophysiology--Congresses. 5. Stauch, Martin--Congresses.
I. Steinacker, Jürgen M. II. Ward, Susan A. III. International
Symposium on the Physiology and Pathophysiology of Exercise
Tolerance (1994 : Ulm, Germany)
   [DNLM: 1. Exercise Tolerance--physiology--congresses.
2. Exercise--physiology--congresses.   QT 255 P582 1996]
QP301.P576 1996
612'.044--dc21
DNLM/DLC
for Library of Congress   96-39630
                                                           CIP

Proceedings of the International Symposium on the Physiology and Pathophysiology of Exercise Tolerance,
held September 21 – 24, 1994, in Ulm, Germany

ISBN 0-306-45492-0

© 1996 Plenum Press, New York
A Division of Plenum Publishing Corporation
233 Spring Street, New York, N. Y. 10013

PREFACE

Exercise intolerance results when an individual is unable to sustain a particular work rate sufficiently long for the successful completion of the task. It is what, at one extreme, limits an elite athlete from even greater performance and, at the other extreme, what limits an individual with impaired systemic function from maintaining even a mild domiciliary routine.

This recognition was the motivating force for Professor Brian Whipp and Professor Martin Stauch to consider joining forces in bringing together a range of internationally-recognized authorities from both clinical and performance fields to confront the challenges of identifying the limiting causes of exercise intolerance and of designing strategies to extend these limits, both in the patient and the elite athlete. The structure of the Symposium developed over some four years of fruitful collaboration between Professor Stauch and Dr Jürgen Steinacker in the Department of Sports Medicine of the University of Ulm and Professsors Brian Whipp and Susan Ward, first at the University of California at Los Angeles and later at St George's Hospital Medical School in London.

Thus, on September 21, 1994, over 260 scientists from 22 countries assembled at the University of Ulm - the *alma mater* of Albert Einstein - to participate in the international symposium on "The Physiology and Pathophysiology of Exercise Tolerance" and to honor Professor Stauch, upon his retirement as Chairman of the Department of Sports Medicine at the University of Ulm, for his clinical and scientific contributions to cardiology and cardiovascular physiology over a long and fruitful career.

The Symposium, held under the auspices of the German Society of Sports Medicine and The German Cardiac Society, was an official meeting of the *Science and Education Committee of the German Society of Sports Medicine*. We are greatly appreciative of the generous help and advice provided by the members of this committee, especially Professor Hans Rieckert from Kiel in Germany.

We would also like to express our thanks to the sterling efforts of both the Scientific Committee - Gernot Badtke (Potsdam), Niels Secher (Copenhagen), Martin Stauch (Ulm), Jürgen Steinacker (Ulm), and Brian Whipp (London) - and the Organizing Committee - Werner Lormes, Yuefei Liu, Alexandra Opitz-Gress, Susanne Reissnecker, Jürgen Steinacker and Claudia Zeller from Ulm, and Brian Whipp and Susan Ward from London. Finally, the Organizing Committee wishes to express appreciation and thanks to: the organizations without whose generous support the Symposium could not have come to fruition (these are cited overleaf); the Department of Sports Medicine at the University of Ulm and the Department of Physiology at St. George's Hospital Medical School for the support furnished throughout the planning and implementation of the Symposium; and Dr. Yannis Pitsiladis and Ms Maria Dolores Galán from the School of Applied Science at South Bank University who played a key role in the preparation of this volume.

The Symposium delegates were able to take an excursion to the beautiful Allgäu mountains which are close to Ulm and have impressive views of both Switzerland and Austria. These mountains also served as a metaphor for the Symposium: the strain of the task itself and the relief and exhilaration of attaining the summit having, as counterparts for the participants, the strain of preparing their presentations and the exhilaration of the stimulating discussions. All were well aware, however, that what today seem to be mountains are, in reality, only hills in the context of the work which has yet to be done on the mechanisms of exercise tolerance. It is our hope that the perspectives set out in this volume may encourage both a standard of excellence and also serve as a "guide book" for future young scientific and clinical explorers.

Jürgen M. Steinacker, Ulm
Susan A. Ward, London

ACKNOWLEDGMENTS

We wish to express our thanks and appreciation to the following sponsors:

- Deutsche Forschungsgemeinschaft, Bonn - Bad Godesberg, Germany
- Ministerium für Wissenschaft und Kunst, Land Baden-Württemberg, Germany
- The Physiological Society, UK
- Universität Ulm, Ulm, Germany
- Verein zur Förderung der Sportmedizin, Hannover, Germany
- Aircast Europa GmbH, Stephanskirchen, Germany
- Analox Instruments Ltd, London, UK
- ASTRA Chemicals GmbH, Wedel, Germany
- Bayer AG, Leverkusen, Germany
- Boehringer-Mannheim GmbH, Mannheim, Germany
- Butterworth-Heinemann Ltd, Oxford, UK
- Ciba-Geigy GmbH, Wehr, Germany
- Clinical and Scientific Equipment Ltd, Kent, UK
- Marcel Dekker AG, Basel, Switzerland
- Erich Jaeger GmbH & Co KG, Würzburg, Germany
- Gondrom KG, Ulm, Germany
- Gödecke Parke-Davis, Freiburg, Germany
- Hans Rudolph Inc, Kansas City, USA
- Human Kinetics Publishers (Europe) Ltd, Leeds, UK
- Innovision A/S, Odense, Denmark
- Knoll AG, Ludwigshafen, Germany
- Med Graphics GmbH, Düsseldorf, Germany
- Quinton Europe, Hoofddorp, Netherlands
- Radiometer Deutschland GmbH, Willich, Germany
- Schiller Medizintechnik GmbH, Germany
- Schlag GmbH, Bergisch-Gladbach, Germany
- Schwarz-Pharma-AG, Monheim, Germany
- Sensormedics, Yorba Linda, USA
- Waverly Europe Ltd, London, UK

We also wish to thank The United States Air Force, European Office of Aerospace Research and Development and The United States Navy Office of Naval Research, Europe, for their contribution to the success of this conference.

CONTENTS

Part 3. SYSTEMIC LIMITATION TO MAXIMUM EXERCISE IN HEALTHY SUBJECTS

Part 4. PATHOPHYSIOLOGY OF EXERCISE INTOLERANCE

Part 5. SPORTS-SPECIFIC LIMITATIONS TO EXERCISE IN HEALTH AND DISEASE

The Physiology and
Pathophysiology of
Exercise Tolerance

PART 1. INTRODUCTION

GREETINGS ADDRESS OF THE GERMAN SOCIETY OF CARDIOLOGY IN HONOR OF PROFESSOR MARTIN STAUCH

Translated by Professor Eberhard Bassenge

Eberhard Bassenge

Institut für Angewandte Physiologie
Universität Freiburg
Germany

It is indeed a great pleasure to present this address to Professor Martin Stauch on the occasion of the international symposium "Physiology and Pathophysiology of Exercise Tolerance."

This symposium should be scientifically fruitful, rewarding, and memorable. Indeed, perhaps it will be as memorable as the highly successful Congress of the German Society of Cardiology that you, Professor Stauch, organized as Chairman in October 1983 and at which a broad spectrum of new scientific and therapeutic issues in cardiology were raised and comprehensively addressed. At that time, however, there was much less emphasis on exercise physiology in cardiology, as exercise was not yet considered to be the effective diagnostic and therapeutic modality that it has now become—and about which there will be much discussion at this symposium.

This change in "center of gravity" has much to do with the dynamic shifting in your primary focus and efforts throughout your career (almost "sinusoidally") in response to wide-ranging external *stimuli* and perhaps also by *endogeneous forces*—in addition to various *"agonists"* and possibly even by some *"partial antagonists"*—that originated from your local surroundings. In a manner owing as much to your mental endurance as to the more usual *"receptor-mediated"* manner reflecting fruitful stimulation from supporting colleagues, you slowly but progressively underwent a transition from cardiology and angiology to exercise physiology and sports medicine.

You embody the various essential requirements of "high-performance medicine"—a reflection of your excellent and broad scientific background that ranges from the physiology of the isolated working heart and the modulation of diastolic wall stress on ventricular contraction, through analyses of pulmonary function, to your particular interests in internal medicine, surgery, anesthesiology, angiology, nuclear medicine, cardiology, and finally, sports medicine. These multifaceted spectra of potential *adaptive* mechanisms to physical and mental stress have brought you into the desirable position of keeping the circulation, the organ blood supply and the *"non-receptor-mediated"* interactions with your colleagues at optimal performance levels, both with regard to the therapeutic management of your patients and to the various scientific projects in which you and your co-workers have been engaged. Thus, we hope that this scientific seed and solid foundation will thrive and continue to be productive in the immediate and the more-distant future.

As we are all impressed by your excellent performance, condition and activity—no doubt stimulated and supported by your family—we all expect you in the near future to assume the role of philosopher, consultant and seasoned advisor for your former department and colleagues as you continue to anticipate and recognize future developments in both cardiology and exercise physiology. Like the *"exercise-induced"* augmented expression of constitutive endothelial nitric oxide synthase, which supresses undesirable, excessive platelet and leukocyte activation, low-density lipoprotein incorporation and subsequent formation of atheroma (in short, which suppresses atherosclerosis and similar events), we are confident that you will preserve your cardiovascular system and that of your former department in a state of long-lasting youth.

These are my best wishes and also those of our Society. We are all becoming increasingly convinced of the benefits that can accrue from the pursuit of exercise physiology which, facilitated by the impetus that we hope this symposium and its proceedings will confer, is likely to endure as one of the "gold standards" in modern preventive medicine.

PROFESSOR MARTIN STAUCH

His Career

Jürgen M. Steinacker

Abteilung Sport-und Leistungsmedizin
Medizinische Klinik und Poliklinik
Universität Ulm, D-89070 Ulm
Germany

Professor Dr. Martin Stauch was born on March 8, 1928 in Berlin. In 1949 he began his medical studies at the University of Cologne and then subsequently at the University of Frankfurt. The award of a Fulbright Fellowship in 1955 provided the opportunity to undertake a two-year Fellowship in Internal Medicine, first at Ford Sanders Presbyterian Hospital, teaching hospital of the University of Tennessee at Knoxville, and then Colorado State Hospital at the University of Colorado in Pueblo. He returned to Germany to take up a Fellowship of the Academy of Sciences in Mainz. Martin Stauch's career in internal medicine and cardiology was launched at the University of Frankfurt Medical Center with Professor Hoff and Professor Heinecker. In 1967, he was appointed Professor of Cardiology and Medicine at the University of Ulm and headed the Division of Cardiology, Angiology, and Pulmonology until 1989. He served as Professor of Medicine and Head of the Department of Sports Medicine from 1989 until his retirement in 1994.

It was at the University of Frankfurt that Martin Stauch took his first substantial steps in cardiological research, under the guidance of Professor Karl Wezler (a former Fellow of Otto Frank) in the Institute of Physiology. For example, using isolated frog heart muscle preparations, he demonstrated that spontaneous ventricular flutter could be generated by shortening the duration of the action potential, which led in turn to tetanic contraction (1).

In addition, by examining the effects of different external loads on myocardial cell contractility, he discovered the phenomenon of electro-mechanical feedback (2). Furthermore, again using the isolated heart, he was able to show that changes in afterload lead directly to changes in diastolic relaxation (3).

Martin Stauch was the first to demonstrate, in 1975, the therapeutic effects of mononitrates on angina pectoris (4)—to the amused interest of colleagues who now, 20 years later, routinely prescribe mononitrates in their clinical practice! In addition, his experimental work on hamsters with hereditary cardiomyopathy led to a new concept of treating chronic hypertrophic cardiomyopathy by means of calcium antagonists (5).

Over the last two decades, Martin Stauch has conducted a range of non-invasive investigations of cardiac function, using techniques in the newly emerging field of nuclear medicine that he has developed in collaboration with Adam and his colleagues. Many of these studies have addressed pharmacological means of improving cardiac function in the chronically ischemic heart and in congestive heart failure (6,7). "Imaging the heart" became an early passion of Martin Stauch, and is reflected in his promotion of new echocardiographic investigations of mitral insufficiency employing color Doppler flow techniques (8).

The other major area in which Martin Stauch has made significant contributions is that of exercise testing. In his hands, this approach has proved an important tool for analyzing cardiac and metabolic function in patients and athletes for: diagnostic purposes, the control of therapy and rehabilitation, and the design and monitoring of training programs. He pioneered the use of upper-body exercise testing with the design, in 1989, of an innovative fixed-seat rowing ergometer that enables metabolic and cardiac function to be assessed over a wide range of exercise intensities in patients with peripheral occlusive disease (9), independently of peripheral blood flow limitations (10).

Present-day high-performance medicine also incorporates "care." Apart from his clinical involvements, Martin Stauch has amply demonstrated this quality through his dedicated participation in five training camps of the sucessful German Junior National Rowing Team 1990–1994 (11).

Martin Stauch has always been a caring physician. He has always taught his students and fellows that arriving at a diagnosis should involve first listening to the patient, then examining the patient carefully and with respect, and only after these stages are complete should investigative procedures be employed—and then always in a rigorous pathophysiological context.

REFERENCES

1. K. Wezler and M. Stauch. Spontane Superposition und Tetanus am geschädigten Froschherzen. *Arch. Kreislaufforschg* **31**:158–180, 1959.
2. M. Stauch. Elektromechanische Beziehungen am isolierten Froschherzen. Das monophasische Aktionspotential bei isotonischer in isometrischer Kontraktion. *Arch. Kreislaufforschg* **49**:1–14, 1966.
3. M. Stauch and H. Frankenberg. Beziehungen zwischen Relaxationsgeschwindigkeit und arteriellem Belastungsdruck am isolierten Warmblhterherzen. *Arch. Kreislaufforschg* **60**:310–327, 1969.
4. M. Stauch, N. Grewe, and H. Nissen. Die Wirkung von 2- und 5-Isosorbidmononitrat auf das Belastungs-EKG von Patienten mit koronarer Herzkrankheit. *Verh. Dtsch. Ges. Kreislaufforsch* **41**:182–184, 1975.
5. K. Lossnitzer, J. Janke, B. Hein, M. Stauch, and A. Fleckenstein. Disturbed myocardial calcium metabolis: A possible pathogenetic factor in the hereditary cardiomyopathy of the syrian hamster. In: *Rec. Advances in Studies on Cardiac structure and metabolism, Vol. 6.* Baltimore: University Park Press, 1975, pp. 207–217.
6. M. Stauch, W. Haerer, G. Rogg-Dussler, H. Sigel, and W.E. Adam. Effects of molsidomine on regional contraction and global function of the left ventricle. *Amer. Heart J.* **109**:653–658, 1985.
7. M. Stauch, G.Grossmann, A. Schmidt, P.Richter, J. Waitzinger, D. Wanjura, W.E. Adam, and W. König. Effect of gallopamil on left ventricular function in regions with and without ischemia. *Europ Heart J.* **8**(Suppl.G.):77–83, 1989.
8. M. Giesler and M.Stauch. Color Doppler determination of regurgitant flow: From proximal isovelocity surface areas to proximal velocity profiles. An in vitro study. *Echocardioraphy* **9**:51–62, 1992.
9. J.M. Steinacker, C. Hühbner, and M. Stauch. Hämodynamik und metabolische Beanspruchung bei Ruderergometrie, Fahrrad- und Laufbandergometrie von ambulanten Patienten nach Herzinfarkt. *Herz/Kreisl* **25**:239–243, 1993.

10. Y. Liu, J.M. Steinacker, and M. Stauch. Transcutaneous oxygen tension and Doppler ankle pressure during upper and lower body exercise in patients with peripheral arterial occlusive disease. *Angiology* **46:**689–698, 1995.
11. J.M. Steinacker, R. Laske, W.D. Hetzel, W. Lormes, Y. Liu, and M. Stauch. Metabolic and hormonal reactions during training in junior oarsmen. *Int J Sports Med* **14:**S24–S28, 1993.

REPRESENTATIVE BOOKS AND REVIEWS

Adam W.E., and Stauch, M. Clinical pharmacology of antianginal drugs: Radionuclide methods. In: *Handbook of Experimental Pharmacology, Vol. 76,* edited by G.V.R. Born, A. Farah, H. Herken, and A.D. Welch. Berlin: Springer Verlag, 1985, pp. 213–253.
Belz G.G. and Stauch, M. *Notfall EKG-Fibel,* 5th edition (1st edition, 1975). Berlin: Springer Verlag,1994.
Stauch, M. (editor) *Farbatlanten der Medizin, Band 1: HERZ* (1st edition edited by F.F.Yonkman). Stuttgart: G. Thieme Verlag, 1990.
Stauch, M. *Kreislaufstillstand und Wiederbelebung*, 6th edition (1st edition, 1967). Stuttgart: G. Thieme Verlag, 1994.

3

THE BEGINNING OF APPLIED PHYSIOLOGY AND SPORTS MEDICINE IN ULM

H. Rieckert

Abteilung Sportsmedizin
Institut für Sport und Sportwissenschaften
Christian-Albrechts-Universität zu Kiel
Germany

1. INTRODUCTION

When and how did sports medicine and applied physiology start at the University of Ulm?

Twenty years ago, on the 6th of June 1974, the Minister of Cultural Affairs of Baden-Württemberg wrote to the University of Ulm indicating that the House of Representatives of Baden-Württemberg had decided to establish sports medicine examination centers throughout Baden-Württemberg. This was to be directed and guided by Professor Keul at the University of Freiburg and would include a sports medicine institute at the University of Ulm, with Professor Rieckert at its head. Two months later, however, Professor Rieckert was offered a professorship at the University of Kiel. His move to Kiel led first to Professor Gebert taking over and then Professor Wodik. A redesigning of the program at Ulm subsequently took place, with Professor Stauch being appointed to head Clinical Sports Medicine.

In 1970, Professor Brecht was appointed as Director of the Institute of Physiology at the University of Ulm. He, Professor Rieckert who was head of the Sports Medicine Research Program, and Professor Pauschinger - the Director of the Institute of Applied Physiology, who worked tirelessly on behalf of the Program - formed the nucleus from which the present program evolved.

Between 1970 and 1974 Professor Rieckert was the team physician of the Swimming Association of Baden-Württemberg. Every year, swimmers were therefore sent by the Association to the University of Ulm for medical and physiological assessment of their swimming fitness. The general line of sports medical research at that time was exercise physiology.

The Physiology and Pathophysiology of Exercise Tolerance
edited by Steinacker and Ward, Plenum Press, New York, 1996

2. EARLY LINES OF RESEARCH

Sports medicine and applied physiological research at Ulm developed out of a research group in the Department of Physiology. There were several central issues of interest for exercise physiology and sports performance:

     a. peripheral circulatory function
     b. capillary filtration, and
     c. athletic activities and physical fitness in school children

2.1. Peripheral Circulatory Function

This research initially focussed on the factors that determine calf blood flow during exercise; that is, the reduction in muscle vascular resistance and the increase in the arterio-venous blood pressure difference across the muscle.

An early experiment illustrating the effect of posture on the arterio-venous pressure difference is shown in Figure 1. Simultaneous recordings were made of the arterial and venous pressures in the feet (A. tibialis posterior and V. dorsalis pedis, respectively), first in a horizontal position and then in the upright position.

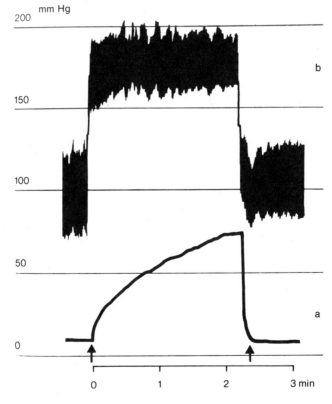

Figure 1. Arterial pressure (b: A. tib. post) and venous pressure (a: V. dors. pedis) in the supine and upright positions. At 0, the subject stood up.

When tilting an individual into the upright position at an angle of about 45°, the arterial blood pressure in the calf rises because of the increased hydrostatic pressure. On the venous side, the valves are closing. This leads to a substantial difference between the arterial and venous blood pressures which, in turn, causes a high blood flow through the calf and foot. The flow of blood through the capillaries and into the veins causes the venous valves to be pushed open. In the upright posture, the pressure in the foot veins is about 90 mmHg. This high pressure enlarges the veins. We found, as a result, a blood volume of 400 ml in both legs when subjects were supine, which increased to 600 ml on standing.

During walking or running, the muscle pump displaces blood against the hydrostatic pressure (Fig. 2). Muscular effort reduces the volume as well as the pressure in the dependent veins. Using telemetry to monitor venous pressure during exercise (e.g. walking and running), we demonstrated that the venous pressure reduction was much greater when the running speed was high. Also, the arterio-venous pressure difference was increased - one of the main factors contributing to the increase in blood flow to the calf. In trained female rowers, the maximum blood flow during rowing was higher than for untrained subjects, and the venous pooling in the calf was twice as high.

Interestingly, we were able to demonstrate that competitive swimmers had a higher maximal oxygen uptake (measured using a closed and an open system: Fig. 3) while swimming in a jet stream than for cycle ergometer exercise (Fig. 4). However, after comparable efforts, the recovery of the blood flow in the arms back to normal was faster in water than on the land.

2.2. Capillary Filtration

At the University of Ulm, we also began to record the capillary filtration and muscle blood flow at rest and during activity. This was found to be influenced by the general condition of the athlete and the intensity of the effort. Endurance-trained athletes evidenced a lower rate of capillary filtration (recorded with a plethysmograph and compared with isotopes) during running than did untrained persons at the same speed. The rate of capillary filtration was proposed to be influenced by the intramuscular accumulation of metabolites (such as lactate) that occurs with anaerobic glycolysis (which was less in the athletes) and

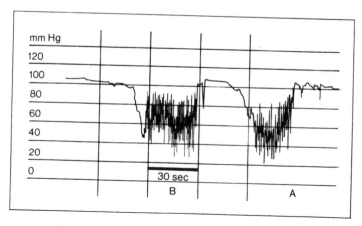

Figure 2. Venous pressure in a foot vein when walking (A) and when running (B). (The trace starts from the right.)

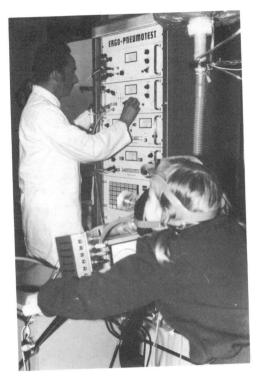

Figure 3. Recording of oxygen uptake.

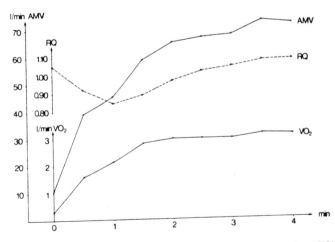

Figure 4. Pulmonary ventilation (AMV), oxygen uptake (VO2), and respiratory quotient (RQ) during backstroke in an 18-year-old girl.

therefore creates an osmotic potential difference tending to draw fluid from the blood into the extravascular space.

2.3. Athletic Activity in School Children

These investigations addressed the intensity of effort in activities such as gymnastics, games, swimming and track and field. Using telemetric measurements, we found that one single period of sports per day over a 5-year span increased the maximal oxygen uptake compared to the non-treatment classes; i.e. from 46.3 ± 5.1 ml/min/kg body weight to 52.4 ± 7.5 ml/min/kg body weight.

3. CONCLUSION

Today, the University of Ulm is an important center of clinical sports medicine. Much excellent scientific research has been conducted here and has provided the impetus for Ulm now being recognized both within Germany and beyond as a center of excellence in sports medicine and applied physiology.

4

PREVENTATIVE CARDIOLOGY AND PHYSICAL ACTIVITY

An Overview

W. Hollmann

Institute for Cardiology and Sports Medicine
German Sport University
Cologne, Germany

1. INTRODUCTION

Scientifically exact judgements on the importance to health of increased physical activity in the context of exercise and sports can be achieved in the following ways:

a. Experimental studies on lack of exercise on the one hand, and on qualitatively and quantitatively differing physical training programs on the other.
b. Clinical practical experience on exercise lack, physical training and sports.
c. Animal experiments.
d. Epidemiological studies.

Only the first aspect can be addressed comprehensively here.

2. EFFECTS OF LACK OF EXERCISE (PROLONGED BED REST)

After the end of World War II, technical developments strongly influenced our lifestyle. New devices and automation reduced the caloric consumption within most occupations and also in our leisure time. In West Germany between 1950 and 1990, the daily caloric consumption for men was diminished by approximately 1700 J, and for women by 1200 J (19a). It seems for the first time in the history of mankind that the minimal level of caloric consumption fell below that boundary which has to be exceeded in order to preserve optimal physiologic conditions according to the biological rule: performance capacity and structure of the human organism depend on the genotype and the quality and quantity of muscular strain (20-22,24).

We felt it important, therefore, to obtain information on the clinical effects of exercise lack. For example, a 9-day-period of bed rest provoked a 20% reduction of maximal oxygen uptake in a group of healthy physical education students. The heart volume, determined in the supine position by x-ray investigations in two different planes, was reduced by 10%. Heart rate, ventilatory minute volume and blood lactate levels were significantly higher for a given submaximal ergometer load than before the bed rest. All these findings can be considered disadvantages to fitness and health (22, 24).

Significant increases in resting heart rate were observed after a 14-day period of absolute bed rest (27). Correspondingly, the myocardial O2-requirement increased, predisposing to risk for patients with coronary insufficiency. The blood volume, and that of the erythrocytes, was significantly reduced. In studies with orthopedic patients, a period of physical inactivity of more than 2 months provoked a reaction similar to diabetes, as demonstrated with the Staub-Traugott blood sugar test. The pathological function curve normalized after 1-4 weeks of gymnastic exercise (Fig. 1) (22). Today we know that these findings belong to the so-called metabolic syndrome. In this condition, the receptor sensitivity for insulin decreases, which is why the insulin level in the blood increases. At the same time, a decrease of HDL and a rise of LDL and VLDL cholesterol, in conjunction with an increase of the arterial blood pressure, can be observed. All these changes imply increased risk for the development of arteriosclerosis and coronary heart insufficiency. A few weeks of exercise to increase general aerobic dynamic endurance normalizes these responses, however (1,5,6,8,9,11-13,17-19,22,24,35,37,38).

Saltin *et al.* demonstrated the same effects and, in addition, described an increase in urinary calcium excretion as an indication of bone demineralization (32). It appears that gravitational stress on the bones is essential for normal bone growth and for the conservation of a normal condition.

Summarizing, it can be stated that lack of exercise has many detrimental effects for health.

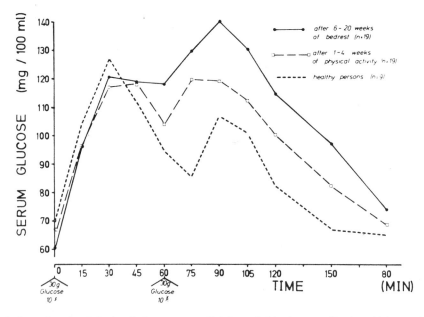

Figure 1. Several weeks of absolute bed rest causes a disturbance in blood sugar utilization which resembles diabetes. After a 1-4 week reactivation, the curve returns to normal (22).

3. RECOMMENDATIONS FOR ENDURANCE TRAINING

The important question is: What is the most suitable quality and quantity of physical exercise to induce health-promoting effects? Optimal sporting events are those which grant a maximum of adaptations desirable for health with a minimum of stress for the exercising person. This can be examined by simultaneously measuring hemodynamic factors (heart rate and arterial blood pressure) and the rate of anaerobic metabolism (arterial lactate level). The greater the oxygen intake for a submaximal work rate and the smaller the simultaneous mean arterial blood pressure and arterial lactate level, the more desirable is that event. The results of investigations propose that recommendable sports are fast walking, uphill walking, slow running (jogging), cycling, swimming, long-distance skiing, and rowing (Figs. 2 and 3). Not so effective for the cardiovascular system, but including other kinds of motor strain and therefore especially desirable for healthy older persons, are games like tennis, hockey, and basketball. The only prerequisite is that the person has to be healthy.

Our recommendations are:

- frequency: 3 x weekly;
- Duration: 30-90 min (depending on the level of physical activity);
- intensity: according to the "rule of thumb" that the heart rate at maximal exercise heart rate should be (180 - years of age) (during a continuous effort) (22).

4. EFFECTS OF AEROBIC DYNAMIC ENDURANCE TRAINING

4.1. Skeletal Muscle

The following adaptations occur with endurance training:

a. an increase in number and size of mitochondria (9,16,34);
b. a rise in the activity of some aerobic and anaerobic enzymes (16);
c. an augmentation of myoglobin levels (9);
d. an increase in intramuscular glycogen content (7,9);
e. an increase in the number of capillaries and of the capillary surface area (9,34).

There are very highly significant negative correlations between: the mitochondrial volume and the arterial lactate level, the number of capillaries per muscle fiber and the arterial lactate level, the capillary surface area and the arterial lactate level, and the capillary surface area at a given submaximal work rate (34) (Fig. 4).

The significance of these peripheral adaptations for the heart can be clearly demonstrated by endurance training conducted with only one leg while sitting on a cycle ergometer. After training 3 times weekly for 30 min per training session for 6 weeks, the heart rate at a given submaximal exercise load declines significantly when exercise is performed with the endurance-trained leg, as does the ventilatory minute volume. In contrast, the same individual does not show any significant changes of heart rate and ventilatory minute volume, compared with pre-training values, when using the non-trained leg (22). This result is consistent with a reduction of the myocardial oxygen demand causing a greater security against, for instance, an imbalance between oxygen supply and demand of the myocardium (Fig. 5).

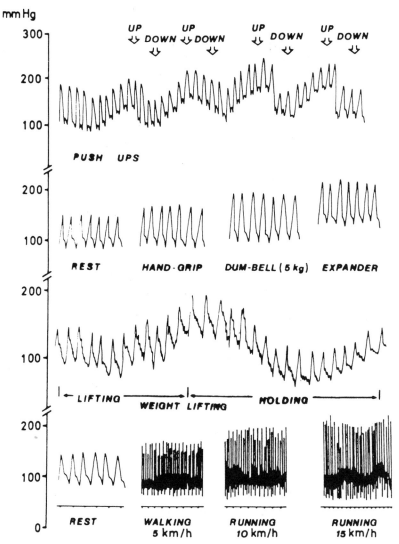

Figure 2. Arterial blood pressure response (A. brachialis) during various static and dynamic muscle demands, and while walking at 5 km/h, and running at 10 and 15 km/h (31). Walking and running are the only kinds of exercise which do not result in a rise of diastolic blood pressure.

The myocardial oxygen demand can be determined non-invasively from the product of heart rate and systolic blood pressure. If the heart rate decreases for a given submaximal work rate, the oxygen demand of the heart muscle is also lowered. The reason for the reduced heart rate after training may be a smaller peripheral sympathetic drive caused by an improved local aerobic capacity combined with an attenuated lactate production (Fig. 6).

4.2. Heart

The main adaptations in response to endurance training are:

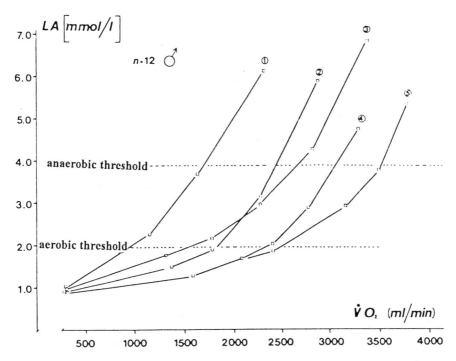

Figure 3. The arterial blood lactate level during an incremental exercise test as a function of submaximal oxygen uptake. 1 = arm-crank exercise, 2 = cycle ergometry in a supine posistion, 3 = cycle ergometry in a seated position, 4 = step test, 5 = running on a treadmill. The lower the lactate level at a given oxygen uptake, the more appropriate is the exercise mode for preventative and rehabilitative cardiology.

a. reduction of the heart rate at rest and for submaximal exercise loads;
b. lengthening of the diastolic period;
c. acceleration of diastolic relaxation;
d. increase of the stroke volume;
e. decreased catecholamine release;
f. diminution of the systolic blood pressure;
g. stabilization of myocardial electrical activity
h. decrease of the peripheral resistance (1,2,9,22,23,31).

Most of these adaptations contribute to the reduction of the myocardial oxygen demand. The prolonged diastolic period causes a lengthening of the intramuscular phase of blood supply. The practical consequences can be explained in the following scheme. If an untrained coronary patient can master a load intensity of 75 watts without complaint, then this individual may reach a higher work rate (e.g. 125 watts) without complaint after training. The difference between 75 and 125 watts functions practically as a safety zone that prevents a disparity between the myocardial oxygen demand and oxygen supply.

4.3. Blood

There are numerous influences of endurance training on blood. The most important one may be that the rigidity of the erythrocyte membrane declines. Therefore, the plastic-

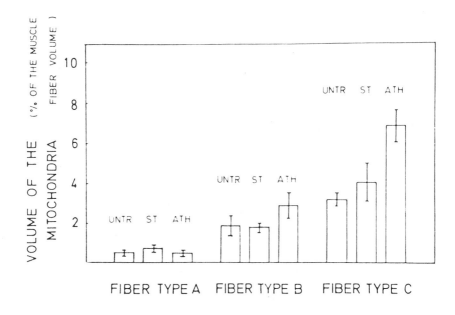

UNTR = untrained persons
ST   = students of physical education without endurance training
ATH = endurance-trained persons

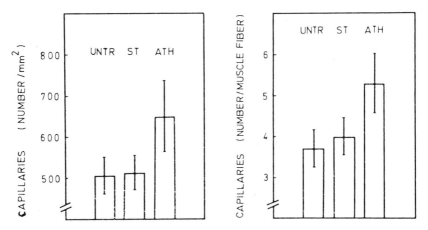

Figure 4. Above: percentage proportion by volume of mitochondria in 3 muscle fiber types in normal subjects, sports students, and endurance-trained persons. Below: capillary number and capillary surface area in m. vastus lat. for the same subjects (n = 19) (34).

ity of erythrocytes increases and the flow properties are improved within the capillaries. At the same time, factors such as the adhesiveness and aggregability of the platelets decline, which opposes thrombosis. In addition, the fibrinogen content is reduced (10,22).

The practical significance of these changes for the prevention of ischemic heart diseases is unknown. Today it is supposed that 98% of all heart infarctions originate from mi-

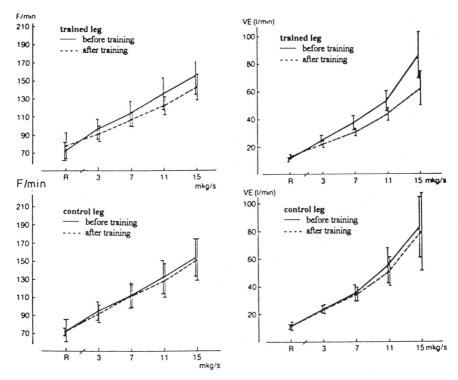

Figure 5. Heart rate before and after a one-legged endurance training on the cycle ergometer performed by the endurance-trained leg (upper panel) and the untrained leg (lower panel). The right panels demonstrate the ventilatory minute volume during different load intensities when using the endurance-trained or the untrained leg before and after training.

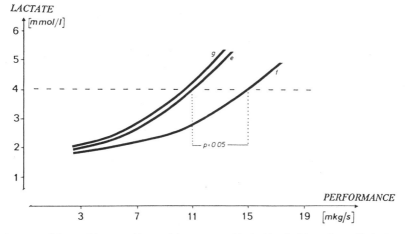

Figure 6. Increase of the aerobic-anaerobic transition, measured in the blood of the v. femoralis, in the endurance-trained leg (f). No change was evident in the untrained left leg (g, e).

crothrombosis. The effects of endurance training on the blood could therefore be very important.

4.4. Lipid Metabolism

We performed cross-sectional and longitudinal investigations on the effects of endurance training in 2 groups of 353 and 24 male subjects between 18 and 82 years of age. In another group of 47 male subjects, the changes of the plasma lipoproteins were observed during the first hours and days after a 3-hour run (11-13). The most important results are described below.

The concentrations of beta-lipoprotein, total, and LDL-cholesterol were reduced in the blood of endurance-trained athletes. At the same time, the concentrations of alpha-lipoproteins and of HDL-cholesterol were increased (Fig. 7).

Elite rowers had a direct correlation between total and HDL-cholesterol and an inverse correlation between triglycerides and HDL-cholesterol. The quotient HDL2/HDL3 was higher in endurance-trained than in untrained individuals. One day after a 3-hour run, the concentrations of total and VLDL-triglycerides and of the total VDL- and LDL-cholesterol were significantly reduced. The ratios of non-esterised cholesterol/total cholesterol and of non-esterised/esterised cholesterol increased significantly 3 hours after this run. One and two days later, these ratios were significantly reduced. The HDL-cholesterol level was significantly higher one day after the 3-hour run than before. One day later, an increase of HDL3-cholesterol ($p < 0.001$) was seen, and 2 days after the run an increase of HDL2-cholesterol ($p < 0.05$) was also evident. The apolipoprotein A-I/A-2 ratio was also significantly increased 2 and 4 days after the 3-hr run ($p < 0.01$). The activity of lecithin cholesterol acyltransferase was significantly elevated 3 hours after the run ($p < 0.01$). Two days later, the activity of this enzyme was decreased relative to resting values ($p < 0.05$). The increased HDL-cholesterol, in particular the HDL2-cholesterol, that results from endurance training may be important for its anti-atherosclerotic effect (4-6,11-13,17-19,25,38). It is possible that changes of plasma proteins after endurance training are the sum of delayed reactions after repeated intense endurance strains (7).

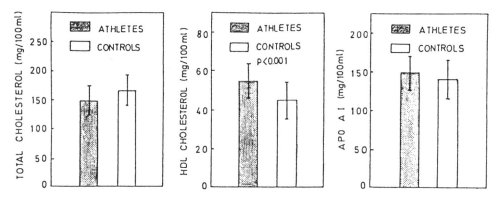

Figure 7. Total cholesterol, HDL, and ApoA-1 in untrained and endurance-trained males (n = 377) (22).

4.5. Hormones

Following a 3-hour run, no significant changes of growth hormone (GH), cortisol, and glucagon were evident (12). Noradrenaline levels during submaximal exercise were significantly reduced in endurance-trained individuals. As catecholamines have been proposed to be "oxygen-robbers", this would suggest that the oxygen demand of the heart muscle at a given work rate is decreased as a consequence of the reduced catecholamine levels in the cell.

Besides the decreased noradrenaline and adrenaline levels in the blood at a particular submaximal exercise load, there is also a reduction of the insulin level. This change is the consequence of an increased sensitivity of the insulin receptors in the muscle cells. As a high insulin level in the blood can favor atherosclerotic developments, this effect of training may be of specific benefit (9,12,24).

5. AGING PROCESS

The maximal oxygen uptake decreases after the 30th year of age by about 8% per decade (Fig. 8) (23,28,29,31,33,36,40,41,51,54). In sufficiently loadable individuals, a training effect can be demonstrated even after the 90th year of age (Figs. 8 and 9).

Physical training, especially endurance training, also improves the mood (Fig. 10). The effect depends on 3 different mechanisms: increased production of endorphins (3), augmentation of serotonin production in the brain because of an increased amount of tryptophan at the blood-brain-barrier, and/or an increase of the insulin level during the recov-

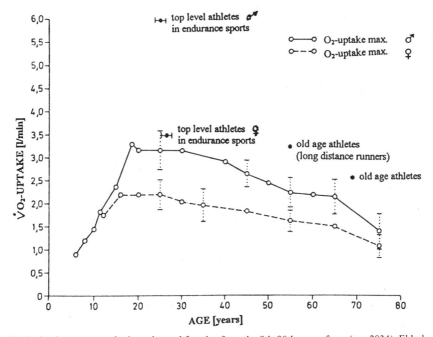

Figure 8. Maximal oxygen uptake in males and females from the 8th-80th year of age (n = 2834). Elderly endurance-trained individuals have marked higher values than untrained elderly individuals (22).

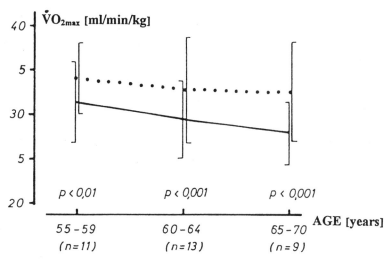

Figure 9. Maximal oxygen uptake (ml/kg·min-1) after a 10-week endurance training program in healthy males who were untrained for at least 3 decades. The trainability of old and elderly persons is qualitatively unchanged. (Before ____, after .....) (36).

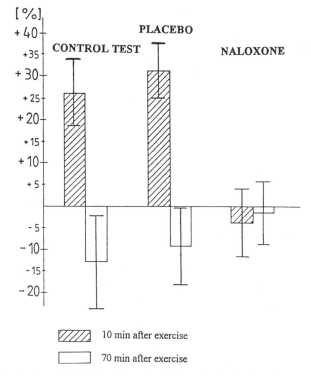

Figure 10. Reduction of pain sensitivity in healthy male subjects after maximal cycle ergometer exercise in a control test (left), with a placebo, and with naloxone blockade (double blind experiment measured the electrical pain sensitivity of tooth pulp) (3).

ery from exercise with the same effect (9,24). At the same time, regional cerebral blood flow rises during dynamic exercise about 20% or 30% (Fig. 11).

6. DECLARATION OF THE WORLD CONSENSUS CONGRESS (TORONTO, 1992)

"Sedentary living is associated with a high incidence of coronary heart disease. This observation is supported by numerous prospective studies from Europe and North America based on groups of apparently healthy individuals who were followed for fatal and non-fatal coronary heart disease for up to 20 years. The majority of studies show an inverse relationship of coronary heart disease rates across physical activity levels. There is near unanimity that physical activity provides some protection against coronary heart disease in studies that have used good or excellent epidemiological methods, including valid and reliable assessment of physical activity and comprehensive surveillance of disease end points. There is an approximate doubling in risk of coronary heart disease when the least active individuals are compared with their most active peers. The association of inactivity to coronary heart disease is not due solely to the confounding influences of other favorable lifestyle behaviors." And elsewhere: "The relative risk for coronary heart disease in the least active compared to most active is approximately 2.0. When fitness is the expo-

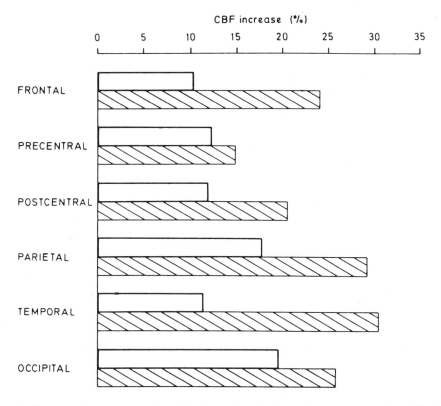

Figure 11. Blood supply to various regions of the human brain during cycle ergometer exercise at 25 watts and 100 watts

sure variable, relative risks as high as 8.0 are seen when comparing least fit with most fit individuals." - "The relationship between fitness and cardiovascular disease mortality show similar relative risks in men and women when individuals with low and high fitness levels are compared." - "Low levels of activity or fitness precede the development of coronary heart disease in healthy individuals. Results are consistent within and across populations, and the epidemiological findings are plausible and coherent with the results from clinical and experimental investigations." "A similar situation is to be seen comparing cardiac patients with physical training in the rehabilitation and without a training program. Several meta-analyses of randomized trials of cardiac rehabilitation involving thousands of patients have demonstrated a 20% reduction of risk for total mortality, and a 25% reduction in the risk for fatal re-infarction" (9).

7. CONCLUSIONS

The physician has to recommend sports and physical training:

    a. *in childhood and adolescence*: in order to acquire an optimal development of organs and of the musculo-skeletal system;

    b. *in adults*: to counteract cardiovascular and metabolic diseases;

    c. *in the elderly and old*: to ameliorate loss of physical and mental performance capacity in the aging process.

For a modern physician, sports and physical training belong to our daily life like our daily dental hygiene.

8. ACKNOWLEDGMENTS

Supported by: Institute for Sports Sciences of the Federal Republic of Germany, Cologne; Krupp-von-Bohlen-und Halbach-Grant, Essen; Eckloff-Winterstein-Grant, Bad Kissingen; and Oertel-Grant, Mülheim/Ruhr, Germany.

9. REFERENCES

1. American College of Sports Medicine. The recommended quantity and quality of exercise for developing and maintaining cardiorespiratory and muscular fitness in healthy adults. *Med. Sci. Sports Exerc.* 265-274, 1990.
2. American Heart Association. A position statement for health professionals by the Committee on Exercise and Cardiac Rehabilitation of the Council on Clinical Cardiology. *Circulation* **81(1)**:396-398, 1990.
3. Arentz, T., K. De Meirleir, and W. Hollmann. Endogenous opioid peptides during bicycle ergometer exercise (German). *Dt. Z. Sportmed.* **37(7)**:210-218, 1986.
4. Assmann, G. Lipidmetabolism and atherosclerosis (German). Stuttgart-New York: Schattauer, 1982.
5. Berg, A., J. Johns, N. Baumstark, M. Kreutz, and J. Keul. Changes in HDL-subfractions after a single extended episode of physical exercise. *Atherosclerosis* **47**:231, 1983.
6. Berg, A., M. Halle, M.W. Baumstark, and J. Keul. The significance of lipoproteins for the pathogenesis of coronary heart disease (German). *Dt. Ärzteblatt* **91(12)**:822-830, 1994.
7. Bergström, J. Muscle electrolytes in man, determined by neutron activation analysis on needle biopsy specimens. A study on normal subjects, kidney patients, and patients with chronic diarrhoea. *Scand. J. Clin. Lab. Invest.* **14(68)**, 1962.

8. Bijnen, F.C.H., C.J. Caspersen, and W.L. Mosterd. Physical inactivity as a risk factor for coronary heart disease: A WHO and International Society and Federation of Cardiology position statement. *Bull. WHO* **72(1)**:1-4, 1994.

9. Bouchard, C., R.J. Shephard, and Th. Stephens (eds.). Physical Activity, Fitness, and Health. International Proceedings and Consensus Statement. Champaign, Ill, USA: Human Kinetics, 1994.

10. Broustet, J.P., M. Boisseau, J. Bouloumie, J.P. Emerian, E. Series, and H. Bricaud. The effects of acute exercise and physical training on platelet function in patients with coronary artery disease. *Cardiac Rehab.* **9(2)**:28, 1978.

11. Dufaux, B., G. Assmann, A. Mader, and W. Hollmann. Nephelometric determination of apolipoprotein A-I in endurance-trained athletes. Spring Meeting, Dresden: The European Atherosclerosis Group, 1979.

12. Dufaux, B., G. Assmann, U. Order, A. Hoederath, and W. Hollmann. Plasmalipoproteins, hormones and energy substrates during the first days after prolonged exercise. *Int. J. Sports Med.* **2**:256, 1981.

13. Dufaux, B., H. Liesen, R. Rost, H. Heck, and W. Hollmann. Effects of endurance training on lipoproteins with special consideration of HDL in young and old persons (German) *Dt. Z. Sportmed.* **30**:123, 1979.

14. Erikssen, J. Physical fitness and coronary heart disease morbidity and mortality. *Acta Med. Scand.* **711**:189-192, 1986.

15. Fox, S.M., and W.L. Haskell. Physical activity and health maintenance. *J. Rehab.* **32**:89, 1966.

16. Gollnick, P.D., L.A. Bertorci, T.B. Kelso, E.H. Witt, and D.R. Hodgson. The effect of high intensity exercise on the respiratory capacity of skeletal muscle. *Pflügers Arch.* **415**:405, 1990.

17. Gordon, T., W.P. Castelli, M.C. Hjortland, W.D. Kannel, and T.R. Dawber. High density lipoprotein as a protective factor against coronary heart disease: Framingham Study. *Am. J. Med.* **62**:707, 1977.

18. Harris, S.S., C.J. Caspersen, G.H. DeFriese et al. Physical activity counseling for healthy adults as a primary preventive intervention in the clinical setting: Report for the U.S. preventive services task force. *J. Amer. Med. Assoc.* **261**:3588-3598, 1989.

19. Haskell, W.L., H.J. Montoye, D. Orenstein. Physical activity and exercise to achieve health-related physical fitness components. *Public Health Rep.* **100**:202, 1985.

19a. Hettinger, Th. Unpublished results.

20. Hollmann, W. Effects of exercise and physical training on heart, circulation and respiration (German). Darmstadt: Steinkopff, 1959.

21. Hollmann, W., H. Liesen, R. Rost, H. Heck, J. Satomi. Preventive cardiology: Lack of exercise and physical training - an epidemiological and experimentally based overview (German). *Z. Kardiol.* **74**:46, 1985.

22. Hollmann, W., R. Rost, B. Dufaux, and H. Liesen. Preventive and rehabilitative cardiology and physical training (German). Stuttgart: Hippokrates, 1983 (1st ed, 1965).

23. Hollmann, W., R. Rost, H. Liesen, B. Dufaux, H. Heck, and A. Mader. Assessment of different forms of physical activity with respect to preventive and rehabilitative cardiology. *Int. J. Sports Med.* **2**:67, 1981.

24. Hollmann, W., Th. Hettinger. Textbook of sports medicine (German). Stuttgart-New York: Schattauer, 1990.

25. Kannel, W.B., A. Belanger, R. D'Agostino, and I. Israel. Physical activity and physical demand on the job and risk of cardiovascular disease and death: The Framingham Study. *Am. Heart J.* **112**:820, 1986.

26. Liesen, H., and W. Hollmann. Endurance and metabolism (German). Schorndorf: Hofmann, 1981.

27. Miller, P.B., R.L. Johnson, and L.E. Lamb. Effects of 4 weeks of absolute bed rest on circulatory functions in man. *Aerospace Med.* **35**:1194, 1964.

28. Morris, J.N., M.G. Everitt, and A.M. Semmence. Exercise and coronary heart disease. In: *Exercise: Benefits, limits and adaptations,* edited by D. Macleod, R. Maughan, M. Nimmo, T. Reilly, C.E. Williams and F.N. Williams. London: Spon Publishers, 1987.

29. Paffenbarger, R.S., A.L. Wing, and R.T. Hyde. Physical activity as an index of heart attack risk in college alumni. *Am. J. Epidemiol.* **108**:161, 1978.

30. Paffenbarger, R.S., R.T. Hyde, A.L. Wing, and C.C. Hsieh. Physical activity, all-cause mortality, and longevity of college alumni. *N. Engl. J. Med.* **314**:605, 1986.

31. Rost, R., and F. Webering (eds.). Cardiology and sport (German). Köln: Deutscher Ärzteverlag, 1987.

32. Saltin, B., G. Blomquist, J. Mitchell, R.L. Johnsen, K. Wildenthal, and C.B. Chapman. Response to exercise after bed rest and after training. *Circulation* **37/38(7)**:1-78, 1968.

33. Saltin, B., L. Hartley, A. Kilbom, and I. Åstrand. Physical training in sedentary middle-aged and older men. *Scand. J. Clin. Lab. Invest.* **24**:323-334, 1969.

34. Schön, F.A., W. Hollmann, H. Liesen, and E. Waterloh. Electromicroscopical findings in m. vastus lat. of untrained, endurance-trained persons and marathon runners and the relations to the relative maximal oxygen uptake and lactate production (German). *Dt. Z. Sportmed.* **31(12)**:343, 1980.

35. Sedgwick, A.W., J.R. Brother, A. Hood, A. Harris-Davidson, R.E. Taplin, and D.B. Thomas. Long-term effects of physical training program on risk factors of coronary heart disease in otherwise sedentary men. *Brit. Med. J.* **1**:7, 1980.
36. Suominen, H., E. Heikkinen, H. Liesen, D. Michel, and W. Hollmann. Effects of 8 weeks' endurance training on skeletal muscle metabolism in 56-70-year old sedentary men. *Europ. J. Appl. Physiol.* **37**:173, 1977.
37. Weidemann, H., and K. Meyer. Textbook of exercise therapy for cardiac patients (German). Darmstadt: Steinkopff, 1991.
38. Wood, P.D., W. Haskell, H. Klein, S. Lewis, M.P. Stein, and J.W. Farquhar. The distribution of plasma lipoproteins in middle-aged male runners. *Metabolism* **25**:1249, 1976.

PART 2. THE PHYSIOLOGICAL BASIS OF MUSCULAR FATIGUE

5

MECHANISM OF FATIGUE IN SMALL MUSCLE GROUPS

N. Maassen

Sports and Exercise Physiology
Medical School
Konstanty Gutschow Strasse 8
30625 Hannover
Germany

1. INTRODUCTION

The issue of this paper is peripheral fatigue, in particular, fatigue due to muscular impairment during dynamic exercise with small muscle groups. Small muscle groups in this sense are of such a size that the change in concentration of circulating catecholamines even during exhausting exercise is negligible. Thus a small muscle group is, in a way, comparable to a muscle exercising *in vitro*. Another similarity is the large distribution volume for substances leaving the muscle. In humans, forearm or calf muscle exercise can be regarded as exercise with small muscle groups.

Fatigue can be defined in different ways: "Inability to maintain the expected force and power output" (6) or "A transient loss of work capacity resulting from preceding work regardless of whether or not the current performance is affected" (14). The latter definition regards fatigue as a continous process starting from the beginning of exercise. In general, fatigue may occur at different sites. At the muscular level, fatigue can occur by impairing the following processes: e.g. cross bridge cycling, Ca^{++} pumping, electromechanical coupling, or electrical transmission.

The following factors are discussed in terms of their ability to cause fatigue at one or more of these sites: decrease in pH (13); depletion of ATP (7) and creatine phosphate (CrP) stores (11); increases in inorganic phosphate (in particular H_2PO_4: Ref. 3), NH_3 and IMP (4); Ca^{++} (18); and oxygen lack (12). The evidence for an involvement of these factors is derived mainly from *in vitro* studies. The role of these factors *in vivo* is still a matter of debate.

The Physiology and Pathophysiology of Exercise Tolerance
edited by Steinacker and Ward, Plenum Press, New York, 1996

2. OXYGEN PRESSURE

Oxygen pressure is proposed to influence many of the factors described above because of its effect on shifting energy metabolism towards a greater reliance on anaerobic metabolism. There is little qustion that hypoxia reduces performance during exercise with a large muscle group. If cycling exercise (incremental test) is performed while 12.5% oxygen is breathed, maximum power is reduced by about 30%. But hypoxia of the same degree does not reduce performance in exhaustive incremental exercise with the forearm muscles (dynamic handgrip exercise; 24 contractions/min), in spite of a reduction of oxygen saturation of as much as 30% in venous blood from the working muscle. There are also no associated changes in blood flow (venous occlusion plethysmography) or in the increase of blood [Lac] or decrease in blood pH. Neither does hyperoxia confer any beneficial effect (Table 1). Thus, under these exercise conditions, oxygen seems not to be a limiting factor.

3. INTRACELLULAR PH AND METABOLITES

The role of intracellular metabolites in fatigue is also conflicting. Using NMR, the role of pH and phosphates can be deduced. pH is argued to influence Ca^{++}-pumping, cross-bridge cycling and the fluxes through metabolic pathways. Comparing patients with heart failure with normal subjects, Arnolda et al. (1) found lower pH values at exhaustion in the patient group. In contrast, in hypertensive patients, higher pH values were measured at exhaustion than for normals (i.e. 6.61 and 6.44, respectively: Ref. 5). Wong et al. (19) found no correlation between pH and endurance time (time to exhaustion: 5 to 25 minutes), while Taylor et al. (16) showed that the pH at exhaustion can differ in a given subject using the same muscle group on different occasions.

Our own NMR spectroscopy measurements (4.7 tesla magnet, 20 sec time resolution) under similar exercise conditions comparing forearm exercise to that of the calf muscles showed only a small fall in muscle pH (from 7.0 to 6.85) during normoxia. This does not seem sufficient to exert the effects on glycolysis or on cross bridge cycling that might be expected to lead to exhaustion. In these experiments, phosphocreatine levels decreased to only 43% of initial values, ruling out depletion of the phosphocreatine store as a probable cause for fatigue. An increase in inorganic phosphate is also unlikely to be involved, especially as the pH change is small. Therefore, the diprotonated phosphate does not increase to any large extent.

Table 1. Forearm blood flow, power output, and muscle venous blood composition at the end of exercise

		HYPO	NORM	HYPER
SO_2	%	21.0 ± 7.7	32.1 ± 5.2	33.3 ± 4.9
pH		7.21 ± 0.02	7.19 ± 0.01	7.18 ± 0.01
$\dot{Q}$	ml/100ml/min	26.8 ± 7.2	33.2 ± 7.2	30.8 ± 3.3
POWER	watt	1.64 ± 0.19	1.62 ± 0.24	1.57 ± 0.20

The difference in SO_2 between hypoxia (Hypo) and normoxia (Norm) is significant ($p<0.001$); all others are not.

No correlation between [Lac] or phosphate levels and fatigue was found by de Haan *et al.* (4) after exercise of high intensity. They proposed an NH_3- or IMP-related effect leading to fatigue. On the other hand, patients with McArdle's disease are able to work with much higher NH_3 and IMP concentrations than normal subjects.

For each of the listed "fatigue-inducing" factors, conditions can be found where their behavior does not correlate with fatigue. Fatigue is therefore regarded as a multifactorial phenomenon, implying that the individual effects are in a way additive. On the other hand, however, it is possible that a relevant factor has not yet been considered. Evidence against these factors can be derived from Bigland Ritchie *et al.* (2), who demonstrated a progressive decline in the ability to maintain maximum force from the beginning of exercise. At this time, none of the factors considered here could contribute significantly because either their changes are very small at this time or they occur in the opposite direction.

4. BLOOD FLOW, NOT OXYGEN SUPPLY, LIMITS PERFORMANCE

That there has to be a factor causing fatigue on the muscle level is shown in the following simple experiment. If forearm exercise (implemented as described above; n = 6) is performed in the upright posture, maximum power is reduced by 13% compared to that attained in the supine posture. If the exercise is performed in the upright posture while hypoxic gas is breathed (12.5% O_2), no influence on [Lac] and no further decrease in power can be seen (Fig. 1). Blood flow rather than oxygen supply therefore appears to be a limit-

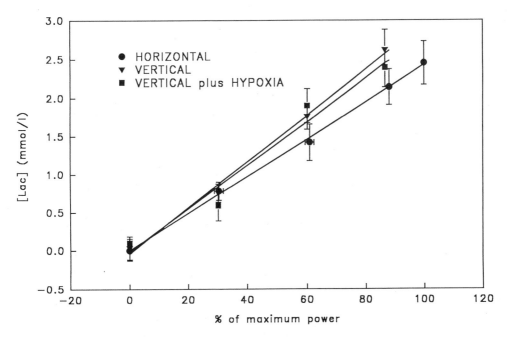

Figure 1. The muscle venous-arterial (V-A) differences in [Lac] for arm exercise with different postures. The results are expressed relative to the maximum power in the supine posture. Maximum power is reduced by about 13% in the upright posture, but is not further reduced by hypoxia (12.5% O_2).

ing factor. As the flow is reduced in the upright posture, relative to the supine posture, a possible mechanism may involve the trapping of some mediator(s) within the muscle during exercise.

One possible factor might be potassium. Immediately with the start of exercise K+ leaves the muscle cells and accumulates in the interstitial space (8,9,17). As a consequence, the resting membrane potential (RMP) should decrease and the propagation of the action potential should be impaired (10). Juel measured intracellular and interstitial [K+] and membrane potential on a rat muscle fibre: RMP decreased from 70 to 58 mV and, in parallel, power output decreased (10). Thus, an impairment of membrane excitation might contribute to fatigue.

5. IS POTASSIUM RELATED TO FATIGUE IN SMALL MUSCLE GROUPS?

We investigated short-term exhaustive forearm exercise which leads to exhaustion within 2 min (n = 10) and incremental exhaustive exercise leading to fatigue within about 10 min (n = 15). Forearm blood flow was measured by venous occlusion plethysmography. Hemoglobin (Hb), hematocrit (Hct), plasma [Na+] and [K+] were determined in arterialized venous blood and in venous blood from the working muscles. [Na+] and [K+] increased in muscle venous blood, but remained constant in the arterialized blood. As Na+ is shifted into the working muscle cell (15), the [Na+] increase is due to a shift of water into the muscle cells. Because there is no gradient in [Na+] between plasma and the interstitial space (17), the increase in [Na+] gives a relative estimate of the water gain by the muscle cells. From plasma flow through the muscle (derived from Hct and blood flow measurements) and the arterio-venous (A-V) difference in [K+] across the muscle, K+ loss from the muscle can be estimated assuming a bone mass of 1/3 of the volume where blood flow is measured. From the data of Hirche et al. (8), we assumed a K+ gradient between interstitial space and blood plasma of about 2.5 mmol/l for the intense exercise and of 2.0 mmol/l for the incremental exercise. The intracellular water space was assumed to be 660 ml/kg and the intracellular concentration of K+ 160 mmol/l. The calculated changes are shown in Table 2.

According to these changes, the resting membrane potential decreases with a resulting impairment of action potential propagation (10). As the calculated resting potential

Table 2. Estimated membrane potential after exhausting exercise.

Duration		Rest	Short 2 min	Incremental 10 min
K^+_{pl}	mmol/kg H_2O	4.4	7.0	6.2
K^+_{int}	mmol/kg H_2O	4.4	9.5	8.2
$\dot{Q}$	ml/100ml/min	1.5	20.0	26.0
K^+-loss	mmol		0.7	2
H_2O_i	ml/kg	660	680	695
K^+_i	mmol/kg H_2O	160	157	147
MP	mV	96	75	78

$\dot{Q}$ is the mean blood flow during the whole exercising time.

represents a mean potential of the muscle group as a whole, the RMP in some individual fibres should be even lower. Thus these fibres cannot be excited and therefore cannot develop force. This mechanism might work at both high and low exercise intensities. During high-intensity exercise, the accumulation of K+ in the interstitial space predominates, while during long-lasting low-intensity exercise K+ loss from, and water shift into, the muscle also come into play. This proposal fits very well to data obtained by muscle biopsy. Sjögaard (15) calculated a decreasing membrane potential (MP) from measurements of intracellular and extracellular [K+] in exercising human quadriceps muscle, suggesting that this mechanism might also play a role in large muscle groups. The importance of this mechanism is underscored by the fact that the first adaptation to training by chronic stimulation of muscle is an increase in Na+/K+-ATPase activity (Pette, this volume).

6. ACKNOWLEDGMENT

NMR spectroscopy was performed with Professor D. Leibfritz and Dr. H. Koch, FBII, Chemistry, University of Bremen, Germany.

7. REFERENCES

1. Arnolda, L., M. Conway, M. Dolecki, H. Sharif, B. Rajagopalan, J.G.G. Ledingham, P. Sleight, and G.K. Radda. Skeletal muscle metabolism in heart failure: A 31P nuclear magnetic resonance spectroscopy study of leg muscle. *Clin. Sci.* **79**: 583-589, 1990.
2. Bigland-Ritchie, B., E. Cafarelli, and N.K. Vollestad. Fatigue of submaximal static contractions. *Acta Physiol. Scand.* **128** (Suppl 556): 137-148,1986.
3. Cooke, R., and E. Pate. The inhibition of muscle contraction by the products of ATP hydrolysis. In: *Biochemistry of Exercise, VII*, edited by A.W. Taylor, P.D. Gollnick, H.J. Green, C.D. Ianuzzo, E.G. Noble, G. Metivier, and J.R. Sutton. Champaign, Il, USA: Human Kinetics, 1990, pp. 59-72.
4. De Haan, A. High-energy phosphates and fatigue during repeated dynamic contractions of rat muscle. *Exp. Physiol.* **75**: 851-854, 1990.
5. Dudley, C.R.K., D.J. Taylor, L.L. NG, G.J. Kemp, P.J. Ratcliffe, G.K. Radda, and J.G.G. Ledingham. Evidence for abnormal Na+/H+ antiport activity detected by phosphorus nuclear magnetic resonance spectroscopy in exercising skeletal muscle of patients with essential hypertension. *Clin. Sci.* **79**: 491-497, 1990.
6. Edwards, R.H.T. Human muscle function and fatigue. In: *Human Muscle Fatigue: Physiological Mechanisms*, edited by R. Porter and J. Whelan. London: Pitman Medical, 1981, pp. 1-18.
7. Ferenczi, M.A, Y.E. Goldman, and R.M. Simmons. The dependence of force and shortening velocity on substrate concentration in skinned muscle fibres from Rana temporaria. *J. Physiol. (Lond.)* 350: 519-543, 1984.
8. Hirche, H., E. Schumacher, and H. Hagemann. Extracellular K+ balance of the gastrocnemius muscle of the dog during exercise. *Plügers Arch.* **387**: 231-237, 1980.
9. Juel, C. Potassium and sodium shifts during in vitro isometric muscle contraction, and the time course of the ion-gradient recovery. *Pflügers Arch.* **406**: 458-463, 1986.
10. Juel, C. Muscle action potential propagation velocity changes during activity. *Musc. nerve* **11**: 714- 719, 1988.
11. Katz, A., K. Sahlin, and J. Henriksson. Muscle ATP turnover rate during isometric contraction in humans. *J. Appl. Physiol.* **60** (6): 1839-1842, 1986.
12. Moritani, T., M. Muro, and A. Kijima. Electromechanical changes during electrically induced and maximal voluntary contractions: electrophysiologic responses of different muscle fibre types during sustained contractions. *Exp. Neurol.* **88**: 471-483, 1984.
13. Sahlin, K., L. Edstroem, and H. Sjoeholm. Fatigue and phosphocreatine depletion during carbon dioxide-induced acidosis in rat muscle. *Am. J. Physiol.* **245**: 15-20, 1983.
14. Simonson, E., and P. Weiser. *Physiological Aspects and Physiological Correlates of Work Capacity and Fatigue.* Springfield, Il, USA: Charles C. Thomas, 1976.

15. Sjögaard, G. Exercise-induced muscle fatigue: The significance of potassium. *Acta physiol. Scand.* **140** (Suppl 593), 1990.

16. Taylor, D.J., P. Styles, M. Matthews, D.A. Arnold, D.G. Gadian, P. Bore, and G.K. Radda. Energetics of human muscle: Exercise-induced ATP Depletion. *Magn. reson. med.* **3**: 44-54, 1986.

17. Tibes, U., E. Haberkorn-Butendeich, and F. Hammersen. Effect of contraction on lymphatic, venous, and tissue electrolytes and Metabolites in rabbit skeletal muscle. *Pflügers Arch.* **368**: 195-202, 1977.

18. Westerblad, H., and D.G. Allen. The contribution of [Ca 2+]i to the slowing of relaxation in fatigued single fibres from mouse skeletal muscle. *J. Appl. Physiol.* **468**: 729-740, 1993.

19. Wong, R., N. Davies, D. Marshall, P. Allen, G. Zhu, G. Lopaschuk, and T. Montague. Metabolism of normal skeletal muscle during dynamic exercise to clinical fatigue: In vivo assessment by nuclear magnetic resonance spectroscopy. *Can. J. Cardiol.* **6** (9):391-395, 1990.

6

FACTORS CONTRIBUTING TO ENHANCED FATIGUE-RESISTANCE IN LOW-FREQUENCY STIMULATED MUSCLE

Dirk Pette

Faculty of Biology
University of Konstanz
D-78434 Konstanz, Germany

1. INTRODUCTION

The ability for sustained contractile activity of skeletal muscle is generally assumed to correlate with a high capacity of aerobic-oxidative energy metabolism. This notion was derived from the observation that muscle fibres differing in their mitochondrial enzyme activities display distinct fatigue properties (4,7,18,22). Motor units composed of fast-twitch glycolytic (FG) fibres are fast-fatiguable, whereas motor units composed of so-called fast-twitch oxidative (FOG) or slow-twitch oxidative (SO) fibres are less fatigable or resistant to fatigue, respectively. Additional evidence in support of this notion has emerged from studies on fast-twitch muscles exposed to chronic electrical stimulation (for review see (25)). A major effect of maximally forced contractile activity by chronic low-frequency stimulation (CLFS) is that stimulated muscles display pronounced increases in enzyme activities of terminal substrate oxidation (Fig. 1) and become non-fatigable (24,25).

As mirrored by the high mitochondrial contents of fatigue-resistant muscles, an overall relationship exists between endurance and aerobic-oxidative capacity. However, evidence has accumulated that additional factors contribute to the ability for sustained performance. For example, long-term electrical stimulation was shown to enhance fatigue-resistance of canine latissimus dorsi muscle; however, no increases were observed in in citrate synthase activity, a marker of aerobic-oxidative metabolism (21).

Our studies on fast-twitch tibialis anterior (TA) muscles of rat and rabbit have shown that elevations in citrate synthase activity induced by CLFS do not strictly parallel improvements in fatigue-resistance (29). Our observations indicated that CLFS-induced elevations in aerobic-oxidative capacity may not be the only prerequisite for attaining enhanced resistance to fatigue. Thus, citrate synthase activity was only slightly elevated during the first week of CLFS while fatigue-resistance was markedly enhanced. Conversely, increases in citrate synthase activity exceeded by far the increment in fatigue-resistance observed during the second week of CLFS. Finally, CLFS for periods longer than 30 days

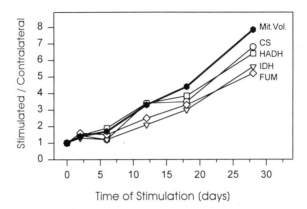

Figure 1. Time course of changes in volume density of total mitochondria and in mitochondrial enzyme activities of the superficial portion of rabbit tibialis anterior muscle in response to CLFS (10 Hz). Modified from Reichmann et al. (26).

did not lead to further improvements in fatigue-resistance, although citrate synthase activity continued to rise (Fig. 2). Taken together, these results suggested that although an overall correlation exists between aerobic-oxidative capacity and fatigue-resistance, this may not be the only factor for endurance performance. In this regard, we became interested in characterizing early adaptive responses that might explain enhanced fatigue-resistance prior to increases in mitochondrial enzyme activities and other parameters of aerobic-oxidative capacity.

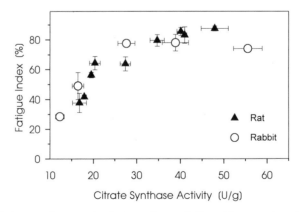

Figure 2. Relationship between increases in citrate synthase activity and increases in resistance to fatigue (expressed as fatigue index) of low-frequency (10 Hz) stimulated tibialis anterior muscles of rat and rabbit. Values represent means ± S.E.M. Modified from Simoneau et al. (29).

2. ENHANCED GLUCOSE UPTAKE AND GLUCOSE PHOSPHORYLATION

Availability of glycogen is essential for sustained performance (1). Low-frequency stimulation leads to a rapid glycogen depletion in extensor digitorum longus (EDL) and TA muscles of the rabbit (10, 20). Hence, in FG fibres free glucose becomes the main fuel of energy metabolism after the first hour of CLFS. As a consequence of enhanced glucose uptake, the concentration of free glucose rises more than two-fold in rabbit TA muscle as early as 15 min after the onset of stimulation (10). A study on rat TA muscle revealed that CLFS rapidly raises the capacity for glucose uptake by translocation of the insulin-sensitive glucose transporter GLUT4 from its intracellular pool into the sarcolemma. A pronounced increase of sarcolemmal GLUT4 was observed 30 min after the onset of CLFS and corresponded to the decrease of GLUT4 in the intracellular membranes. Following this immediate response, low-frequency stimulation enhanced transcriptional and translational activities leading to approximately two-fold elevations of the total amount of GLUT4 within a few days (10). A single-fibre study on low-frequency stimulated rabbit TA muscle showed that the total amount of GLUT4 had increased 2.5-fold in FG fibres after 1 day of CLFS, reaching levels 14-fold above control after 3 weeks of CLFS (17). As indicated by the steep rise of free glucose during CLFS, a major limiting step of glucose utilization seems to be its phosphorylation, not its uptake. This suggestion is supported by the effects of CLFS on hexokinase II, the major isoform of hexokinase in muscle tissue.

As shown for TA muscles of rabbit and rat, low-frequency stimulation has multiple effects on hexokinase. It affects the intracellular distribution of hexokinase and greatly increases its total cellular concentration. Hexokinase exists in skeletal muscle as soluble and mitochondrially bound activities (28). Its mitochondrially bound fraction rises steeply after the onset of stimulation, most likely resulting in an increase of its catalytic activity (23,30). In addition, enhanced transcriptional and translational activities result in a steep rise in total cellular hexokinase (12, 13, 26, 30). In rat TA, the level of messenger RNA encoding hexokinase II rises more than five-fold three hours after the onset of stimulation, reaches a maximum (approximately 30-fold elevation) at 12 hours and then declines to a stable level five-fold above control (Fig. 3). Hexokinase synthesis increases in a similar

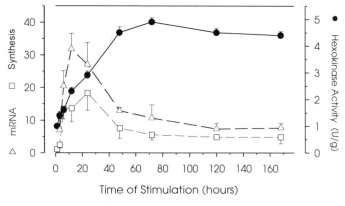

Figure 3. Time-dependent changes in mRNA content, relative synthesis rate and activity of hexokinase II in low-frequency (10 Hz) stimulated rat tibialis anterior muscle. Values (means ± S.E.M.) in the stimulated muscle have been referred to the values of the unstimulated, contralateral muscles. Data from Hofmann and Pette (13).

manner. It reaches a maximum (approximately 20-fold elevation) at 24 hours; thereafter it decays and stabilizes at a five-fold elevated level. The rise in hexokinase activity, which corresponds to an increase in hexokinase II protein (30), is significant at 12 hours and reaches a maximum after two days. Interestingly, cessation of stimulation is followed by a steep decline in hexokinase synthesis (13). These findings show that the capacity for glucose phosphorylation in skeletal muscle is rapidly adjusted to specific functional demands. It may be speculated that improved fatigue-resistance during the first days of CLFS relates to enhanced capacities of glucose uptake and phosphorylation. Obviously, these changes precede by far the rise in enzyme activities of fatty acid oxidation and terminal substrate oxidation.

3. CHANGES IN Na+,K+-ATPase CONTENT

During enhanced contractile activity, skeletal muscle loses K+. Therefore, a tight control of the active Na+-K+ transport is essential for the maintenance of optimum muscle function (6). In fact, it has been shown that low-frequency stimulation leads to decreases in [K+] and increases in [Na+] in rat fast-twitch muscle soon after the onset of stimulation (8). An efflux of K+ and its accumulation in the T-tubule impairs excitation-contraction coupling and, therefore, impairs contractile activity. These changes, as well as our observation of steep decreases in force output of rabbit TA muscle within the first minutes after the onset of low-frequency stimulation (5, 10), have drawn our attention to the effect of CLFS on Na+,K+-ATPase. In rabbit EDL, we observed a two-fold increase in the content of Na+,K+-ATPase with stimulation periods exceeding two days (9). This increase, which seems to coincide with the recovery of force output, precedes the elevation of citrate synthase activity (Fig. 4). It also suggests that the capacity of Na+,K+-ATPase in fast-twitch muscle is insufficient for maintaining ionic homeostasis during forced contractile activity. Therefore, the elevation of the sarcolemmal Na+,K+-ATPase seems to be an important adaptive response contributing to enhanced fatigue-resistance.

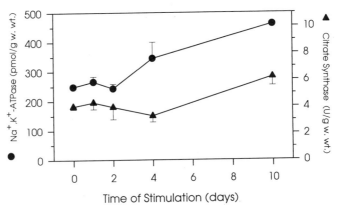

Figure 4. Time course of changes in the concentration of Na+,K+-ATPase and in citrate synthase activity of low frequency (10 Hz) stimulated extensor digitorum longus muscle of rabbit. Values are means ± S.E.M. Data from Green et al. (9).

4. INCREASES IN PERFUSION, CAPILLARIZATION, AND AEROBIC-OXIDATIVE CAPACITY

Increased blood flow and capillarization are important factors for enhanced oxygen delivery and fuel supply. Similarly, they are important for the removal of metabolic end-products such as lactate. Perfusion and capillary density have been shown to be markedly increased in low-frequency stimulated muscles (3,14,15). Initial increases in capillary density were recorded as early as four days after the onset of CLFS. An additional factor contributing to an enhanced exchange of fuels and metabolic products is a pronounced expansion of the extracellular, interstitial space around the fibres in low-frequency stimulated muscle (12). As indicated by a five to six-fold increase in extracellular albumin content during the first week (11), the expansion of the extracellular space represents an early adaptive response to CLFS. An expansion of the extracellular space would also serve to mitigate the depolarizing effects of an increase in extracellular [K+].

Increases in myoglobin content follow a much slower time course, similar to that of fatty acid binding protein (16). Maximum levels of these two proteins are reached in rat TA muscles after stimulation periods of 3-4 weeks (16) and, therefore, represent relatively late adaptive responses. They coincide with the increases in mitochondrial content and elevations in enzyme activities of fatty acid oxidation, the citric acid cycle and the respiratory chain (25-27).

5. FAST-TO-SLOW TRANSITIONS IN MYOSIN ISOFORMS

The response of the contractile apparatus to the persistently increased contractile activity induced by CLFS consists of sequential transitions in myofibrillar protein isoforms of both the thick and thin filaments (19,25). At the level of the thick filament, transitions in myosin heavy chain (MHC) isoforms follow the order MHCIIb $\rightarrow$ MHCIId $\rightarrow$ MHCIIa $\rightarrow$ MHCI. However, these transitions in MHC isoform expression occur at a time when resistance-to-fatigue has been attained. Recent studies of Bottinelli and coworkers (2) on single muscle fibres, identified according to their MHC isoform composition, revealed that specific ATPase activities and energy costs for isometric tension of these isoforms decrease in the order MHCIIb > MHCIIx(d) > MHCIIa > MHCI. It may be speculated, therefore, that the exchange of fast with less fast and, ultimately, slow MHC isoforms contributes to a more economical use of energy for contractile activity under conditions of sustained performance.

6. CONCLUSIONS

Fast-twitch, fast-fatigable muscle fibres attain enhanced fatigue-resistance when subjected to CLFS. Increases in the capacities for glucose uptake and phosphorylation are among the earliest changes observed during the first days of CLFS. Elevations in the amount of sarcolemmal Na+,K+-ATPase, counteracting the loss of K+ during forced contractile activity, also contribute to enhanced fatigue-resistance during an early phase of adaptation. Enhanced perfusion and increased capillarization represent additional factors contributing to reduced fatigability in response to sustained performance. Increases in mitochondrial volume density and enzyme activities of fatty acid oxidation and aerobic-oxidative metabolic pathways further contribute to enhanced fatigue-resistance. These alterations, however, represent late changes when compared to the early adaptive responses that occur within a time frame of hours and days.

7. REFERENCES

1. Bergström, J., L. Hermansen, E. Hultman, and B. Saltin. Diet, muscle glycogen and physical performance. *Acta Physiol. Scand.* **71**: 140-150, 1967.
2. Bottinelli, R., M. Canepari, C. Reggiani, and G. J. M. Stienen. Myofibrillar ATPase activity during isometric contraction and isomyosin composition in rat single skinned muscle fibres. *J. Physiol. (Lond.)* **481**: 663-675, 1994.
3. Brown, M. D., M. A. Cotter, O. Hudlická, and G. Vrbová. The effects of different patterns of muscle activity on capillary density, mechanical properties and structure of slow and fast rabbit muscles. *Pflügers Arch. Europ. J. Physiol.* **361**: 241-250, 1976.
4. Burke, R. E., D. N. Levine, F. E. Zajac, P. Tsairis, and W. K. Engel. Mammalian motor units: Physiological-histochemical correlation in three types in cat gastrocnemius. *Science* **174**: 709-712, 1971.
5. Cadefau, J. A., J. Parra, R. Cusso, G. Heine, and D. Pette. Responses of fatigable and fatigue-resistant fibers of rabbit muscle to low-frequency stimulation. *Pflügers Arch. Europ. J. Physiol.* **424**: 529-537, 1993.
6. Clausen, T. Regulation of active Na+-K+ transport in skeletal muscle. *Physiol. Rev.* **66**: 542-580, 1986.
7. Edström, L. and E. Kugelberg. Histochemical composition, distribution of fibres and fatiguability of single motor units. Anterior tibial muscle of the rat. *J. Neurol. Neurosurg. Psychiatry* **31**: 424-433, 1968.
8. Everts, M. E., T. Lomo, and T. Clausen. Changes in K+, Na+ and calcium contents during in vivo stimulation of rat skeletal muscle. *Acta Physiol. Scand.* **147**: 357-368, 1993.
9. Green, H. J., M. Ball-Burnett, E. R. Chin, L. Dux, and D. Pette. Time dependent increases in Na+,K+-ATPase concentration of low-frequency stimulated rabbit muscle. *FEBS Lett.* **310**: 129-131, 1992.
10. Green, H. J., S. Düsterhöft, L. Dux, and D. Pette. Metabolite patterns related to exhaustion, recovery, and transformation of chronically stimulated rabbit fast-twitch muscle. *Pflügers Arch. Europ. J. Physiol.* **420**: 359-366, 1992.
11. Heilig, A. and D. Pette. Albumin in rabbit skeletal muscle. Origin, distribution and regulation by contractile activity. *Eur. J. Biochem.* **171**: 503-508, 1988.
12. Henriksson, J., M. M.-Y. Chi, C. S. Hintz, D. A. Young, K. K. Kaiser, S. Salmons, and O. H. Lowry. Chronic stimulation of mammalian muscle: changes in enzymes of six metabolic pathways. *Am. J. Physiol.* **251**: C614-C632, 1986.
13. Hofmann, S. and D. Pette. Low-frequency stimulation of rat fast-twitch muscle enhances the expression of hexokinase II and both the translocation and expression of glucose transporter 4 (GLUT-4). *Eur. J. Biochem.* **219**: 307-315, 1994.
14. Hudlická, O., L. Dodd, E. M. Renkin, and S. D. Gray. Early changes in fiber profile and capillary density in long-term stimulated muscles. *Am. J. Physiol.* **243**: H528-H535, 1982.
15. Hudlická, O. and S. Price. The role of blood flow and. *Pflügers Arch. Europ. J. Physiol.* **417**: 67-72, 1990.
16. Kaufmann, M., J.-A. Simoneau, J. H. Veerkamp, and D. Pette. Electrostimulation-induced increases in fatty acid-binding protein and myoglobin in rat fast-twitch muscle and comparison with tissue levels in heart. *FEBS Lett.* **245**: 181-184, 1989.
17. Kong, X. M., J. Manchester, S. Salmons, and J. C. Lawrence. Glucose transporters in single skeletal muscle fibers - Relationship to hexokinase and regulation by contractile activity. *J. Biol. Chem.* **269**: 12963-12967, 1994.
18. Kugelberg, E. and B. Lindegren. Transmission and contraction fatigue of rat motor units in relation to succinate dehydrogenase activity of motor unit fibres. *J. Physiol. (Lond.)* **288**: 285-300, 1979.
19. Leeuw, T. and D. Pette. Coordinate changes in the expression of troponin subunit and myosin heavy chain isoforms during fast-to-slow transition of low-frequency stimulated rabbit muscle. *Eur. J. Biochem.* **213**: 1039-1046, 1993.
20. Maier, A. and D. Pette. The time course of glycogen depletion in single fibers of chronically stimulated rabbit fast-twitch muscle. *Pflügers Arch. Europ. J. Physiol.* **408**: 338-342, 1987.
21. Mayne, C. N., W. A. Anderson, R. L. Hammond, B. R. Eisenberg, L. W. Stephenson, and S. Salmons. Correlates of fatigue resistance in canine skeletal muscle stimulated electrically for up to one year. *Am. J. Physiol.* **261**: C259-C270, 1991.

22. Nemeth, P. M., D. Pette, and G. Vrbová. Comparison of enzyme activities among single muscle fibres within defined motor units. *J.Physiol.(Lond.)* **311**: 489-495, 1981.

23. Parra, J. and D. Pette. Effects of low-frequency stimulation on soluble and structure-bound activities of hexokinase and phosphofructokinase in rat fast-twitch muscle. *Biochim. Biophys. Acta* **1251**: *154-160*, 1995.

24. Peckham, P. H., J. T. Mortimer, and J. P. van der Meulen. Physiologic and metabolic changes in white muscle of cat following induced exercise. *Brain Res.* **50**: 424-429, 1973.

25. Pette, D. and G. Vrbová. Adaptation of mammalian skeletal muscle fibers to chronic electrical stimulation. *Rev. Physiol. Biochem. Pharmacol.* **120**: 116-202, 1992.

26. Reichmann, H., H. Hoppeler, O. Mathieu-Costello, F. von Bergen, and D. Pette. Biochemical and ultrastructural changes of skeletal muscle mitochondria after chronic electrical stimulation in rabbits. *Pflügers Arch. Europ. J. Physiol.* **404**: 1-9, 1985.

27. Reichmann, H., R. Wasl, J.-A. Simoneau, and D. Pette. Enzyme activities of fatty acid oxidation and the respiratory chain in chronically stimulated fast-twitch muscle of the rabbit. *Pflügers Arch. Europ. J. Physiol.* **418**: 572-574, 1991.

28. Rose, I. A. and J. V. B. Warms. Mitochondrial hexokinase. Release, rebinding, and location. *J. Biol. Chem.* **242**: 1635-1645, 1967.

29. Simoneau, J.-A., M. Kaufmann, and D. Pette. Asynchronous increases in oxidative capacity and resistance to fatigue of electrostimulated muscles of rat and rabbit. *J.Physiol.(Lond.)* **460**: 573-580, 1993.

30. Weber, F. E. and D. Pette. Changes in free and bound forms and total amount of hexokinase isozyme II of rat muscle in response to contractile activity. *Eur. J. Biochem.* **191**: 85-90, 1990.

7

ENERGY METABOLISM AND MUSCLE FATIGUE DURING EXERCISE

Kent Sahlin

Department of Physiology and Pharmacology
Physiology III, Karolinska Institute
Department of Sport and Health Sciences
Stockholm University Collegue of Physical Education and Sports
Lidingövägen 1, Box 5626, S-114 86 Stockholm
Sweden

1. INTRODUCTION

The physiology of exercise is a matter of transforming energy stored in the form of chemical compounds into mechanical energy. Hydrolysis of ATP is the immediate energy source but since the store of ATP is rather small it has to be replenished continuously. The breakdown of ATP to ADP and the rephosphorylation of ADP back to ATP constitutes the ATP-ADP cycle by which the energy consuming processes are coupled to the energy yielding processes (Fig. 1).

The transition from rest to exercise involves a drastic increase in energy demand. The rate of ATP utilization can increase more than 100 times and corresponds to a utilization of the whole muscle store of ATP in about 2-3 sec. Despite large fluctuations in energy demand, muscle [ATP] remains practically constant and demonstrates a remarkable precision of adjusting the rate of the ATP-generating processes to the demand. The control involves an intricate interplay between both feed-forward and feed-back mechanisms.

The energy requirement during sustained exercise is met almost exclusively through oxidative or aerobic processes which involve combustion of fuels with oxygen within mitochondria. The metabolic end-products (CO_2 and H_2O) can easily be handled by the organism. Regeneration of ATP can also be achieved through anaerobic processes: breakdown of phosphocreatine (PCr) or glycolysis. The term anaerobic denotes that these processes can regenerate ATP in the absence of oxygen. Nevertheless, the anaerobic processes will supply ATP also during aerobic conditions, albeit at a low rate compared with the oxidative processes. The anaerobic processes will, in contrast to the aerobic processes, produce waste products (e.g. H+ and inorganic phosphate). During conditions of high rates of anaerobic energy utilization or/and ischemia, waste products will accumulate in

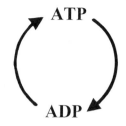

Energy yielding processes:
Anaerobic:
 -PCr breakdown
 -Glycolysis
Aerobic:
 -CHO oxidation
 - fat oxidation

Energy demanding processes:
-muscle contraction
-anabolic processes
-transport processes

Figure 1. The ATP-ADP cycle.

the working muscle and may impair both the contraction process and the energetic processes.

There are two inherent limits of the energetic processes: the maximum rate (power) and the amount of ATP (capacity) that can be produced (9). The power and the capacity vary drastically between the different energetic processes. Factors such as training status of the subject, nutritional factors and muscle mass are of additional importance. Observed peak values of power and capacity in human skeletal muscle are shown in Figure 2. The limitations of the metabolic processes will set an upper bound for the energy production and thus the intensity and the duration at which exercise can be performed. Breakdown of PCr is the energy source which can sustain the highest rate of ATP production but has, on the other hand, the lowest capacity of ATP production. PCr will thus be important for short bursts of high intensity exercise. In contrast, combustion of fat has a low power but since fat is present abundantly the capacity of fat oxidation is not considered to be limiting. Fat will consequently be the dominating fuel during prolonged exercise at low intensities and exercise duration will in this case not be limited by metabolic factors. The relative exercise intensity ($\dot{V}O_2$max) is an important determinant of to the extent to which the various energetic processes are recruited. This will influence the relationship between aerobic/anaerobic processes of energy production and carbohydrate (CHO) and fat oxidation. Other factors of importance are the availability of oxygen, availability of fuels, and hormonal changes.

2. LIMITATIONS IN POWER AND CAPACITY OF THE ENERGETIC PROCESSES

2.1. Breakdown of PCr

The maximum rate of PCr breakdown presented in Figure 2 (120 mmol ATP/kg muscle/min) was observed during short term (1.3 s) electrical stimulation of m. quadriceps femoris during isometric conditions (6). The rate of PCr breakdown during 10s cycling was lower (71 mmol ATP/kg muscle/min: Ref. 8). It is not clear whether the difference is due to the type of activity or to the difference in exercise duration. The maximum value is close to the Vmax of myosin ATPase activity measured *in vitro* (for references, see Ref. 10). It is therefore possible that power under these conditions is limited by the ATP demand rather than by the rate of ATP regeneration through the creatine kinase reaction.

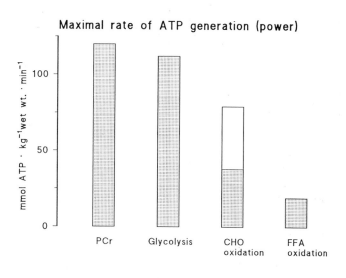

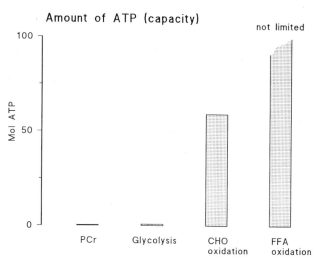

Figure 2. Power and capacity of the energy yielding processes in human skeletal muscle. Values of power are based on observed values during the following conditions: PCr breakdown, 1.3s electrical stimulation (6); glycolysis, 10s cycling (8); CHO oxidation (filled bar), calculated from O_2 extraction during two-leg cycling assuming that 72% of $\dot{V}O_2$max (4 l/min) is utilized by a working muscle mass of 20 kg; CHO oxidation (unfilled bar), one-leg knee extension (2); FFA oxidation, assumed to be 50% of that of CHO oxidation (see text). Values of capacity have been derived from muscle content of PCr, glycogen (80 mmol/kg), maximal muscle lactate accumulation and a working muscle of 20 kg. The amount of ATP that can be produced from oxidation of FFA is not limited, hence the stippled bar is incomplete.

From thermodynamic considerations, one would expect that the maximum rate of PCr breakdown decreases when the PCr content of the muscle decreases. Availability of PCr may therefore be a limiting factor of power output also before the muscle content of PCr is totally depleted.

The amount of energy that can be produced from PCr is limited by the muscle content, which is about 19 mmol/kg wet wt in human skeletal muscle. Muscle content of PCr in fast-twitch fibers is about 15% higher than in slow-twitch fibers (13). With the maximal rate of PCr breakdown depicted in Figure 2, one would expect complete depletion in about 10 s. However, the contributions to ATP resynthesis from other energy sources and decreased energy expenditure (fatigue) will prolong this time.

2.2. Glycolysis

The maximum reported power of glycolysis in human muscle is 112 mmol/kg wet wt/min and has been observed in m. quadriceps femoris after short-term (10s) cycling (8). A lower value (71-73 mmol/kg wet wt/min: Refs. 1 & 6) was observed after voluntary and electrically stimulated isometric contraction. Although most reports demonstrate that glycolysis contributes to ATP generation already at the onset of exercise, there appears to be a lag of a few seconds before the maximum rate of glycolysis is achieved. The rate of glycolysis is controlled to a large extent by the activity of two enzymes: glycogen phosphorylase (GP) and phosphofructokinase (PFK). Both GP and PFK catalyze nonequilibrium reactions and exhibit a complex and diverse mode of control. The Vmax of GP and PFK are close to the observed maximum rate of glycolysis *in vivo* (for references, see Ref. 3) and may therefore set the limit of maximum glycolytic power. The activities of PFK and GP are reduced by acidosis. The product of glycolysis (lactic acid) can therefore reduce the rate of glycolysis through feed-back inhibition. This could be regarded as a safety mechanism by which cellular damage due to excessive lactic acid accumulation is prevented. Both the power and the capacity of glycolysis (i.e. amount of lactic acid produced) may therefore be limited by product accumulation (i.e. H+). Factors such as muscle buffering capacity and transmembrane transport of H+ are likely to modulate the glycolytic response. Evidence exists that the contraction process can be directly impaired by acidosis through an effect on the contractile proteins (5). The restraint on glycolysis would, in this case, be determined by the ATP demand rather than by the activities of glycolytic enzymes. Also, glycogen depletion may theoretically limit the maximal power of glycolysis. However, the affinity of phosphorylase for glycogen is very high (low Km), and a number of studies have shown that a low pre-exercise glycogen concentration neither influences the rate of glycogenolysis nor the rate of blood lactate increase (for references, see Ref. 3).

2.3. Aerobic Processes

During two-leg exercise, muscle O_2 utilization may increase 50-fold to, for example, 140 ml/kg muscle/min which corresponds to a rate of ATP generation of 38 mmol/kg muscle/min (Fig. 2). It is generally agreed that the major determinant of whole body maximal aerobic power ($\dot{V}O_2$max) is cardiac output, which sets an upper limit of O_2 delivery. It has been estimated that exercise with a muscle mass of 10 kg is sufficient to tax the maximal cardiac output in a sedentary subject (2). The maximal aerobic power of the muscle tissue is therefore not utilized during two-leg exercise, where the working muscle mass is about 20 kg or more. However, during exercise with small muscle groups, the rate of aerobic energy production may be limited by peripheral factors (e.g. mitochondrial density or O_2 diffusion). Thus, during one-leg knee extension, muscle O_2 utilization can increase to 300 ml/min/kg muscle (2), corresponding to a power of 79 mmol ATP/kg muscle/min which is 70% of the maximum power of glycolysis.

The amount of energy that can be produced by CHO oxidation is limited by the CHO stores. Cycling or running at intensities between 60-80 % of $\dot{V}O_2$max can normally proceed for 1-2 h before exhaustion is reached, and coincide with, depletion of the glycogen stores in the working muscle.

Several lines of evidence suggest that oxidation of fatty acids cannot procede at the same rate as for CHO oxidation. First, it has been shown that tissue homogenates and isolated mitochondria cannot oxidize palmitate at the same rate as pyruvate (7, 14). Furthermore, it is known that ultradistance running, which results in a depletion of the body storage of carbohydrates and a switch to fat oxidation (4), causes a decline in the maximum attainable power output to the equivalent of about 50% of $\dot{V}O_2$max. The reason for the lower maximum rate of aerobic ATP formation from fat is under debate. Combustion of fatty acids may be limited by factors such as: the rate of fuel transport from the fat depots to the site(s) of utilization, the enzyme activities involved in the degradation of fatty acids, and the higher O_2 requirements for ATP formation compared with CHO oxidation. The maximal rate of fat oxidation may also be limited by the rate of NADH production in the tricarboxylic acid cycle (TCA), since the level of tricarboxylic acid cycle intermediates (TCAI) are dependent on pyruvate concentration and consequently on the availability of CHO (11).

2.4. Influence of Muscle Mass

As discussed above, the power of anaerobic energy production appears to be limited by the activities of key enzymes in glycolysis, and also creatine kinase or/and myosin ATPase activities. As the total enzyme activities are proportional to muscle mass, an increased working muscle mass will therefore result in a proportional increase in the maximum rate of anaerobic energy production. In contrast, the rate of aerobic energy production is limited to a major extent by cardiac output. An increase in muscle mass will therefore not increase aerobic power. The capacities of both aerobic and anaerobic energy production are limited by intrinsic muscular factors such as the amount of glycogen, the amount of PCr, and the volume available within which to distribute interfering metabolic end-products. During both aerobic and anaerobic conditions, a large working muscle mass will therefore be of advantage for the amount of energy and thus the amount of work that can be produced. However, an increased muscle mass may not necessarily be of advantage for performance since energy expenditure may increase because of the increase in body weight.

From Figure 2, it is evident that the main limitation to energy release through the aerobic processes is the low rate, whereas anaerobic processes are limited by the amount of ATP that can be produced. This is parallelled by expression of aerobic training status (i.e. $\dot{V}O_2$max) as ml oxygen consumed per min and per kg body weight (i.e. a measure of the rate of ATP production) and anaerobic training status (maximal oxygen deficit) as ml O2 per kg body weight (i.e. a measure of the amount of ATP that can be produced).

3. MUSCLE ENERGETICS AND MUSCLE FATIGUE

The hypothesis that muscle fatigue is caused by failure of the energetic processes to generate ATP at a sufficient rate is classic. The evidence for this hypothesis is that interventions which increase the power (i.e. aerobic training) or capacity (i.e. CHO loading, creatine supplementation, glucose supplementation) of the energetic processes result in increased performance and delayed onset of fatigue. Similarly, factors that impair the energetic processes (i.e. depletion of muscle glycogen, intracellular acidosis, hypoxic conditions, reduced muscle blood flow) have a negative influence on performance. The evidence is however circumstantial and a direct cause-and-effect relationship remains to be established. It has been argued that since muscle [ATP] remains practically unchanged during exhaustive exercise, energetic causes of fatigue are unlikely. This line of argument may, however, be too simplistic since temporal and spatial gradients of adenine nucleotides may exist in the contracting muscle. Furthermore, the fatigue mechanism may be related to increased levels of the products of ATP hydrolysis (i.e. ADP, AMP or Pi) rather than to decreases in [ATP] *per se*. A small decrease in [ATP] will cause relatively large increases in [ADP] and [AMP], because of their much lower concentrations. Muscle fatigue is generally associated with increased catabolism of adenine nucleotides and signifies a condition of energetic stress (10). This lends further support for the hypothesis that muscle fatigue under many conditions is caused by energetic deficiency.

4. CONCLUSIONS

The present paper has focused on the limitations of aerobic and anaerobic energetic processes and how these may be related to muscle fatigue. Metabolic factors are likely to play an important role in physical performance *in vivo* but there is no doubt that conditions exist where fatigue cannot simply be explained by metabolic changes. Considering the diversity and complexity of exercise, this is perhaps not unexpected.

5. ACKNOWLEDGMENTS

Financial support from the Swedish Medical Research Council (grant 8671) and the Swedish Research Council of Sports Medicine is gratefully acknowledged.

6. REFERENCES

1. Ahlborg, B., J. Bergström, L.-G. Ekelund, G. Guarnieri, R.C. Harris, E. Hultman, and L.-O. Nordesjö. Muscle metabolism during isometric exercise performed at constant force. *J. Appl. Physiol.* **33**: 224-228, 1972.
2. Andersen, P., R.P. Adams, G. Sjögaard, A. Thorboe, and B. Saltin. Dynamic knee extension as model for study of isolated exercising muscle in humans. *J. Appl. Physiol.* **59**: 1647-1653, 1985.
3. Connett, R.J., and Sahlin K. Control of glycolysis and glycogen metabolism. In: *Handbook of Physiology: Integration of Motor, Circulatory, Respiratory and Metabolic Control during Exercise*, edited by L.B. Rowell and J.T. Shepherd. Bethesda, MD, USA: The American PhysiologicalSociety, In press.
4. Davies, C.T.M., and M.W. Thompson. Aerobic performance of female marathon and male ultramarathon athletes. *Eur. J. Appl. Physiol.* **41**: 233-245, 1979.
5. Fitts, R.H. Cellular mechanisms of muscular fatigue. *Physiol. Rev.* **74**:49-94, 1994.
6. Hultman, E., and H. Sjöholm. Substrate availability. *Biochem. Exer.* **13**:63-75, 1983.
7. Ivy, J.L., R.T. Withers, P.J. Van Handel, D.H. Elgers, and D.L. Costill. Muscle respiratory capacity and fiber type as determinants of lactate threshold. *J. Appl.Physiol.* **48**: 523-527, 1980.
8. Jones, N.K., N. McCartney, T. Graham, L.L. Spriet, J.M. Kowalchuk, G.J.F. Heigenhauser, and J.M. Sutton. Muscle performance and metabolism in maximal isokinetic cycling at slow and fast speeds. *J. Appl. Physiol.* **59**: 132-136, 1985.
9. McGilvery, R.W. The use of fuels for muscular work. In: *Metabolic Adaptation to Prolonged Physical Exercise*, edited by H. Howald and J.R. Poortmans. BAsel, Switzerland: Birkhäuser, 1973, pp. 12-30.

10. Sahlin, K., and S. Broberg. Adenine nucleotide depletion in human muscle during exercise: causality and significance of AMP deamination. *Int. J. Sports Med.* **II**:S62-S67, 1990.

11. Sahlin, K., A. Katz, and S. Broberg. Tricarboxylic acid cycle intermediates in human muscle during prolonged exercise. *Am. J. Physiol.* **259**: C834-C841, 1990.

12. Saltin B., and Gollnick P.D. Skeletal muscle adaptability: significance for metabolism and performance. In: *Handbook of Physiology. Skeletal Muscle*, edited by L.D. Peachey, R.H. Adrian and S.R. Geiger. Bethesda, MD, USA: American Physiological Society, Bethesda, 1983, pp. 555-631.

13. Söderlund, K., and E. Hultman. ATP and phosphocreatine changes in single human muscle fibers after intense electrical stimulation. *Am. J. Physiol.* **261**: E737-E741, 1991.

14. Wibom, R., and E. Hultman. ATP production rate in mitochondria isolated from microsamples of human muscle. *Am. J. Physiol.* **259**: E204-E209, 1990.

8

DIFFERENCES IN CONCENTRICALLY AND ECCENTRICALLY DETERMINED LOCAL MUSCLE ENDURANCE OF THE SHOULDER MUSCULATURE WITH DIFFERENT TRAINING STATUS

F. Mayer, T. Horstmann, L. Yin, A. Niess, and H.-H. Dickhuth

Medical Clinic and Policlinic
Department of Sports Medicine
Hölderlinstr.11, D-72074 Tübingen
Germany

1. INTRODUCTION

Past investigations have shown that greater force occurs under eccentric stress of a muscle group than in isometric and concentric modes of exercise (1,2,5). Additional passive elastic forces are seen as the cause (2,5). Most of the available studies are concerned with large muscle groups such as knee extensors and flexors (3,9). Isokinetic strength measurements of smaller muscle groups like the shoulder musculature show only small differences in eccentric strength development compared to isometric exercises in untrained male and female individuals (6,7). It is also known that higher maximum strength capacity is evolved during eccentric stress of the shoulder musculature in sports with mainly shoulder stress (7). The decrease in maximum torque with increasing movement velocity under concentric exercise conditions can be demonstrated (2,4,6).

Determinations of local muscle endurance of the upper thigh musculature showed less fatigue under eccentric exercise compared to concentric exercise forms (6). A lower metabolic stress during eccentric exercise due to additional passive elastic forces is seen as responsible for this (3; unpublished observations). It is unclear whether similar differences exist between these various types of work with respect to local muscle endurance during shoulder stress and if there are any diffferences in dependence on the different training status.

2. METHODS

We examined local muscle endurance of the shoulder musculature in a total of 117 subjects of various training levels and different athletic exercise types. The subjects were divided into 5 groups (Table 1).

On the one hand, differentiation was made by the different movement possibilities (flexion *flex*, extension *ext*, abduction *abd*, adduction *add*, external rotation *ero*, internal rotation *iro*) and on the other hand by eccentric or concentric working modes. The dominant and non-dominant shoulders were measured in all cases. Measurements of local muscle endurance were made concentrically at a movement velocity of 180o/s and eccentrically at 60o/s, so that the achieved work was comparable in both forms of stress during the complete measurement (unpublished observations). The LIDO-ACTIVE equipment manufactured by Loredan was used for isokinetic strength measurements.

Measurements were made in a 1-minute test for all shoulder movements, with as many repeats as possible being made in the defined time under maximum arbitrary exertion. Measurements of flexion/extension and external rotation/internal rotation were made supine, and measurement of abduction/adduction with the subject on his/her side. The final measurement value for local endurance performance capacity of the shoulder musculature was a quotient calculated from the arithmetic mean of the maximum torque of the

Table 1. Anthropometric data of the subjects

mean (± SD)	[n]	age [years]	height [cm]	weight [kg]
tennis players (male)	25	21.2 ±3.5	180.2 ±3.5	71.7 ±6.2
gymnasts (male)	16	23.5 ±0.8	176.1 ±6.1	70.2 ±1.6
volleyball players (female)	14	23.0 ±3.5	177.5 ±6.7	67.5 ±7.7
untrained individuals (male)	32	25.7 ±3.8	181.2 ±5.4	72.1 ±2.6
untrained individuals (female)	30	24.6 ±4.9	173.4 ±5.1	61.2 ±6.8

first 5 repeats and the arithmetic mean of the last 5 repeats (Fig. 1). Differences in means were checked using multi-factor variance analysis with repeated measurements ($\alpha = 0.05$).

3. RESULTS

The results show a higher local endurance performance capacity under eccentric exercise compared to concentric forms of exercise ($p<0.05$). The differences apply to all groups, independent of athletic exertion and training status (Fig. 2).

A tendency for lower endurance performance capacity of high-performance gymnasts compared to the other groups was seen. No significant differences could be observed between the various groups ($p>0.05$). Determination of maximum torque showed significant differences between the various athletic groups compared to untrained subjects ($p<0.05$, Fig. 3).

No difference in the extent of local muscle endurance was evident when comparing the various movements in the shoulder joint (Fig. 4).

The different training status also showed no differences between the individual shoulder movements. The difference between eccentrically and concentrically determined local endurance performance capacity appeared more pronounced in the male than in the female subjects ($p>0,05$). Comparison of the dominant and non-dominant shoulders showed no difference in these results ($p>0.05$). This was found to apply independent of the shoulder movement in question for all subjects of both sexes.

4. DISCUSSION

It is concluded from the results of this investigation that lower local muscle fatigue of the shoulder musculature can be expected under conditions of eccentric exercise. This endurance performance capacity, which is higher than in concentric exercise, is independent of the training status of the shoulder musculature and of the type of sport involved. A lower metabolic stress of the working musculature seems to be responsible for

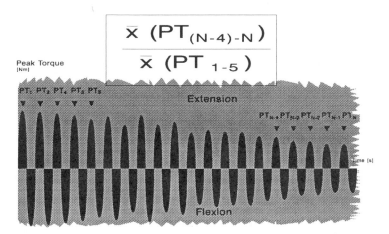

Figure 1. Calculation of an endurance-quotient as a measure of the local muscle endurance of the shoulder musculature (e.g. flexion/extension movement).

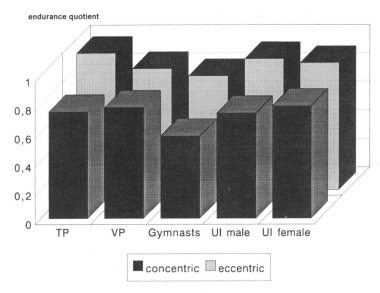

Figure 2. Local muscle endurance of the extension movement of the dominant shoulder joint in different types of sport in comparison to untrained male and female individuals.

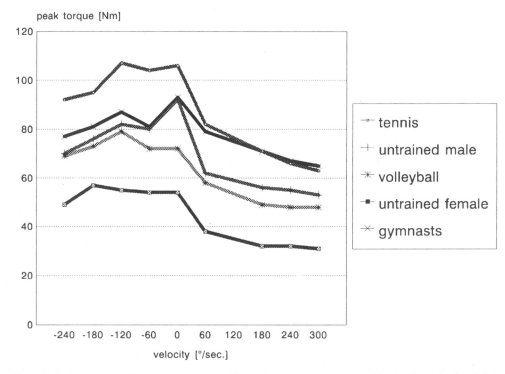

Figure 3. Peak torque in different movement velocities of the extension movement of the dominant shoulder joint in different types of sport in comparison to untrained individuals.

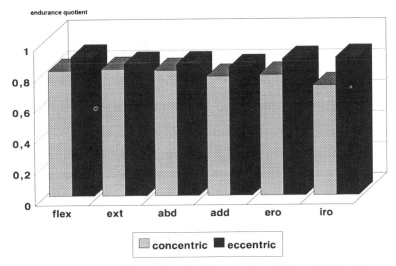

Figure 4. Local muscle endurance of the dominant shoulder musculature in female high-performance volleyball players.

the lower eccentric fatigue (3). Based on series elastic components of the musculature, additional passive elastic forces arise during eccentric stress, which lead overall to lower metabolic stress and lower electrical activation compared to concentric stress over the same exercise time (1,2,3,5). The fatigue in the working muscle group is thus lower.

In considering the method used to determine local endurance performance capacity in this investigation, it is apparent that the two measurements were performed at different movement velocities and thus with a varying number of repeats during the same exercise time. Our own as yet unpublished studies showed that the work performed with this format during the 1-minute stress phase is practically the same for both exercise forms, so that comparability of the two measurements can be assumed, even though the number of repeats is different.

The slight differences between the individual groups show that local muscle endurance apparently plays a minor role in evaluating performance capacity of the shoulder musculature. Past investigations concerning isokinetic strength measurements have shown that parameters of local muscle endurance give further information in addition to peak torque estimates only in special situations (4,6).

Unlike determination of maximum torque, no differences could be determined even in the types of sports examined in which the shoulder is exercised. The tendency for lower endurance values among high-performance gymnasts in concentric measurements may be attributable to the isometric stresses which are more likely to occur in this type of sport and to an increased reliance on coordinative elements. However, no certain conclusion can be drawn on this from the results of the present study.

In summary, it can be concluded that, independent of the training status and type of sports, local muscle endurance of the shoulder musculature is of less importance compared to maximum strength development of the shoulder musculature. As observed in earlier studies on large muscle groups, lower fatigue levels can be observed in the shoulder musculature under eccentric exercise than under conditions of concentric exercise.

5. REFERENCES

1. Asmussen, E. Positive and negative muscular work. *Acta Physiol. Scand.* **28**: 364-382, 1952.
2. Gülch, R.W., P. Fuchs, A. Geist, M. Eisold, and H. C. Heitkamp. Eccentric and posteccentric contractile behaviour of sceletal muscle: a comparative study in frog single fibers and in humans. *Eur. J. Appl. Physiol.* **63**: 323-329, 1991.
3. Horstmann, T., F. Mayer, J. Fischer, J. Maschmann, K. Röcker, and H. H. Dickhuth. The cardiocirculatory reaction to isokinetic exercises in dependance on the form of exercise and age. *Int. J. Sports Med.* **15** (Suppl): 50-55, 1994.
4. Kannus, P. Isokinetic evaluation of muscular performance: Implications for muscle testing and rehabilitation. *Int. J. Sports Med.* **15** (Suppl): 11-18, 1994.
5. Komi, P.V., and E. R. Buskirk. Effect of eccentric and concentric muscle conditioning on tension and electrical activity of human muscle. *Ergonomics* **15**: 417-434, 1972.
6. Mayer, F., T. Horstmann, W. Küsswetter, and H. H. Dickhuth. Isokinetics - a review. *Dt. Z. Sportmed.* **45**: 7/8:272-287, 1994.
7. Mayer, F., T. Horstmann, H. Kachel, H. H. Dickhuth, and W. Küsswetter. Isokinetic maximum strength development in the shoulder joint in high-performance tennis players compared to normal persons. *Int. J. Sports Med.* **15**: 378, 1994.
8. Mayer, F., T. Horstmann, K. Röcker, H. C. Heitkamp, and H. H. Dickhuth. Normal values of isokinetic maximum strength, the strength/velocity curve, and the angle at peak torque of all degrees of freedom in the shoulder. *Int. J. Sports Med.* **15** (Suppl): 19-25, 1994.
9. Prietto, C.A., and V. J. Caiozzo. The in vivo force-velocity relationship of the knee flexors and extensors. *Am. J. Sports Med.* **17**: 607-611, 1989.

9

INACTIVITY ALTERS STRUCTURAL AND FUNCTIONAL PROPERTIES OF THE NEUROMUSCULAR JUNCTION

Y. S. Prakash,[1] Hirofumi Miyata,[2] Wen-Zhi Zhan,[1] and Gary C. Sieck[1]

[1] Departments of Anesthesiology, and Physiology and Biophysics
Mayo Clinic, Rochester, Minnesota
[2] Department of Exercise Physiology
Yamaguchi University, Japan

1. INTRODUCTION

The diaphragm (DIAm) is the most important inspiratory muscle involved in mammalian ventilation, and accordingly it has a high duty cycle (fraction of time active versus inactive; ~40% for the rat DIAm (7, 8)), compared to limb muscles (~2% for the soleus muscle to ~15% for the extensor digitorum longus muscle (5)). This unique activation pattern of the DIAm might make it particularly susceptible to inactivity.

As in other skeletal muscles, motor units comprise the essential elements of neuromotor control of the DIAm (14). Motor units in the DIAm, which are composed of different fiber types, can be classified into four different types based on differences in contractile and fatigue properties (3): slow (S) units comprising type I fibers; fast fatigue-resistant (FR) units comprising type IIa fibers; fast fatigue-intermediate (FInt) units comprising type IIx fibers; and fast fatigable (FF) units comprising type IIx/IIb fibers (15, 17).

The neuromuscular junction (NMJ) is the sole communicative link between a motoneuron and the fibers it innervates. A key feature of neuromotor control is the functional match between motoneuron and muscle fiber properties that is mediated by neurotrophic and myotrophic influences. Therefore, it is likely that the structural and functional properties of the NMJ also match those of motor units, and are influenced by alterations in neurotrophic and myotrophic interactions.

In the present study, morphological adaptations of NMJs on type-identified muscle fibers of the rat DIAm were examined after two weeks of inactivity induced by: (a) hemisection of the spinal cord at C_2, thereby removing descending inspiratory drive to the phrenic motoneuron pool (spinal isolation; SI); or (b) tetrodotoxin (TTX) blockade of axonal propagation. The effect of inactivity on neuromuscular transmission was also evaluated.

2. METHODS

Male Sprague-Dawley rats (250-300 g, 12 weeks of age) were divided into four groups: spinal cord hemisected (spinal isolation, SI; n=6), sham spinal hemisected (Sham SI; n=6), tetrodotoxin-blocked (TTX; n=6), and sham tetrodotoxin block (Sham TTX; n=6). Animals were anesthetized by i.m. administration of ketamine (60 mg/kg) and xylazine (2.5 mg/kg). Surgical procedures were carried out in an aseptic environment. Detailed descriptions of the surgical procedures have been recently published (8).

2.1. Spinal Cord Hemisection

The cervical spinal cord was exposed at C2 and the right half of the cord was sectioned from the dorsal root to the ventral root such that only the ventral and lateral funiculi were cut, while the dorsal funiculus was preserved. Using this surgical procedure phrenic motoneurons were inactivated without significant limb motor deficits. During surgery, DIAm electromyographic (EMG) and phrenic nerve activities were recorded prior to, and immediately following, spinal cord section. The continued efficacy of SI was confirmed by the absence of inspiratory-related EMG activity at various times during the two week survival period. The absence of both EMG and phrenic nerve activities was also confirmed immediately before the terminal experiment. The Sham-SI animals were exposed to all surgical procedures except sectioning of the spinal cord.

2.2. TTX Blockade of Action Potential Propagation

The right phrenic nerve was exposed in the lower neck and a silastic cuff was loosely placed around the nerve. A short length of polyethylene tubing connected the nerve cuff to a miniosmotic pump that continously perfused a 0.0125% solution of TTX in saline (pH 7.4) onto the nerve at 0.5 μl/hr over a two week period. In the Sham TTX animals, the phrenic nerve was superperfused with saline. The efficacy of TTX in blocking axonal action potential propagation was verified by recording right phrenic nerve activity proximal and distal to the cuff, before and after TTX superfusion. The TTX block was also evaluated by stimulating the phrenic nerve proximal to the block and verifying the absence of evoked nerve action potentials distal to the block. Inspiratory-related EMG activity of the right DIAm was absent while activity of the left side continued. A number of verification procedures provided assurance that there was minimal, if any, nerve damage resulting from the TTX cuff (8).

2.3. Tissue Extraction

At the end of the two week period, animals were re-anesthetized, and the right midcostal region of the DIAm muscle was quickly excised and cut into 1-1.5 cm wide strips (4 samples per animal). Three samples were used for immunocytochemistry, and one for assessment of neuromuscular transmission.

2.4. Three-Color Immunocytochemistry and Confocal Imaging

The immunocytochemical technique used in the present study has also been recently described, and is only summarized here (10, 11). The resting (excised) muscle length for each strip was measured, and the strips were stretched and pinned to 1.5 times this length (an approximation to the optimal length for force production (9)). First, motor endplates were labelled by incubating tissue samples in 5-10 µg/ml tetramethylrhodamine α-bungarotoxin in phosphate buffer. The tissue samples were then thoroughly washed in buffer and fixed in 2% paraformaldehyde. Following fixation, the samples were blocked in 4% normal donkey serum and then incubated in a primary solution of 0.6 µg/ml rabbit anti-protein gene product (to label axons and nerve terminals (18)) and 0.25 µg/ml mouse antibody to fast myosin heavy chain (MHC) isoforms. The samples were then washed and incubated in a secondary solution containing 15 µg/ml DTAF-conjugated donkey anti-rabbit IgG and 15 µg/ml Cy5-conjugated donkey anti-mouse IgG. The triple-labelled tissue samples were finally washed, blotted dry, mounted, and coverslipped with immersion oil for immediate confocal imaging.

A Bio-Rad MRC500 confocal system equipped with an Ar-Kr laser (emitting at 488, 568 and 647 nm), and a 40X objective lens, was used to obtain optical sections of the triple-labelled muscle fibers. Based on previously published validation techniques for confocal imaging, the step size for optical sectioning was set at 0.8 µm (12, 13). Nerve terminals, endplates and muscle fibers were sequentially imaged in three-dimensions (3D) while ensuring image registration between different wavelengths (10). Two-dimensional (2D) projections of nerve terminals and endplates were obtained by superposing optical sections. Muscle fibers were classified as type I or type II based on their immunoreactivity patterns for anti-fast MHC, and the clear presence of both labelled sarcomeres and unlabelled myonuclei in immunopositive fibers.

The digitized images were analyzed using a comprehensive image manipulation and analysis software package (ANALYZE). Planar areas of nerve terminals and endplates were measured from the 2D projection images, and normalized to muscle fiber diameter. The extent of overlap between nerve terminal and motor endplate was estimated by digitally subtracting a binary image of the nerve terminal from that of the endplate (10). The surface areas of nerve terminals and motor endplates, and endplate gutter depth, were measured from three-dimensional (3D) reconstructions obtained using a voxel-gradient shading algorithm. The pattern of arborization of nerve terminals and motor endplates was expressed as the total number of branches and total branch length.

2.6. Assessment of Neuromuscular Transmission Failure

A DIAm strip, together with the phrenic nerve, was mounted vertically in a glass chamber containing Ringers' solution, aerated with 95% O_2-5% CO_2, and maintained at a pH of 7.4 and a temperature of 26°C. The central tendon of the muscle was attached to a calibrated force transducer, and the other end of the muscle to a micromanipulator. Muscle length was adjusted to optimal fiber length. The phrenic nerve was repetitively stimulated at 40 Hz in 330 ms duration trains repeated each s using a suction electrode at supramaximal intensity. At 15 s intervals, the muscle was directly stimulated using field stimulation at supramaximal intensity. The extent of neuromuscular transmission failure (NF) during repetitive stimulation was estimated by the difference in forces evoked by nerve versus direct muscle stimulation, using the following formula:

$$NF = (NS-MS)/1-MS$$

where NS is the decrement in force during nerve stimulation and MS is the decrement in force during muscle stimulation.

3. RESULTS

3.1. Morphometric Analysis

DIAm fiber diameters in Sham SI, Sham TTX and SI animals were comparable (Table 1). In contrast, with TTX, there was significant hypertrophy of type I fibers, while type II fibers atrophied ($P<0.05$).

In all animals, diameters of phrenic axons innervating type I fibers were significantly smaller than those innervating type II fibers ($P<0.05$; Table 1). Axon diameters were not affected by either SI or TTX. However, compared to corresponding sham controls, there was a significantly greater number of axon collaterals in both SI and TTX animals, with the number being greatest in TTX animals ($18.3\pm0.9\%$ of axons in SI and $27\pm1.2\%$ of axons in TTX; $P<0.05$).

The planar (2D) areas of both nerve terminals and motor endplates (collectively referred to as NMJs from here forward) were not different between sham groups and did not differ between type I and II fibers (Table 1). In the SI animals, the planar area of NMJs on type II fibers increased by ~50% compared to sham SI ($P<0.05$), while the planar areas of NMJs on type I fibers were unchanged. In contrast, in the TTX animals, the planar area of type II fiber NMJs was only ~10% larger than corresponding sham controls, while the planar area of type I fiber NMJs was unchanged.

When normalized for fiber diameter, in both sham groups, the planar areas of NMJs innervating type I fibers were ~50% larger than those innervating type II fibers ($P<0.05$; Table 1). In SI animals, the normalized planar areas of type II fiber NMJs were larger than that for type I fiber NMJs ($P<0.05$). In contrast, the normalized planar area of type I fiber

Table 1. Effect of muscle inactivity on the structure of diaphragm neuromuscular junctions

	Fiber type	Control	SI	TTX
Muscle Fiber Diameter (μm)	I	27.7±1.3	30.6±1.5	39.5±2.1*
	II	51.8±2.4†	51.6±2.7†	34.1±4.6*
Axon Diameter (μm)	I	1.9±0.4	1.9±0.3	2.0±0.4
	II	3.4±0.7†	3.6±0.6†	3.3±0.6†
2D Nerve Terminal Area (μm^2)	I	375±20	346±32	351±24
	II	349±33	515±42*†	385±36
Normalized Terminal Area (μm)	I	14.3±0.8	11.1±0.7	11.9±0.8
	II	9.8±1.1†	15.0±0.9*†	13.5±1.1*
2D Endplate Area (μm2)	I	380±23	369±34	363±21
	II	431±36	569±48*†	446±36
Normalized Endplate Area (μm)	I	14.6±0.9	12.1±0.7	12.3±1.0
	II	11.8±1.1†	16.2±1.1*†	15.6±1.2*
Endplate Surface Area (μm3)	I	4,520±475	4,271±243	4,291±381
	II	2,654±474†	3,777±234*†	2,905±461†
Number of Terminal Branches	I	6±1	6±1	8±1*
	II	15±4†	22±2*†	24±2*†
Total Branch Length (μm)	I	100±8	117±14	97±10
	II	120±7†	187±21*†	129±11†
% Overlap	I	95.5±1.7	93.9±2.0	91.8±1.9
	II	81.1±2.1†	90.9±1.9*	86.3±2.0*†

Values are means±SE. Control data are pooled from Sham SI and Sham TTX animals (no significant difference between the two groups; $P<0.01$). * indicates significant difference from control ($P<0.05$). † indicates significant difference between type I and II ($P<0.05$).

NMJs in the TTX animals did not change, while type II fiber NMJ area increased significantly and was comparable to that of the SI group (P<0.05). Thus, the greater extent of type II fiber atrophy in the TTX animals amplified the changes in NMJ size.

The structure of NMJs was far more elaborate on type II fibers, compared to type I fibers, in all experimental groups (Fig. 1). In sham animals, the total number of terminal branches was ~2.5-fold greater for type II fiber NMJs compared to type I fibers (P<0.05; Table 1). Following SI, the number of branches of type II fiber NMJs increased significantly (P<0.05), while the number of branches of type I fiber NMJs remained the same. In contrast, the number of terminal branches of NMJs on both type I and II fibers increased in the TTX group (P<0.05), compared to the corresponding sham group.

In all groups, the 3D surface area of motor endplates on type I fibers was significantly greater than those on type II fibers (P<0.05; Table 1). After SI, the surface areas of type II fiber endplates increased by ~45% compared to corresponding sham controls (P<0.05), while the surface area of endplates on type I fibers was unchanged. In contrast, surface areas of both type I and II fiber endplates in the TTX group were not different from the corresponding sham group. In both sham groups, the primary gutter depths of motor endplates on type I fibers were significantly greater than those of type II fiber endplates (P<0.05). After SI or TTX, there was no significant change in primary gutter depth of motor endplates on either type I or II muscle fibers.

In sham animals, there was a significant phenotypic difference in the extent of overlap between nerve terminals and motor endplates, with type I fiber NMJs displaying ~95% overlap, while type II fiber NMJs showed only ~80% overlap (P<0.05; Table 1). After SI, the extent of overlap of pre- and postsynaptic elements of type II fiber NMJs increased significantly to ~91% (P<0.05), while type I fiber NMJs continued to display nearly complete overlap. As a result, there was no difference in the extent of overlap of nerve terminal and motor endplate between type I and II fibers after SI. The extent of overlap in type II fiber NMJs in the TTX group was also greater than corresponding sham controls (P<0.05), but significantly less than that observed in type II fiber NMJs of the SI group.

3.2. Neuromuscular Transmission Failure

There was no difference in NF between sham control groups and therefore these results were combined (Fig. 2). In the SI animals, the extent of NF was significantly reduced compared to sham controls. In contrast, in TTX animals the extent of NF was markedly increased.

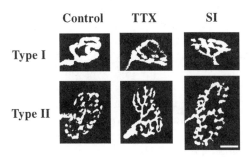

Figure 1. Fiber-type differences in the structure of DIAm nerve terminals of control, SI and TTX animals. Type I fibers typically have NMJs which are smaller and less elaborate than those on type II fibers. Note the selective enlargement of type II fiber NMJs in the SI animals. Scale bar is 10 μm.

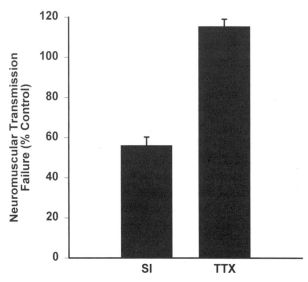

Figure 2. Effect of DIAm muscle inactivity on the extent of NF. In Sham SI and Sham TTX controls, the extent of NF was 73.2±2.0% and 71.3±3.1% of total fatigue. Note the significant decrease in NF with SI, and the significant increase with TTX (P<0.05).

4. DISCUSSION

The present study has demonstrated that two weeks of DIAm muscle inactivity induced by spinal cord hemisection (SI) resulted in significant expansion of the NMJ on type II muscle fibers and an improvement in neuromuscular transmission. In contrast, two weeks of DIAm inactivity induced by TTX blockade of axonal action potential propagation caused little expansion of the NMJ but a marked worsening of neuromuscular transmission. These two models of DIAm inactivity differ in their effects on phrenic motoneuron activity. Following spinal cord hemisection, descending inspiratory drive to phrenic motoneurons is removed, causing the motoneurons to become inactive. Thus, in the SI animals there was a match between phrenic motoneuron and DIAm inactivity. In contrast, in the TTX animals, phrenic motoneuron activity actually increases (8). Therefore, in the TTX animals there was a mismatch between phrenic motoneuron activity and DIAm inactivity. It is likely that the production of neurotrophic factors is affected by the match between phrenic motoneuron and DIAm activity. However, whether neurotrophic factors are involved and, if so, which one(s) remains unresolved.

In the present study, we limited our analysis to a comparison of NMJs on type I and II fibers. The results indicated phenotypic differences in NMJ adaptations to SI-induced DIAm inactivity, with only type II fiber NMJs expanding and displaying increased complexity. In a recent study, we extended our immunocytochemical labelling of MHC isoform expression to include identification of type II fiber subgroups, i.e., IIa, IIx and IIb fibers (10, 11). We found that both nerve terminals and endplates on type I and IIa fibers were smaller and less complex than NMJs on type IIx and IIb fibers. Therefore, type IIa fiber NMJs resemble those on type I fibers, while NMJs on type IIx and IIb fibers are similar. It is likely that the morphological adaptations of type II fiber NMJs following SI were more pronounced on type IIx and IIb fibers.

The phenotypic differences in DIAm NMJs may reflect differences in motor unit activation histories. Based on the forces generated by different DIAm motor unit types, it is likely that most ventilatory behaviors of the DIAm can be achieved by the recruitment of only fatigue resistant (S and FR) units comprising type I and IIa fibers (16). Thus, in the normal DIAm, type I and IIa fibers are frequently activated, while type IIx and IIb fibers are rarely recruited. The similarity in junctional architecture between type I and IIa fibers, on one hand, and type IIx and IIb fibers on the other, may reflect similarities in activation histories. However, this possible relationship between NMJ morphology and muscle fiber activation history appears paradoxical when considering that NMJ adaptations to inactivity were observed only in type II fibers. If type IIx and IIb fibers are relatively inactive to begin with, it would seem more reasonable that they would show less adaptation to imposed inactivity, while the more active type I and IIa fibers should be more susceptible. Therefore, motoneuron and muscle inactivity *per se* does not appear to be the driving force for NMJ adaptations.

The observation of nerve terminal sprouting at type II fiber NMJs in the present study, represented by the increased number of branches, appears to be in general agreement with previous reports where phenotypic differences in NMJ adaptations to muscle inactivity have also been observed (for a review, see ref. (1)). Accordingly, it has been suggested that inactive muscle fibers release a growth factor that induces NMJ adaptations (2, 4). In the present study, we found increased terminal sprouting following inactivity induced by both SI and TTX. However, unlike other morphological adaptations of pre- and postsynaptic elements of the NMJ, terminal sprouting was considerably more prevalent in the TTX group compared to the SI group, even though the inactivity induced in these two models was the same. It is possible that the differential expression of some neurotrophic substance, most likely to be dependent on the level of motoneuron activity (i.e., expressed more abundantly in phrenic motoneurons of SI animals), is involved in inhibiting the terminal sprouting induced by DIAm inactivity. The differences in NMJ adaptations between SI and TTX models would suggest that the balance or match between motoneuron and muscle fiber activities, rather than muscle fiber or motoneuron inactivity *per se*, determines how NMJs adapt to muscle inactivity.

Phenotypic differences in nerve terminal and endplate morphology may partially explain the progressively increasing susceptibility of type I, IIa, IIx and IIb muscle fibers of the rat DIAm muscle to neuromuscular transmission failure (6). In the present study, we found that neuromuscular transmission was significantly improved in the SI model. It is possible that the expansion of the NMJ on type IIx and IIb fibers was primarily involved in this improvement of neuromuscular transmission. Since there were no changes in muscle fiber size with SI, the expansion of the NMJ relative to fiber size is conducive to such an improvement in neuromuscular transmission. In contrast to these adaptations following SI, TTX blockade resulted in an increased susceptibility to neuromuscular transmission failure. Since the NMJs on type II fibers displayed a modest increase in size following TTX blockade, while type II fibers atrophied, neuromuscular transmission should have improved. This discrepancy would suggest that with TTX blockade, there was a depletion of acetylcholine at the nerve terminal, and/or a desensitization of the acetylcholine receptor.

5. ACKNOWLEDGMENTS

This work was supported by National Heart, Lung and Blood Institute Grants HL34817 and HL37680. YSP is supported by a fellowship from Abbott Laboratories. The authors thank Dr. Kenneth Smithson for his helpful comments.

6. REFERENCES

1. Deschenes, M. R., C. M. Maresh, J. F. Crivello, L. E. Armstrong, W. J. Kraemer, and J. Covault. The effects of exercise training of different intensities on neuromuscular junction morphology. *J. Neurocytol.* **22**:603-615, 1993.
2. Dohrmann, U., D. Edgar, M. Sendtner, and H. Thoenen. Muscle derived factors that support and promote fiber outgrowth from embryonic chick spinal motor neurons in culture. *Dev. Biol.* **118**:209-221, 1986.
3. Fournier, M. and G. C. Sieck. Mechanical properties of muscle units in the cat diaphragm. *J. Neurophysiol.* **59**:1055-66, 1988.
4. Gurney, M., B. Appattoff, and S. Heinrich. Suppression of terminal axon sprouting at the neuromuscular junction by monoclonal antibodies against a muscle-derived antigen of 56,000 daltons. *J. Cell Biol.* **102**:2264-2272, 1986.
5. Hernsbergen, E. and D. Kernell. Daily duration of activity in ankle muscles of cats. In *XXXIInd IUPS,* Glasgow, 1993.
6. Johnson, B. D. and G. C. Sieck. Differential susceptibility of diaphragm muscle fibers to neuromuscular transmission failure. *J. Appl. Physiol.* **75**:341-348, 1993.
7. Miyata, H., W. Z. Zhan, Y. S. Prakash, and G. C. Sieck. Influence of inactivity on contribution of neuromuscular transmission failure to diaphragm fatigue. *Med. Sports Exercise* **26**:S167, 1994.
8. Miyata, H., W. Z. Zhan, Y. S. Prakash, and G. C. Sieck. Myoneural interactions affect muscle adaptations to inactivity. *J. Appl. Physiol.* **79**:1640-1649, 1995.
9. Prakash, Y. S., M. Fournier, and G. C. Sieck. Effects of prenatal undernutrition on the developing rat diaphragm. *J. Appl. Physiol.* **75**:1044-1052, 1993.
10. Prakash, Y. S., S. M. Miller, M. Huang, and G. C. Sieck. Morphology of diaphragm neuromuscular junctions on different fibre types. *J. Neurocytol.* **25**:88-100, 1996.
11. Prakash, Y. S., S. M. Miller, and G. C. Sieck. Three-color confocal imaging of the neuromuscular junction. In *Fourth IBRO World Congress of Neuroscience,* Kyoto, Japan, 1995.
12. Prakash, Y. S., K. G. Smithson, and G. C. Sieck. Application of the Cavalieri principle in volume estimation using laser confocal microscopy. *NeuroImage* **1**:325-333, 1994.
13. Prakash, Y. S., K. G. Smithson, and G. C. Sieck. Measurements of motoneuron somal volumes using laser confocal microscopy: Comparisons with shape-based stereological estimations. *NeuroImage* **1**:95-107, 1993.
14. Sherrington, C. S. Some functional problems attaching to convergence. *Proc. R. Soc. Lond. (Biol)* **105**:332-362, 1929.
15. Sieck, G. C. Neural control of the inspiratory pump. *Nips.* **6**:260-264, 1991.
16. Sieck, G. C. and M. Fournier. Diaphragm motor unit recruitment during ventilatory and nonventilatory behaviors. *J. Appl. Physiol.* **66**:2539-45, 1989.
17. Sieck, G. C., M. Fournier, and J. G. Enad. Fiber type composition of muscle units in the cat diaphragm. *Neurosci. Lett.* **97**:29-34, 1989.
18. Thompson, R. J., J. F. Doran, P. Jackson, A. P. Shillon, and J. Rode. PGP 9.5- a new marker for vertebrate neurons and neuroendocrine cells. *Brain Res.* **278**:224-228, 1983.

10

EFFECT OF PASSIVE AND ACTIVE RECOVERY ON PCR KINETICS

Takayoshi Yoshida and Hiroshi Watari

Exercise Physiology Laboratory
Faculty of Health and Sports Sciences
Osaka University, Toyonaka
Osaka 560
Japan

1. INTRODUCTION

Recently, [31]phosphorus-nuclear magnetic resonance spectroscopy (^{31}P-MRS) has been used as a noninvasive technique to measure changes in the concentrations of adenosine 5′-triphosphate (ATP), phosphocreatine (PCr) and inorganic phosphate (Pi), as well as the intramuscular pH, both during and after exercise. Splitting of the Pi peak into two has been observed during exercise and is attributable to two different pH distributions in exercising muscle (high pH and low pH) (1, 3, 5, 8, 11, 15). Previously, we reported that the two split Pi peaks showed different time courses at the onset of exercise and during recovery (16, 19, 22); high-pH Pi increased promptly at the onset of exercise and disappeared rapidly after exercise, while the low-pH Pi peak increased gradually after a delay of approximately 60 sec at the onset of exercise and decreased over a longer period after exercise was stopped.

The slow disappearance of the low-pH Pi peak during passive recovery may be a manifestation of the suppression of glycolysis at low pH resulting from accumulation of lactate in glycolytic muscle fibers. In this context, it is noteworthy that Hermansen and Stensvold (4) and McLellan and Skinner (7) showed a more rapid clearance of blood lactate during active than passive recovery. However, less attention has been paid to the effects of active and passive recovery on intramuscular pH.

In the present study, we therefore compared intramuscular pH profiles during active and passive recovery, using ^{31}P-MRS to determine the splitting patterns of Pi peaks.

2. MATERIALS AND METHODS

2.1. Subjects

Six male long-distance runners participated in the study. The approval of the National Institute of Physiological Sciences Human Ethics Committee and the informed written consent of each subject were obtained. The average age of the subjects was 19.8 years (± 0.4 SEM), height 168.3 cm (± 1.6 SEM), mass 54.2 kg (± 1.4 SEM), and maximal oxygen uptake 70.8 ml.kg^{-1}.min^{-1} (± 1.9 SEM).

2.2. Exercise Procedure

The lever of a wooden exercise ergometer, specially designed for use inside the bore of a magnet, was connected to the leg of each subject while he lay in a prone position (16–22). The work rate was determined from the weight lifted by the leg using a pulley, the flexion rate (set at 50 times per min following a metronome), and the vertical displacement of the weight.

The exercise tests consisted of submaximal exercise with passive recovery (Passive recovery test) and active recovery (Active recovery test)(Fig. 1):

1. Passive recovery test involved an exercise test consisting of 2 min of exercise at 60%max intensity, followed by 2 min of passive recovery. The 2 min of exercise was then repeated and followed by 5 min of passive recovery.
2. Active recovery test involved an exercise test consisting of 2-min unloaded warm-up, then 2 min of exercise at 60%max, followed by 2 min of unloaded exercise (active recovery). The cycle of 2-min exercise and 2-min active recovery was then repeated, and followed by a resting recovery period of 5 min.

2.3. Data Collection

NMR acquisition. Intramuscular pH and changes in the levels of phosphorus metabolites during the rest-exercise-recovery sequence were determined with very short time resolution (interval of 5 sec) by ^{31}P-MRS. ^{31}P-MRS data were obtained using a 67-cm bore 2.1-Tesla superconducting magnet and a spectrometer (EX90, JEOL, Japan). The subject's entire body was placed inside the bore of the magnet and a surface coil (8 cm in diameter) was placed over the center of the biceps femoris muscle of the left thigh to obtain ^{31}P-MRS signals. After the subject was placed in the magnet, magnetic field homogeneity was optimized by shimming on a proton signal of the biceps flexor muscles. ^{31}P-MRS data were collected with an optimal pulse width at a pulse rate of one per 0.416 sec throughout the experiment. ^{31}P-MRS data for 12 scans were averaged to produce a single spectrum, so each represented the data from a 5.0 sec period.

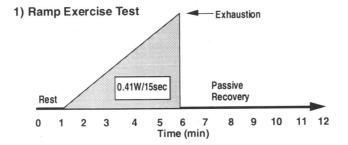

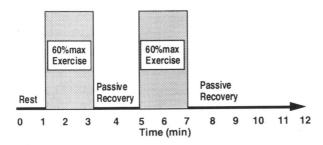

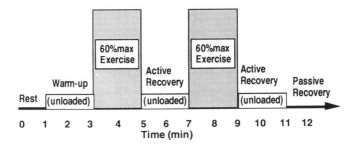

Figure 1. Schematic illustration showing exercise protocols: 1) Ramp incremental exercise test, 2) Passive recovery test and 3) Active recovery test.

2.4. Data Analysis

The relative areas under the Pi and PCr peaks of the spectrum were determined by integration. Intramuscular pH was determined using the chemical shift of Pi relative to PCr according to the following variant of the Henderson-Hasselbalch equation:

$$pH = 6.73 + \log10((a-3.275)/(5.685-a))$$

where a is the chemical shift from Pi to PCr. When the Pi peak was split, the Pi peaks were separated by a model fitting procedure using the Lorentzen curve with the least mean squares method, and pH data were determined from the center of the curve.

2.5. Statistical Analysis

Results are expressed as the means ± SEM. Differences between mean values obtained for passive and active recovery were compared by Student's t-test. A probability value of less than 0.05 was considered to indicate significance.

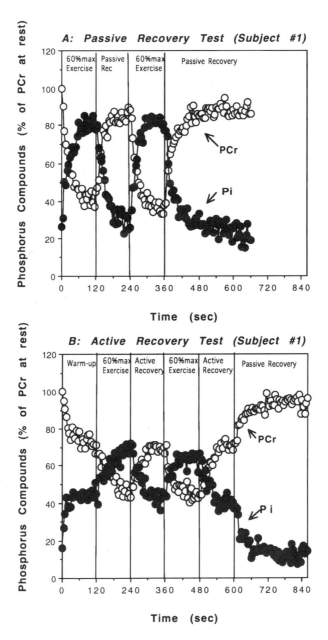

Figure 2. An example of Pi and PCr changes during Passive (A) and Active recovery tests (B) in subject #1.

3. RESULTS

Figure 2 shows examples of the time courses of changes in PCr and Pi during Passive and Active recovery tests for one subject (subject #1). The changes in PCr and Pi values were expressed relative to resting PCr as 100%. During exercise, group mean values of PCr decreased to 47.5 ± 2.3% for the Passive recovery test and 49.0 ± 5.0% for the Active recovery test, and the Pi level increased to 71.9 ± 3.9 % for the Passive recovery test and 64.1 ± 3.5% for the Active recovery test. PCr and Pi changes were reciprocal. The reproducibility of PCr and Pi changes during both types of exercise tests was good (0.92 < r < 0.96; P < 0.001). In the Passive recovery test, the PCr recovery was very rapid at first, followed by very slow kinetics. In the Active recovery test, however, no tailing of the recovery was observed. The kinetics of active recovery appeared a little slower (τ =30.2 ± 3.0 sec and 32.1 ± 2.5 sec) than those of passive recovery in the Passive recovery test (25.3 ± 2.6 sec and 25.9 ± 2.1 sec), but it is interesting that the kinetics of passive recovery were similar between the two tests (25.9 ± 2.1 sec in the Passive recovery test and 27.7 ± 3.2 sec in the Active recovery test).

Splitting of Pi peaks was apparent during exercise in both the Passive and Active recovery tests. It is interesting to note that the high-pH Pi peak disappeared more rapidly during passive recovery in the Passive recovery test than the low-pH Pi peak, with the latter peak remaining at the acidified site after 3 min; i.e. the low-pH Pi peak was displayed as a tail during passive recovery. In the Active recovery test, the high-pH Pi peak decreased slowly but remained at a low level throughout the active recovery, while the low-pH Pi peak recovered more quickly and disappeared within 2 min.

4. DISCUSSION

The splitting of the Pi peak during exercise was considered to reflect data acquisition from active and inactive muscle groups, because its detection was dependent on the size of the surface coil (5). Another possibility that has been considered is compartmentalization of Pi between intracellular and interstitial spaces. However, recent studies have indicated that splitting of the Pi peak during exercise is due to differences in energy metabolism between oxidative and glycolytic muscle fibers (1, 10, 15, 20, 22). Skeletal muscle fibers of vertebrates have been classified into two main types based on their characteristic features as well as functions, i.e. type I and type II by staining for myofibrillar ATPase, slow twitch (ST) and fast twitch (FT) muscle fibers by contractile kinetics, and oxidative and glycolytic fibers by dominant energy metabolism. There are many reports in several species of animals demonstrating that the characteristics of a muscle fiber correspond to its contractile function and activity of energy metabolism (13, 14). Thus, type I muscle fibers have slow contractile kinetics and high mitochondrial oxidative activity, while type II muscle fibers are fast and glycolytic. However, the activities of oxidative enzymes in human FT muscle enzymes have been shown to be greatly enhanced by endurance training, and the absolute level of these enzymes can surpass that in ST muscle fibers of untrained subjects (13). Moreover, MRS cannot distinguish between type I and type II muscle fibers, nor between ST and FT muscle fibers, but can provide information on a metabolic classification. Therefore, the present study is confined to metabolic features of muscle fibers; i.e. oxidative and glycolytic fibers.

Although there appeared to be at least two Pi peaks during exercise, only a small Pi peak was seen in the high pH region at rest (ca. 7.04 pH units). This may indicate that each fiber type has the same pH value. However, the distribution of Pi was not even, being mostly contained in oxidative muscle fibers, as demonstrated in experimental animals by earlier investigations (8). At the onset of exercise with light intensity, the height of the Pi peak increased a little without shifting. This implies that when work intensity is low, energy is supplied mainly by oxidative phosphorylation and low amounts of lactate are produced. It has also been shown, using glycogen depletion, that oxidative muscle fibers are recruited first during low intensity exercise, and that glycolytic muscle fibers are subsequently recruited with increasing workload (13). In practice, during 60%max exercise, an additional Pi signal appeared at a slightly lower ppm, and shifted gradually to 3.8 ppm relative to PCr (ca. 6.1 pH units) with increases in exercise intensity indicating the recruitment of glycolytic muscle fibers and thereby production of large amounts of lactate. Subsequently, when the two kinds of fibers are recruited simultaneously during high intensity exercise, at least two different pH distributions can exist in the musculature. Obvious Pi peak splitting was observed during the higher exercise intensity (eg. 60%max exercise), whereas only a single Pi peak was seen at rest, and during unloaded exercise and passive recovery. During 60%max exercise as well as during the active recovery period, the high-pH Pi peak was dominant. In other words, the low-pH Pi peak was actually rather small and did not account for a sizable portion of the total Pi peak. This could be due to the use of endurance trained athletes, who have mostly high oxidative fibers.

It is well-documented that PCr hydrolysis is controlled by myosin ATPase and sarcomere creatine kinase, and that its resynthesis is controlled by aerobic mitochondrial oxidative phosphorylation as well as by anaerobic glycolysis (10). It is not easy to directly estimate the rate of ATP generation in the exercising muscle. However, it is highly likely that the rapid phase of PCr recovery after cessation of exercise reflects oxidative phosphorylation and little consumption of ATP in the oxidative muscle fibers (6, 7, 17, 20, 22). Theoretically, when the oxygen supply is sufficient, the rate of oxidative phosphorylation is a function of the level of free ADP in the oxidative muscle fibers, which in turn is proportional to the Pi/PCr ratio as determined by ^{31}P-MRS (2). Actually, the PCr time constant during passive recovery was 25.3 ± 2.6 sec for the first, and 25.9 ± 2.1 sec for the second recovery of the Passive recovery test (Fig. 2). These values concur with those determined in our previous study, in which long-distance runners showed very fast PCr kinetics in four repeated exercise cycles (20). The fast phase of PCr recovery was followed by very slow kinetics, which may correspond to the very slow disappearance of the low-pH Pi peak during passive recovery. This may be due not only to lower oxidative capacity but also to the complete halt of anaerobic glycogenolysis after cessation of exercise in the glycolytic muscle fibers (12).

Even in the Active recovery test, active recovery was followed by 5 min of passive recovery, and this passive recovery showed similar fast PCr kinetics (τ=27.7 ± 3.2 sec), but very slow kinetics, i.e. tailing, were not observed in the late phase (Fig. 2). This implies that recovery metabolism takes place only in the oxidative muscle fibers during passive recovery after the active recovery period. In accordance with the recovery of PCr, the low-pH Pi, which accumulated in glycolytic muscle fibers during exercise, disappeared during active recovery.

As shown in Fig. 3, the rate of return of the low-pH peak to the resting value was much faster (0.095 ± 0.019 pH units/min) during active recovery than during passive recovery (0.014 ± 0.019 pH units/min) (P<0.01). It has been reported that blood lactate removal following high intensity exercise is enhanced by active recovery compared to that by passive recovery. It is highly likely that the metabolic effects of active versus passive recovery involve circulatory aspects, ie, blood flow is maintained at a higher rate in the exercising muscle during active recovery, thereby preventing lactate accumulation.

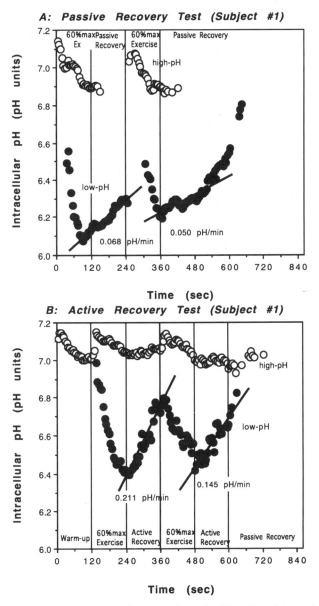

Figure 3. Changes in high and low pH values during repeated exercise (A) and interval exercise (B) in subject #1

In conclusion, during heavy exercise Pi appeared to form two different pH peaks. Probably, the high-pH peak corresponds to oxidative muscle fibers, and the low-pH peak to glycolytic fibers. The low-pH Pi quickly disappeared during active recovery because of enhanced oxidative activity and efficient lactate removal due to the higher rate of blood flow in the exercising muscle. However, the slow disappearance of the low pH Pi peak during passive recovery may be due to halting of glycogenolysis and the low oxidative capacity of resting glycolytic muscle fibers.

6. REFERENCES

1. Achten E., M. Van Cauteren, R. Willem, R. Luypaert, W.J. Malaisse, G. Van Bosch, G. Delanghe, K. De Meirleir, and M. Osteaux. ^{31}P-NMR spectroscopy and the metabolic properties of different muscle fibers. *J. Appl. Physiol.* **68**: 644–649, 1990.

2. Chance, B., J.S. Leigh Jr, J. Kent, and K. McCully. Metabolic control principles and ^{31}P-NMR. *Fed. Proc.* **45**: 2915–2920, 1986.

3. Clark III, B.J., M.A. Acker, K.K. McCully, H.V. Subramanian, R.L. Hammond, S. Salmons, B. Chance, and L.W. Stephenson. In vivo ^{31}P-NMR spectroscopy of chronically stimulated canine skeletal muscle. *Am. J. Physiol.* **254** (*Cell Physiol.* **23**): C258-C266, 1988.

4. Hermansen, L., and I. Stensvold. Production and removal of lactate during exercise in man. *Acta Physiol. Scand.* **86**: 191–201, 1972.

5. Jeneson, J.A.L., M.W. Wesseling, R.W. De Boer, and H.G. Amelink. Peak-splitting of inorganic phosphate during exercise. anatomy or physiology? A MRI-guided ^{31}P MRS study of human forearm muscle. In: *Proceedings of the Eighth Annual Meeting of Society of Magnetic Resonance in Medicine.* Abstract 1030. Society of Magnetic Resonance in Medicine, Berkeley, California, 1989.

6. Kushmerick, M.J., and R.A. Meyer. Chemical changes in rat leg muscle by phosphorus nuclear magnetic resonance. *Am. J. Physiol.* **248** (*Cell Physiol.* **17**): C542-C549, 1985.

7. McLellan, T.M., and J.S. Skinner. Blood lactate removal during active recovery related to the aerobic threshold. *Int. J. Sports Med.* **3**: 224–229, 1982.

8. Meyer, R.A., T.R. Brown, and M.J. Kushmerick. Phosphorus nuclear magnetic resonance of fast- and slow-twitch muscle. *Am. J. Physiol.* **248** (*Cell Physiol.* **17**): C279-C287, 1985.

9. Mizuno, M., L.O. Justesen, J. Bedolla, D.B. Riedman, N.H. Secher, and B. Quistroff. Partial curarization abolishes splitting of the inorganic phosphate peak in ^{31}P-NMR spectroscopy during intense forearm exercise in man. *Acta Physiol. Scand.* **139**: 611–612, 1990.

10. Mole, P.A., R.L. Coulson, J.R. Caton, B.G, Nichols, and T.J. Barstow. In vivo ^{31}P-NMR in human muscle: transient patterns with exercise. *J. Appl. Physiol.* **59**: 101–104, 1985.

11. Park, J.H., R.L. Brown, C.R. Park, K.K. McCully, M. Cohn, J. Hasellgrove, and B. Chance. Functional pools of oxidative and glycolytic fibers in human muscle observed by 31P magnetic resonance spectroscopy during exercise. *Proc. Natl. Acad. Sci. USA* **84**: 8976–8980, 1987.

12. Quistorff, B., L. Johansen, and K. Sahlin. Absence of phosphocreatine resynthesis in human calf muscle during ischemic recovery. *Biochem. J.* **291**: 681–686, 1992.

13. Saltin, B., and P.D. Gollnick. Skeletal muscle adaptability: significance for metabolism and performance. In: *Handbook of physiology. Skeletal Muscle.* Bethesda, MD: Am. Physiol. Soc., 1983, Sect.10, Chapt.19, p.555–631.

14. Tesch, P.A., A.Thorsson, and N. Fujitsuka. Creatine phosphate in fiber types of skeletal muscle before and after exhaustive exercise. *J. Appl. Physiol.* **66**: 1756–1759, 1989.

15. Vandenborne, K., K. McCully, H. Kakihara, M. Prammer, L. Bolinger, J.A. Detre, K. De Meirleie, K. Walter, and B. Chance. Metabolic heterogeneity in human calf muscle during maximal exercise. *Proc. Natl. Acad. Sci. USA*, **88**: 5714–5718, 1991.

16. Yoshida, T., and H. Watari. Noninvasive and continuous determination of energy metabolism during muscular contraction and recovery. *Med. Sci. Sport Exerc.* **37**: 364–373, 1992.

17. Yoshida, T., and H. Watari. Muscle metabolism during repeated exercise studied by ^{31}P-MRS. *Ann. Physiol. Anthrop.* **11**: 241–250, 1992.

18. Yoshida, T., and H. Watari. Effect of hypoxia on muscle metabolite during exercise studied by ^{31}P-MRS. *Med. Sci. Sport Exerc.* **25**(Suppl): S174, 1993.

19. Yoshida, T., and H. Watari. ^{31}P-MRS study of the time course of energy metabolism during exercise and recovery. *Eur. J. Appl. Physiol.* **66**: 494–499, 1993.

20. Yoshida, T., and H. Watari. Metabolic consequences of repeated exercise in long distance runners. *Eur. J. Appl. Physiol.* **67**: 261–265, 1993.

21. Yoshida, T., and H. Watari. Changes in intracellular pH during repeated exercise. *Eur. J. Appl. Physiol.* **67**: 274–278, 1993.

22. Yoshida, T., and H. Watari. Exercise-induced splitting of the inorganic phosphate peak: investigation by time-resolved 31P-nuclear magnetic resonance spectroscopy. *Eur. J. Appl. Physiol.* **69**: 465–73, 1994.

11

PHOSPHOCREATINE AS AN ENERGY STORE AND ENERGY SHUTTLE IN HUMAN SKELETAL MUSCLES

D. Matisone, J. Skards, A. Paeglitis, and V. Dzerve

Latvia Institute of Cardiology
Pilsonu Str. 13, LV-1002, Riga, Latvia

1. INTRODUCTION

The metabolic rate of skeletal muscles during maximal voluntary static contractions can be increased more than 120-fold above the basal rate (5). Phosphocreatine (PCr) in the skeletal muscles serves as an energy (ATP) store at the onset of muscle contraction, the breakdown of which increases the rate of oxidative phosphorylation and stimulates anaerobic glycolysis (1,2). However, little is known about the integration of aerobic and anaerobic pathways of metabolism in skeletal muscles during contraction and ischemia. The aim of this study was to determine the role of PCr in forearm skeletal muscle energetics in two different situations: during voluntary static contractions with different relative forces of maximal voluntary contraction (MVC), and in resting skeletal muscles following forearm arterial occlusion (AO) of different durations.

2. METHODS

The study was carried out on 10 healthy volunteers between the ages of 20 and 40 years, after informed consent was obtained. Forearm static contraction (handgrip) was performed with 5%, 10%, 15%, 30%, 50% and 100% MVC until fatigue. AO lasted 1, 3, 5, 7, 15 and 30 min. Blood flow ($\dot{I}$) and the arterio-venous content differences of O_2 and LA across the forearm skeletal muscles were determined at intervals of 10-15 s at rest, during contraction and recovery, as well as during reactive hyperemia (Rh). Oxygen uptake ($\dot{V}O_2$) and LA release were calculated by the Fick principle. Total ATP turnover ($\Sigma\Delta$ATP), PCr debt and accumulated LA in the muscles were calculated from the VO_2 fast and slow fractions during recovery from contraction and in the Rh phase, according to stochiometric equations (3,4). External work capacity was expressed as tension time (TT): TT = %MVC x time of contraction. The energetic cost of one external work unit (ATP per

TT) was calculated as $\Sigma\Delta ATP/TT$. Capillary filtration coefficient (CFC) was determined from the increase in forearm volume during elevation of venous pressure.

3. RESULTS

Low contraction forces (5% and 10% MVC) are associated with appreciable $\Sigma\Delta ATP$, TT and efficiency of forearm muscle contraction (low ATP/TT). It should be noted that ATP turnover occurs mainly during contraction, and that the ATP equivalent of the O_2 debt paid during recovery does not exceed 6-10% $\Sigma\Delta ATP$. The increase of contraction force from 10% to 15% MVC (only about 5%) evokes a sharp decrease of $\Sigma\Delta ATP$ and TT (more than 2-fold). The ATP equivalent of the O_2 debt increases up to 38% of $\Sigma\Delta ATP$, and the energetic cost of one external work unit also increases. Increasing the contraction force from 15% to 50% MVC evokes a further decrease in $\Sigma\Delta ATP$ and TT. The ATP equivalent of the O_2 debt increases to 75% and 100% of $\Sigma\Delta ATP$ after contraction at 30% and 50% MVC, respectively. These data show that an increase of forearm voluntary contraction force in the range between 10% and 15% MVC leads to a decrease in $\Sigma\Delta ATP$ and TT and to an increase in the O_2 debt (Fig. 1).

Essentially mirror-image response profiles were evident in total oxygen uptake (ΣVO_2) and total lactate production (ΣLA) during contraction at the different work loads.

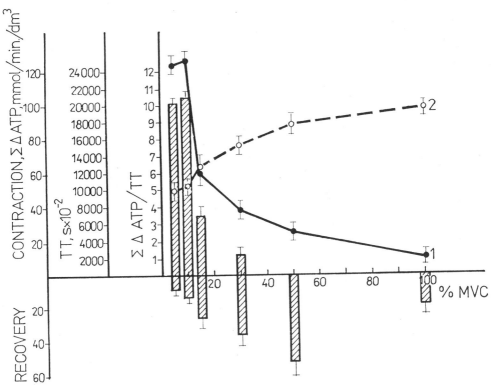

Figure 1. Total ATP turnover during contraction and recovery ($\Sigma\Delta ATP$ - hatched bars), tension time (TT: 1) and energetic cost of one external work unit ($\Sigma\Delta ATP/TT$: 2) as a function of relative force of maximal voluntary contraction (%, MVC).

The curves reflecting the amount of Σ VO_2 and Σ LA during contraction depended on the relative muscle force, intersecting in the range between 10% and 15% MVC. It implies that the change of metabolism from aerobic to anaerobic occurs in this range of contraction force. The total oxygen debt (DO_2) is lowest in the range of low contraction forces (5% and 10% MVC), and increases with the augmentation of contraction force up to 50% MVC. At each contraction force investigated, the dominant fraction in the DO_2 is the slow fraction, which reflects oxidation of accumulated LA and glycogen resynthesis in the skeletal muscles. The magnitude of the fast fraction of the DO_2, which reflects the degree of PCr hydrolysis, increases monotonically with increasing contraction force up to 50% MVC (Fig. 2).

During handgrip to fatigue at 10% MVC, a long-term steady state of $\dot{I}$, $\dot{V}O_2$, LA release and CFC is established. It should be noted that the vasodilation supporting the steady state of $\dot{I}$ during handgrip at 10% MVC is only 40% of the maximal value the muscle can achieve, and that the corresponding CFC does not exceed 60% of its maximal value (Fig. 3).

During handgrip to fatigue at 5% MVC, $\dot{I}$ rises continuously and at the end of contraction it reaches a maximal value. The augmentation of $\dot{V}O_2$ is similar to the increase of $\dot{I}$, but LA release rises exponentially during contraction. Early in the contraction, CFC is significantly greater ($p < 0.001$) than during contraction with 10% MVC, and by the middle of the contraction has already reached a maximum (Fig. 4).

Data on Σ $\dot{V}O_2$, the ATP equivalent of Σ VO_2 and Σ LA in resting forearm muscles during AO from 1 to 30 min shows that the DO_2 which develops in these muscles during ischemia increases proportionally with the duration of AO, and that this DO_2 is completely repaid during reactive hyperemia even after AO of 30 min. Σ LA during AO increases proportionally to that in resting muscles only until the 7th min of AO, when the PCr debt is 3.6 ± 0.2 mmolATP/dm^3. A further increase of AO evokes an exponential increase of Σ LA. This indicates that activation of anaerobic glycolysis occurs only above a definite level of depleted PCr (Fig. 5).

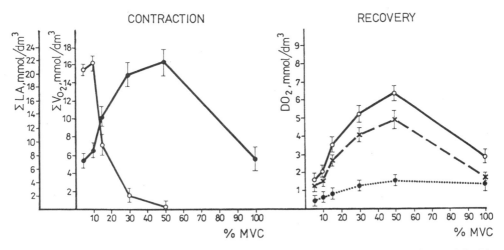

Figure 2. Total O_2 uptake (Σ VO_2 - o) and total lactate production (Σ LA - ·) during contraction, total O_2 debt (DO_2 - o) and its fast (x) and slow (·) fractions of repayment during recovery, for different relative forces of maximal voluntary contraction (%, MVC).

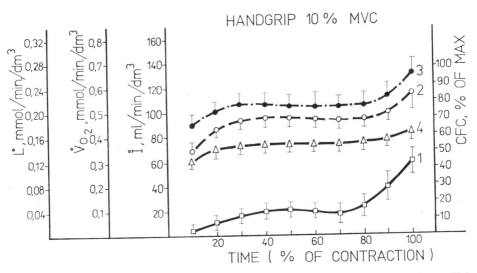

Figure 3. Lactate release (L̇: 1), oxygen uptake (V̇O₂: 2), blood flow (İ: 3) and capillary filtration coefficient (CFC: 4) during voluntary static contraction of forearm muscles at 10% MVC, using a normalized time scale (% of contraction time).

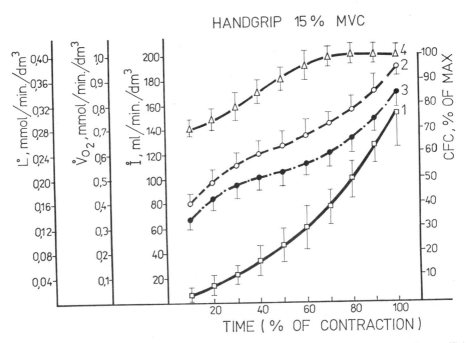

Figure 4. Lactate release (L̇: 1), oxygen uptake (V̇O₂: 2), blood flow (İ: 3) and capillary filtration coefficient (CFC: 4) during voluntary static contraction of forearm muscles at 15% MVC, using a normalized time scale (% of contraction time).

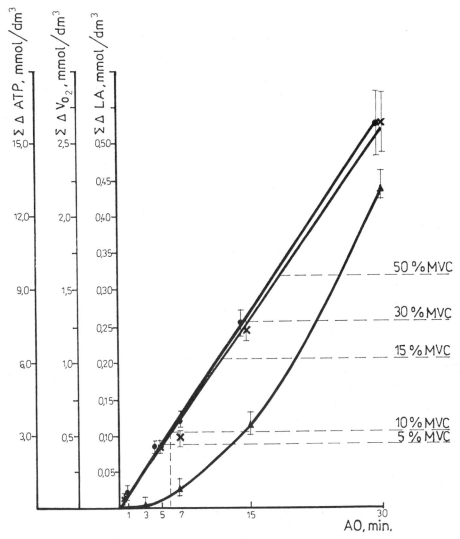

Figure 5. Total oxygen uptake ($\Sigma\ VO_2$ - ·), energetic equivalent of $\Sigma\ VO_2$ ($\Sigma\Delta$ATP - x) and total lactate production (Σ LA - ▲) after different durations of forearm arterial occlusion (AO).

4. DISCUSSION

The obtained data show that PCr acts as an energy store which completely ensures ATP resynthesis in the resting skeletal muscles during ischemia until 7 min AO. The activation of anaerobic glycolysis occurs only above a definite value of PCr debt - 3.6 ± 0.2 mmol ATP/dm³ both in the resting and contracting skeletal muscles. Therefore, two different levels of PCr depletion may be recognized: the first level, up to 3.6 ± 0.2 mmol ATP/dm³, when splitting of PCr evokes an increase in oxidative phosphorylation with a parallel increase in an aerobic glycolysis; and the second level, above this value, when splitting of PCr provokes a further increase in oxidative phosphorylation accompanied by the activation of anaerobic glycolysis and an excess LA production.

Our data show that the fast component of DO_2, reflecting the amount of depleted PCr after exhaustive handgrip with 5% and 10% MVC, corresponds to the PCr depletion in the range of the first level when anaerobic glycolysis is not activated. Such small PCr depletion ensures a long-term aerobic work capacity of the muscles (80 ± 6 min and 42 ± 3 min, respectively). We propose that, within this level of depletion, PCr in skeletal muscle fibers acts as an energy shuttle providing ATP transportation from its generation sites (mitochondria) to its utilization sites (myofibrils).

Results of our previous investigations suggested that, during contraction with low relative forces, exercise hyperemia occurred only at the level of the active muscles fibres and that the recruitment of muscles fibers during contraction is accompanied by redistribution of blood flow to the active muscle fibres by closing precapillary vessels of the fatigued muscle fibres (5,6). During forearm muscle contraction at 15% MVC, the fast component of the DO_2 at the end of contraction exceeds 3.6 ± 0.2 mmol ATP/dm^3, thus evoking a sharp activation of anaerobic glycolysis and excess LA formation. The dynamics of I and CFC during contraction imply that the functional closing of precapillary vessels of the fatigued muscle fibers does not occur. It provokes metabolic and hemodynamic shunting of the active muscle fibers and rapid muscle fatigue.

Our results show that, in any situation which evokes a discrepancy between O_2 demand and O_2 delivery in the skeletal muscles, PCr acts as an energy store to buffer imbalances between energy demand and supply, but that the contribution of aerobic and anaerobic metabolic pathways in the ATP resynthesis depends on the level of PCr depletion.

5. REFERENCES

1. Bessman, S.P. The creatine-creatine phosphate energy shuttle. *Ann. Rev. Biochem.* **54**: 831-862, 1985.
2. Hochachka, P.W. Fuels and pathways as designed systems for support of muscle work. *J. Exp. Biol.* **115**: 149-164, 1985
3. McCann, D.J., P.A. Molé, and J.R. Caton. Phosphocreatine kinetics in humans during exercise and recovery. *Med. Sci. Sports Ex.* **27**: 378-389, 1995.
4. Piiper, J., P.E. Di Prampero, and P. Cerretelli. Relationship between oxygen consumption and anaerobic metabolism in stimulated gastrocemius muscle of the dog. In: *Exercise Bioenergetics and Gas Exchange*, edited by P. Cerretelli and B.J. Whipp. Amsterdam: Elsevier/North Holland, 1980, pp. 35-44.
5. Skards, J., and A. Paeglitis. The energetic supply of the voluntary static contraction of human skeletal muscle. *Izvestia Akademii Nauk Latv. SSR* **403**: 105-122, 1981 (in Russian).
6. Skards, J., A. Paeglitis, D. Matisone, and E. Eglitis. The role of sympathetic vasoconstrictors providing aerobic work capacity of skeletal muscle. *Sechenov. Physiol. J. USSR* **72(9)**: 1275-1283, 1986 (in Russian).

PART 3. SYSTEMIC LIMITATION TO MAXIMUM EXERCISE IN HEALTHY SUBJECTS

12

DOMAINS OF AEROBIC FUNCTION AND THEIR LIMITING PARAMETERS

Brian J. Whipp

Department of Physiology
St. George's Hospital Medical School
London, United Kingdom

1. INTRODUCTION

Aerobic energy transfer during muscular exercise requires that hydrogen atoms be "stripped" out of previously stored substrate molecules, and their component proton and electrons put to work to generate ATP in the mitochondrial electron transport chain. The electron flow is used to supply the redox potential necessary to establish the transmembrane proton gradients which subsequently power the phosphorylation. These reactions require oxygen as the terminal electron transport chain oxidant. Consequently, the ability to sustain muscular exercise is dependent in large part on the body's ability to transport oxygen from the atmosphere to the cytochrome oxidase terminus of the mitochondrial electron transport chain. The time course of pulmonary O_2 uptake ($\dot{V}O_2$) at high work rates should therefore be considered a major index of systemic O_2 transport function. It is perhaps surprising, therefore, how little attention has been paid to the physiological control inferences which may be drawn from the nonsteady-state response profiles of $\dot{V}O_2$. Such determinations are likely to be revealing, as the bulk of the control information regarding a physiological system resides in its transient rather than its steady-state behavior.

Furthermore, from a practical standpoint, as the kinetic parameters of the $\dot{V}O_2$ profile (e.g. gains, time constants, delays) have been shown to differ characteristically at different exercise intensities, their responses (and those of the associated pulmonary gas exchange and acid-base variables) may consequently be used to provide a means of assigning the domains of exercise intensity and the parameters which partition them.

2. MODERATE EXERCISE

This may be considered to reflect the range of work rates which are below the lactate threshold (θL). There seems little justification for subdividing this domain further - for example, into "mild" and "moderate". In this domain, there is little or no sustained metabolic acidemia or increase in arterial blood or muscle lactate levels (others dispute this, however: Refs. 7,12). The $\dot{V}O_2$ kinetics in this domain may be considered to be the "fundamental" response and therefore usefully serve as the frame of reference for the altered profiles that occur at higher intensities.

These "fundamental" kinetics perhaps are best appreciated in the context of square-wave (or constant-load) exercise. The general components, each having different physiological determinants, are: (a) the early, usually rapid, response (phase 1, $\phi 1$); (b) the slower, exponential increase (phase 2, $\phi 2$); and, if attained, (c) the steady state or phase 3 ($\phi 3$) (26).

Pulmonary $\dot{V}O_2$ is determined by: (a) pulmonary blood flow ($\dot{Q}p$) which, in normal subjects, is functionally equal to the cardiac output ($\dot{Q}t$) and (b) the arterial-to-mixed venous O_2 content difference ($CaO_2 - C\bar{v}O_2$). In $\phi 3$, the steady-state increment in $\dot{V}O_2$ equals that of the mean rate of muscle O_2 utilization ($\dot{Q}O_2$). This is not the case for the nonsteady state, however. It is important to recognize that $\dot{Q}O_2$ is not capable of increasing to its new steady-state level immediately following exercise onset. Rather, it increases with a monoexponential time course. The time constant ($\tau\dot{Q}O_2$) of this response reflects its control by the turnover dynamics of the intramuscular high-energy phosphate pool - although the exact mechanisms remain topics of debate (5,8,14). The expression of this $\dot{Q}O_2$ response profile at the lung, however, is delayed as a result of the vascular transit delay between the exercising muscles and the pulmonary capillaries (2). This delay (some 15-20 sec) reflects that period after exercise onset during which alterations of muscle venous composition do not yet influence $C\bar{v}O_2$ (26). Phase 1 is therefore a period of time, and not a pattern of response.

The subsequent $\phi 2\dot{V}O_2$ response has been consistently characterized as a monoexponential function of the form (11,17,25,26) (Fig. 1: "moderate"):

$$\Delta\dot{V}O_2(t) = \Delta\dot{V}O_2(ss) (1 - e^{-(t-\delta)/\tau})$$

where τ is the $\dot{V}O_2$ time constant (2), and δ is a delay term reflecting the muscle-to-lung vascular transit time (26). However, because of the confounding influence of breath-to-breath "noise" (16), it is often necessary to superimpose several replicates of a particular test to establish a sufficiently high signal-to-noise ratio for the acceptable discrimination of these delay features.

In $\phi 2$, the influence of the decreasing muscle venous O_2 content ($Cv(m)O_2$) on $C\bar{v}O_2$ dominates the $\dot{V}O_2$ response, even though $\dot{Q}p$ may still be increasing and is therefore contributory (26). A further influence on the dissociation of $\dot{V}O_2$ from $\dot{Q}O_2$ is that the mixed venous O_2 content which is established at the muscle level at a particular time will be associated with a higher $\dot{Q}p$ by the time it is expressed at the lung. That is, the blood flow will have increased *during* the transit delay.

Although $\tau\dot{V}O_2$ is unlikely to equal $\tau\dot{Q}O_2$ as a consequence, the differences may actually be relatively small. For example, there are two conditions in which the $\phi 2$ $\tau\dot{V}O_2$ will be *exactly* equal to $\tau\dot{Q}O_2$: firstly, if the proportional change in muscle blood flow matches that of $\dot{Q}O_2$ exactly or, alternatively, if muscle blood flow does not change at all - assuming, of course, that there are sufficient local stores to sustain the aerobic energy exchange. In the first case, there would be no $\phi 2$ - the entire response would be $\phi 1$; in the latter, there would be no increase in $\dot{V}O_2$ during $\phi 1$.

The more physiological pattern, however, produces an intermediate response. The mean O_2 concentration in the venous effluent from the contracting muscle ($Cv(m)O_2$) will

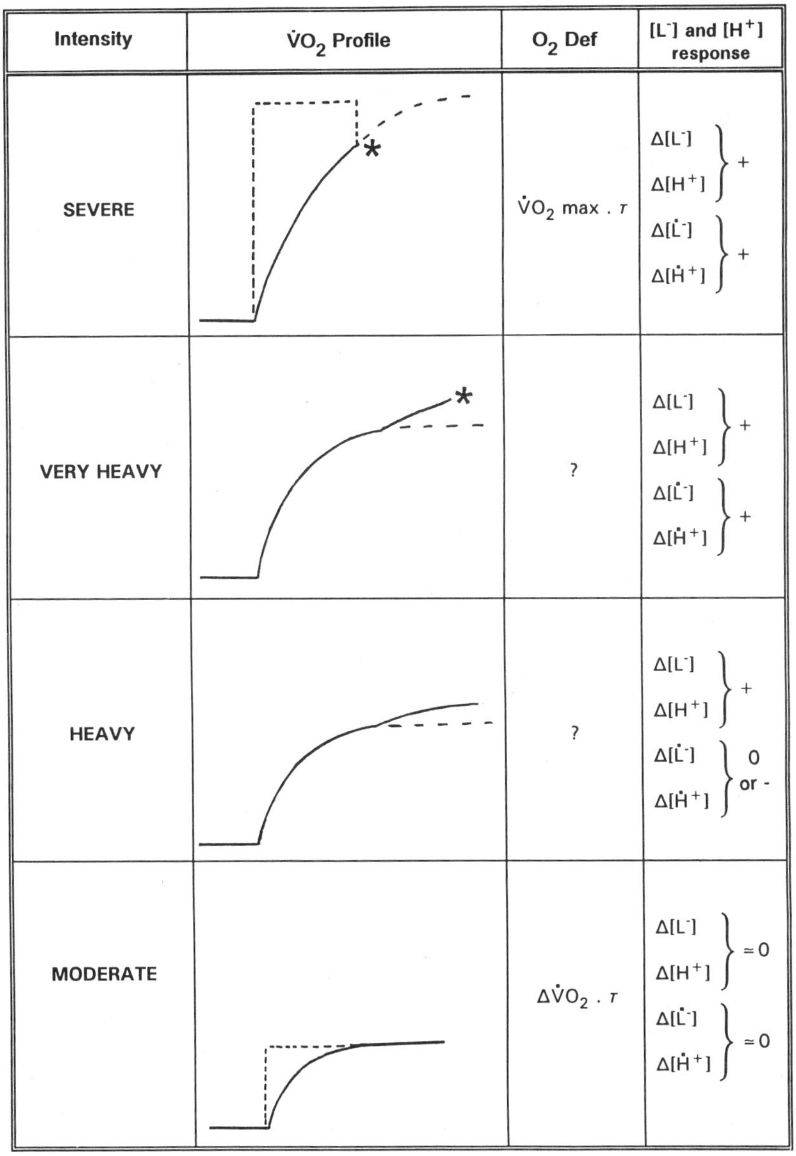

Intensity	$\dot{V}O_2$ Profile	O_2 Def	[L⁻] and [H⁺] response
SEVERE		$\dot{V}O_2$ max . τ	$\left.\begin{array}{l}\Delta[\text{L}^-] \\ \Delta[\text{H}^+]\end{array}\right\} +$ $\left.\begin{array}{l}\Delta[\dot{\text{L}}^-] \\ \Delta[\dot{\text{H}}^+]\end{array}\right\} +$
VERY HEAVY		?	$\left.\begin{array}{l}\Delta[\text{L}^-] \\ \Delta[\text{H}^+]\end{array}\right\} +$ $\left.\begin{array}{l}\Delta[\dot{\text{L}}^-] \\ \Delta[\dot{\text{H}}^+]\end{array}\right\} +$
HEAVY		?	$\left.\begin{array}{l}\Delta[\text{L}^-] \\ \Delta[\text{H}^+]\end{array}\right\} +$ $\left.\begin{array}{l}\Delta[\dot{\text{L}}^-] \\ \Delta[\dot{\text{H}}^+]\end{array}\right\} \begin{array}{l}0 \\ \text{or} -\end{array}$
MODERATE		$\Delta\dot{V}O_2$. τ	$\left.\begin{array}{l}\Delta[\text{L}^-] \\ \Delta[\text{H}^+]\end{array}\right\} \approx 0$ $\left.\begin{array}{l}\Delta[\dot{\text{L}}^-] \\ \Delta[\dot{\text{H}}^+]\end{array}\right\} \approx 0$

Figure 1. Schematic representation of the temporal response of O_2 uptake ($\dot{V}O_2$) to constant-load exercise at different work intensities. *Moderate:* below the lactate threshold (θL); note that there are no sustained increases in [lactate] or [H+] (i.e. both the absolute increments above baseline, Δ[L-] and Δ[H+], and their rates of change, $\Delta[\dot{\text{L}}$-] and $\Delta[\dot{\text{H}}$+], equal zero. *Heavy:* above θL, with $\dot{V}O_2$ reaching a steady state but with a delayed time course; note that there are now sustained increases in [lactate] and [H+] (i.e. Δ[L-] and Δ[H+] are positive) but that their rates of change ($\Delta[\dot{\text{L}}$-] and $\Delta[\dot{\text{H}}$+]) eventually decline back to zero or even become negative. *Very Heavy:* above θL, but with a component of "excess" $\dot{V}O_2$ that leads $\dot{V}O_2$ to attain the maximum $\dot{V}O_2$ ($\dot{V}O_{2max}$) despite the steady state value predicted from a first-order response being sub-maximal; note that there are sustained increases in both [lactate] and [H+] and their rates of change (i.e. Δ[L-] and Δ[H+], and $\Delta[\dot{\text{L}}$-] and $\Delta[\dot{\text{H}}$+] are now all positive). *Severe:* a supra-maximal work rate where fatigue occurs so rapidly that the excess $\dot{V}O_2$ component has not had time to develop discernibly; again, Δ[L-], Δ[H+], $\Delta[\dot{\text{L}}$-] and $\Delta[\dot{\text{H}}$+] are positive. O_2Def represents the calculable O2 deficit, depicted by the area enclosed by the short dashes; * indicates $\dot{V}O_{2max}$.

be determined by the mean change in the ratio between $\dot{Q}O_2$ and the muscle blood flow ($\dot{Q}m$):

$$Cv(m)O_2 = CaO_2 - (\dot{Q}O_2/\dot{Q}m)$$

The exercise-induced change in $\dot{Q}m$ will be reflected virtually instantaneously in an increased $\dot{Q}p$. Consequently, the proportional change of $\dot{Q}p$ will differ from that of $\dot{Q}m$ only as a result of changes in (a) the mean blood flow in the remaining vascular beds - i.e. the "distributive" effect - (the net change is thought to be small at these work rates: Ref. 6); and (b) in the volume of the intervening venous pool - i.e. the "capacitative" effect. In contrast, the influence of the altered $Cv(m)O_2$ on $C\bar{v}O_2$ will necessarily be delayed. The net effect is that the simple mono-exponential increase in $\dot{Q}O_2$ is transformed into a more complex two-component increase in $\dot{V}O_2$, but with a time constant that is likely to be less than 10% different from that of muscle $\dot{Q}O_2$ (see Ref. 2 for discussion).

The size of the oxygen deficit (O_2Def) is dominated by the $\phi2$ $\dot{Q}O_2$ kinetics and may be computed as:

$$O_2\text{Def} = A \cdot \tau'$$

where A is the required steady-state increment in $\dot{V}O_2$, which in this domain equals $\Delta\dot{V}O_2$(ss), and τ' is the "effective" time constant (25) or "mean response time" (17) of the $\dot{V}O_2$ response. It is important to recognize that τ' in this formulation is not the actual $\phi2$ time constant (26); rigorous computation of O_2Def requires that the entire data set from the start of the exercise be included. The appropriate fitting procedure for O_2Def could either utilize a single exponential constrained to start at exercise onset (i.e. $\delta = 0$ in eq. 1) or from the sum of the best exponential-plus-delay to the data. It should be noted that the delay derived from the latter strategy (ie. the "effective" delay) has no physiological meaning - unlike the "real" or physiological delay which is manifest as the $\phi1$-$\phi2$ transition.

In this intensity domain, $\tau\dot{V}O_2$ does not vary appreciably between work rates of different amplitudes (15,24-26). The early transient rise in blood lactate (4) which is not uncommon at these work rates does not seem to influence the response discernibly. Furthermore, the off-transient $\dot{V}O_2$ time constant is not appreciably different from that at the on-transient (9,15), despite lactate having typically returned to resting levels before the start of the recovery phase.

3. HEAVY EXERCISE

The $\dot{V}O_2$ response becomes appreciably more complex above the lactate threshold (3,17,18,21), with both time- and amplitude-based nonlinearities of response. For square-wave exercise below θL on a cycle ergometer, the steady-state increment in $\dot{V}O_2$ ($\Delta\dot{V}O_2$(ss)) increases as a linear function of work rate ($\dot{W}$) with a slope, $\Delta\dot{V}O_2$(ss)/$\Delta\dot{W}$, of ~10 ml · min-1· Watt. At work rates above θL, a steady state of $\dot{V}O_2$ is either delayed or is unattainable. When $\dot{V}O_2$ does eventually stabilize, or when its asymptotic value is estimated, $\Delta\dot{V}O_2$(ss)/$\Delta\dot{W}$ has been shown to be markedly increased; the increase being a function of both the supra-threshold work rate and time: values of 13 ml· min-1· Watt are not uncommon during tests of 10-15 min duration (10,21,22) (Fig. 1: "heavy").

The difference between the "expected" steady-state value (i.e. projected from the sub-θL $\dot{V}O_2$-$\dot{W}$ relationship) and the actual $\dot{V}O_2$ achieved in the quasi-steady state was shown by Whipp & Mahler (24) to be positive for the range of work rates above θL at which the steady-state projection is less than the subject's $\dot{V}O_{2max}$. This additional increment in $\dot{V}O_2$ may, for convenience, be termed "excess" $\dot{V}O_2$ ($\dot{V}O_2(xs)$). This $\dot{V}O_2(xs)$ is a result of a slow component of the $\dot{V}O_2$ kinetics which is *superimposed* upon the early $\dot{V}O_2$ response. Furthermore, this superimposed component is of delayed onset, beginning some minutes into the test (3,17,18,24).

Both Paterson & Whipp (18) and Barstow & Molé (3) were able to demonstrate that not only was this slow component of delayed onset but that the early component of the kinetics remains exponential and projects to a steady-state value that gives the same gain (ie. $\Delta\dot{V}O_2/\Delta\dot{W}$) as for subthreshold exercise.

The presence of the delayed "excess" O_2 uptake component during high-intensity exercise appears therefore to undermine three fundamental assumptions of models of human pulmonary gas exchange. First, the gain term is not constant; that is $\Delta\dot{V}O_2$ is *not* a linear function of the work rate (1). Secondly, the power-duration curve being attributable to a single aerobic term (having a rapid τ of only 10-20 s) is unjustified (27). And, finally, the assumptions inherent in the conventional means of computing the O_2 deficit under these conditions need fundamental reappraisal.

As the additional component is both slow and of delayed onset, its influence is virtually undetectable for constant-load tests in which the subject reaches the maximum $\dot{V}O_2$ in a few minutes. Under these conditions, the O_2 deficit (O_2Def) becomes independent of the work rate. Rather, it can be determined as the product of $\dot{V}O_{2max}$ and τ' (23) (Fig. 1: "severe"). This is only true, however, if there has not been sufficient time for the excess $\dot{V}O_2$ component to develop and 'distort' the mono-exponentiality and if $\tau\dot{V}O_2$ remains unchanged (23).

In contrast, the O_2 deficit may not be rigorously determined from conventional formulations when there is excess $\dot{V}O_2$. This component, being of delayed onset, yields an asymptotic value of $\dot{V}O_2$ which is inappropriate for the early $\dot{V}O_2$ response (23) (Fig. 1: "heavy" and "very heavy").

The off-transient $\dot{V}O_2$ response, however, is often mono-exponential in this domain (18). This may prove important in elucidating the mechanism of $\dot{V}O_2(xs)$. For example, it seems to rule out a significant effect of either the O_2 cost of cardiac and respiratory work or of the $\dot{Q}10$ effect. (Potential mediators of the excess $\dot{V}O_2$ are discussed in Ref. 19.)

While there is a range of supra-θL work rates in which a delayed steady state may eventually be reached (19-22), $\dot{V}O_2$ continues to increase throughout the test at even higher work rates until the maximum $\dot{V}O_2$ is attained (19-21) (Fig. 1: "very heavy"). It is not possible, therefore, for a subject to perform a constant work rate that provides a particular percentage of the $\dot{V}O_{2max}$ at these work rates. This % $\dot{V}O_{2max}$ can only be attained fleetingly, as $\dot{V}O_2$ continues to rise. The highest work rate at which a steady state of $\dot{V}O_2$ can be attained (and hence a sustainable % $\dot{V}O_{2max}$) appears to coincide with the highest work rate at which blood [lactate] and [H+] do not continue to rise throughout the test (19,21) (Fig. 1). This work rate correlates closely with the asymptote of the subject's power-duration curve (19). Above this critical power, the more rapidly the slow component projects $\dot{V}O_2$ towards $\dot{V}O_{2max}$, the shorter will be the tolerable duration of the work rate.

4. CONCLUSIONS

As both the lactate threshold and the critical power represent such highly variable proportions of the maximum $\dot{V}O_2$ in different subjects and also in the same subject at different stages of training, assigning work intensity either as multiples of resting metabolic rate or as percentages of $\dot{V}O_{2max}$ seems no longer justifiable. A more logically-defensible strategy for establishing such intensity domains (for dynamic muscular exercise, at least) might be based upon characteristics of pulmonary gas exchange and acid-base profiles. The advantage of this strategy is that the parameters which partition the intensity domains are physiologically justifiable and experimentally discriminable.

The lactate threshold can be used to establish the transition between the "moderate" domain, in which ventilation and gas exchange attain steady states with first-order kinetics but without sustained metabolic acidemia, and the "heavy" domain, in which ventilation and gas exchange attain steady states with more complex kinetics and with sustained, but not progressive, metabolic acidemia. Above the critical power, however, steady states of ventilation, pulmonary gas exchange and acid-base variables are not attainable. $\dot{V}O_{2max}$ is achieved at all work rates above the critical power, even when the apparent $\dot{V}O_2$ requirement (i.e. from sub-threshold projection) is less than $\dot{V}O_{2max}$. This intensity may be considered to be "very heavy". Work rates for which the projected $\dot{V}O_2$ steady state is above $\dot{V}O_{2max}$ may be considered to be of "severe" intensity.

Such categorization would allow common features of ventilatory, gas exchange and acid-base profiles to characterize a particular exercise intensity. The further complicating factor of how these patterns are perceived by different subjects, however, may also need to be incorporated subsequently into intensity assignments. Similarly, exercise which is limited by impaired systemic function (eg. lung disease) will add an additional layer of complexity.

The not-insignificant flaw in implementing the proposed strategy, however, is that while the lactate threshold and the maximum $\dot{V}O_2$ can be estimated or measured from the results of a single incremental work-rate test, there is currently no similarly-convenient means of determining an individual's critical power. Devising such a test, that is valid and reliable, should be seen as a major experimental-design challenge with potential benefits in both the athletic and clinical exercise milieux.

5. REFERENCES

1. Åstrand, P.-O., and K. Rodahl. *Textbook of Work Physiology*. New York, McGraw-Hill, 1970, p. 284.
2. Barstow, T.J., N. Lamarra, and B.J. Whipp. Modulation of muscle and pulmonary O_2 uptakes by circulatory dynamics during exercise. *J. Appl. Physiol.* **68**: 979-989, 1990.
3. Barstow, T.J., and P.A. Molé. Linear and nonlinear characteristics of oxygen uptake kinetics during heavy exercise. *J. Appl. Physiol.* **71**: 2099-2106, 1991.
4. Cerretelli, P., and P.E. Di Prampero. Gas exchange in exercise. In: *Handbook of Physiology 3. The Respiratory System, vol.IV*, edited by L.E. Fahri and S. M. Tenney. Bethesda, American Physiological Society, 1987, pp. 297-339.
5. Chance, B., J.S. Leigh, Jr., B.J. Clarke, J. Maris, J. Kent, S. Nioka, and D. Smith. Control of oxidative metabolism and oxygen delivery in human skeletal muscle: a steady-state analysis of the work/energy cost transfer function. *Proc. Natl. Acad. Sci.* **82**: 8384-8388, 1985.
6. Clausen, J.P. Circulatory adjustments to dynamic exercise and effect of physical training in normal subjects and in patients with coronary artery disease. *Prog. Cardiovasc. Dis.* **18**: 459-495, 1976.
7. Dennis, S.C., T.D. Noakes, and A.P. Bosch. Ventilation and blood lactate increase exponentially during incremental exercise. *J. Sports Sci.* **10**: 437-449, 1992.

8. Funk, C.I., A. Clark, Jr., and R.J. Connett. A simple model of aerobic metabolism: applications to work transitions in muscle. *Amer. J. Physiol.* **258**: C995-C1005, 1990.

9. Griffiths, T.L., L.C. Henson, and B.J. Whipp. Influence of peripheral chemoreceptors on the dynamics of the exercise hyperpnea in man. *J. Physiol. (Lond.)* **380**: 387-403, 1986.

10. Henson, L.C., D.C. Poole, and B.J. Whipp. Fitness as a determinant of oxygen uptake response to constant-load exercise. *Europ. J. Appl. Physiol.* **59**: 21-28, 1989.

11. Hughson, R.L., and M. Morrissey. Delayed kinetics of respiratory gas exchange in the transition from prior exercise. *J. Appl. Physiol.* **52**: 921-929, 1982.

12. Hughson, R.L., K.H. Weisiger, and G.D. **Swanson**. Blood lactate concentration increases as a continuous function in progressive exercise. *J. Appl. Physiol.* **62**: 1975-1981, 1987.

13. Krogh, A., and J. Lindhard. The regulation of respiration and circulation during the initial stages of muscular work. *J. Physiol. (Lond.)* **47**: 112-136, 1913.

14. Kushmerick, M.J., R.A. Meyer, and T.R. Brown. Regulation of oxygen consumption in fast- and slow-twitch muscle. *Amer. J. Physiol.* **263**: C598-C606, 1992.

15. Lamarra, N., B.J. Whipp, M. Blumenberg, and K. Wasserman. Model-order estimation of cardiorespiratory dynamics during moderate exercise. In: *Modelling and Control of Breathing*, edited by B.J. Whipp and D.M. Wiberg. New York: Elsevier, 1983, pp. 338-345.

16. Lamarra, N., B.J. Whipp, S.A. Ward, and K. Wasserman. Breath-to-breath "noise" and parameter estimation of exercise gas-exchange kinetics. *J. Appl. Physiol.* **62**: 2003-2012, 1987.

17. Linnarsson, D. Dynamics of pulmonary gas exchange and heart rate changes at start and end of exercise. *Acta Physiol. Scand.* (suppl.) **415**: 1-68, 1974.

18. Paterson, D.H., and B.J. Whipp. Asymmetries of oxygen uptake transients at the on- and off-set of heavy exercise in humans. *J. Physiol. (Lond.).* **443**: 575-586, 1991.

19. Poole, D.C., S.A. Ward, G.W. Gardner, and B.J. Whipp. Metabolic and respiratory profile of the upper limit for prolonged exercise in man. *Ergonomics* **31**: 1265-1279, 1988.

20. Poole, D.C., S.A. Ward, and B.J. Whipp. Effect of training on the metabolic and respiratory profile of heavy and severe exercise. *Europ. J. Appl. Physiol.* **59**: 421-429, 1990.

21. Roston, W.L., B.J. Whipp, J.A. Davis, R.M. Effros, and K. Wasserman. Oxygen uptake kinetics and lactate concentration during exercise in man. *Amer. Rev. Resp. Dis.* **135**: 1080-1084, 1987.

22. Whipp, B.J. Dynamics of pulmonary gas exchange. *Circulation* **76**: VI-18 -VI-28, 1987.

23. Whipp, B.J. The slow component of O_2 uptake kinetics during heavy exercise. *Med. Sci. Sports Ex.* **26**: 1319-1326, 1994.

24. Whipp, B.J., and M. Mahler. Dynamics of pulmonary gas exchange during exercise. In: *Pulmonary Gas Exchange, vol. II*, edited by J.B. West. New York, Academic Press, 1980, pp. 33-96.

25. Whipp, B.J., and S.A. Ward. Physiological determinants of pulmonary gas exchange kinetics during exercise. *Med. Sci. Sports Ex.* **22**: 62-71, 1990.

26. Whipp, B.J., S.A. Ward, N. Lamarra, J.A. Davis, and K. Wasserman. Parameters of ventilatory and gas exchange dynamics during exercise. *J. Appl. Physiol.* **52**: 1506-1513, 1982.

27. Wilkie, D.R. Equations describing power input by humans as a function of duration of exercise. In: *Exercise Bioenergetics and Gas Exchange*, edited by P. Cerretelli and B.J. Whipp. Amsterdam: Elsevier, pp. 75-80, 1980.

13

DOES VENTILATION EVER LIMIT HUMAN PERFORMANCE?

Craig A. Harms and Jerome A. Dempsey

John Rankin Laboratory of Pulmonary Medicine
Department of Preventive Medicine
University of Wisconsin-Madison
504 N Walnut Street
Madison, Wisconsin 53705

1. INTRODUCTION

Factors which limit human performance have been of interest to both researchers and athletes for many years. While it is well known that human performance is complex and multifaceted, at least in endurance related activities (e.g. running, cycling, etc), success is related to how well an individual is able to introduce, distribute, and utilize oxygen in the body. Maximal oxygen consumption ($\dot{V}O_{2max}$) is determined by the product of cardiac output and arterio-venous oxygen difference (Fick equation). Specifically, arterial oxygen content is greatly influenced by the effect of pulmonary ventilation and alveolar-capillary diffusion on arterial oxygen pressure (PaO_2) and arterial oxygen saturation (SaO_2). Therefore, a central question in oxygen transport is whether or not the pulmonary system can serve as a weak link in the oxygen transport chain. Traditionally, oxygen transport is believed to be limited by the heart's ability to distribute blood and oxygen throughout the body (11). However, in recent years it has been demonstrated that in highly aerobic individuals, the demand of the exercise for O_2 transport may exceed the capacity of the respiratory system and therefore lead to impairments in pulmonary gas exchange and consequently lead to exercise induced hypoxemia (EIH). Also, high ventilatory work during intense exercise may contribute to exercise limitation via a high oxygen cost of breathing and therefore "steal" blood flow from limb locomotor muscles. It is the intent of this discussion to focus on some of the possibilities in which the pulmonary system could potentially limit human performance in the young healthy adult.

The Physiology and Pathophysiology of Exercise Tolerance
edited by Steinacker and Ward, Plenum Press, New York, 1996

2. LACK OF VENTILATORY LIMITATION IN THE MODERATELY TRAINED

First, it is important to establish that in most healthy young adult males with VO_{2max} < 65 ml/kg/min, the lung and chest wall operate efficiently and easily within their relative capacities at all exercise intensities up to maximum and for long durations of exercise at high intensity. The alveolar to arterial PO_2 difference widens 2-2.5 fold at VO_{2max}, however. Nevertheless, PaO_2, SaO_2, and oxygen content are all well maintained. Also, ventilation at VO_{2max} is less than the maximal available ventilation and its O_2 cost is only about 8-10% of VO_{2max}. Finally, alveolar ventilation (VA) increases out of proportion to increasing VCO_2 or VO_2, arterial PCO_2 falls, and alveolar PO_2 rises at VO_{2max}. Therefore it is safe to conclude that at least in the majority of the healthy population, the pulmonary system does not limit human performance under any exercise conditions. It is likely that maximum cardiac output to working muscle presents the dominant limiting factor to VO_{2max} (11).

3. EXERCISE-INDUCED HYPOXEMIA

In many, but not all athletes with high aerobic capacity (>65 ml/kg/min) alveolar-end capillary oxygen disequilibrium may occur in heavy exercise causing arterial hypoxemia and thereby limiting VO_{2max} (3,4,5,6). During maximal and near maximal exercise in many of these athletes, excessive widening (> 30 mmHg) of the alveolar-arterial oxygen difference (A-a DO_2) occurs, the arterial pressure of oxygen (PaO_2) falls below 75 mmHg, and the oxygen saturation of arterial blood declines below 90%. The following examples demonstrates typical responses of young men with average VO_{2max} and high VO_{2max}:

Note that in untrained Subject 1 the A-a DO_2 widens, but hyperventilation increased PAO_2 to such an extent that PaO_2 was maintained at resting levels. In Subjects 2 & 3 with high VO_{2max}, both have a substantially greater A-a DO_2 than Subject 1 (45 vs 25 mmHg), probably due primarily to greater diffusion limitation (14). Both also have significant arterial hypoxemia and O_2 desaturation, but the most hypoxemic of the two (Subject 3) shows no significant alveolar hyperventilation at maximal exercise. This excessively widened A-a DO_2 is inevitable at high VO_{2max} and is an essential feature of EIH as attested to by the high correlation commonly reported between PaO_2 and A-a DO_2 at maximal exercise (3,5,6). At the same time, severe arterial hypoxemia is achieved in many subjects when the widened A-a DO_2 is *combined with* an "inadequate" hyperventilatory response (Subject 3). In a sense then, the widened A-a DO_2 lowers PaO_2 in the highly trained endurance athlete with high VO_{2max} to the shoulder of the oxygen dissociation curve and the inade-

Table 1. Typical blood and alveolar gases of three young adult men at VO_2max. VO_2max expressed in ml/kg/min; alveolar PO_2, arterial PO_2, A-a DO_2, and $PaCO_2$ expressed in mmHg, SaO_2 expressed in %

Subject	VO_2max	PAO_2	PaO_2	A-a DO_2	SaO_2	$PaCO_2$
1	40	120	95	25	94	30
2	75	120	75	45	90	30
3	75	105	60	45	87	40

quate hyperventilation further drives it down to the steep portion of the curve where small changes in partial pressure result in large changes in O_2 saturation and O_2 content. It is certainly true that inadequate hyperventilation *by itself* will not cause significant EIH, but this claim is of little relevance to heavy exercise. A blunted ventilatory response in the highly trained subjects will also be accompanied by a widened A-a DO_2 and it is the combination of these factors that commonly produces severe EIH.

Exercise-induced hypoxemia is attributed to the relative lack of adaptability of the lung and airways in response to the exercise training stimulus and/or to inborn structural capacities which predispose the subject to a higher VO_{2max} (15). That is, the maximal demand for O_2 transport to locomotor muscles has apparently, through intense training and/or genetic predisposition of these endurance athletes, exceeded the capacity of their pulmonary system to maintain adequate arterial oxygenation. In other words, this once overbuilt pulmonary system (in the least trained subjects) has not shown significant adaptations with increasing fitness and is now *underbuilt* with respect to the demand for O_2 transport in the highly trained athlete. This hypothesis pertains both to the alveolar-capillary diffusion surface area and to the structure of the airways (see below).

4. HYPERVENTILATION OF EXERCISE

Ventilatory requirement must rise with increasing maximum metabolic rate, but what controls hyperventilation in heavy exercise? In general, hyperventilation is determined by the magnitude of the available neuro-chemical stimuli to breathe, the degree of mechanical limitation to flow and volume present in maximum exercise, and the subject's inherent responsiveness (of respiratory motor output) to the combination of mechanical constraints and increasing neurochemical stimuli. It has been reported that at rest, endurance athletes show a sluggish ventilatory response to chemical stimuli (8) and those subjects which demonstrate EIH are less responsive to hypoxic and hypercapnic inspirates (4). However, this may not always be the case during heavy and maximum exercise where some subjects show a marked hyperventilation that helps maintain a high PaO_2 and yet drives them to their mechanical limits for flow, volume, and pressure; whereas others show little or no hyperventilatory response and remain well within their mechanical reserves for ventilatory output yet become hypoxemic (6).

Airway mechanics have been demonstrated to contribute to the arterial hypoxemia of exercise. The upper limits to maximum exercise frequency, tidal volume and ventilation are determined structurally by the lung's resistive and elastic properties and the force-velocity characteristics of the inspiratory muscles. In humans, there is no indication that these mechanical properties of the airways undergo significant change with changing VO_{2max} (6,7). Other data demonstrate that intense whole-body physical training may increase the endurance capacity of respiratory muscles (2,9). Johnson et al (6) have demonstrated that these athletes achieve considerable expiratory flow limitation at maximal exercise even when VO_{2max} and VE_{max} are only 20-30% greater than the untrained subject. To determine if ventilation was in fact mechanically limited, these investigators first increased the inhaled (and end-tidal) PCO_2 to the flow-limited subjects and found that at maximal exercise, ventilation failed to increase with this added stimulus. When a different group of subjects with similar levels of VO_{2max} breathed helium:oxygen (which acts to increase the maximal flow:volume envelope), an immediate and sustained increase in ventilation and reduced $PaCO_2$ were observed (3). Therefore, during intense exercise, mechanical loads might reflexively inhibit and therefore constrain ventilation during high intensity exercise loads which are substantially below those instances when ventilation is not responsive to added stimuli.

5. VENTILATORY LIMITATIONS AND HUMAN PERFORMANCE

Is a reduced ventilatory response to maximal exercise in the endurance athlete beneficial or does it limit performance? On the beneficial side, a reduced ventilatory response spares the work of breathing, blood flow to the respiratory muscles, dyspnea, and avoids a decrease in cerebral blood flow. However, on the negative side, a reduced hyperventilatory response leads to limited increase in PaO_2 during intense exercise, resulting in: a) lower PaO_2 for any given A-a DO_2 (Table 1), and b) a reduced alveolar to capillary diffusion gradient and therefore a slowed rate of O_2 equilibrium along the pulmonary capillary (14). Also, there is reduced compensation for metabolic acidosis which occurs during intense exercise. Therefore, it is apparent that the effects of hypoxemia certainly outweigh any benefits which may arise from a low hyperventilatory response during intense exercise.

What are the performance consequences of these limits to diffusion or airflow? First, when the arterial oxygen desaturation (SaO_2 <92%) that was observed in some athletes during intense exercise was prevented by the subjects breathing an inspirate of 26% O_2, $\dot{V}O_{2max}$ increased significantly and in approximate proportion to the magnitude of the original hypoxemia (10). The overall extent of the influence of this limiting factor is thus probably not more than 10-15% of $\dot{V}O_{2max}$ (based on the level of desaturation demonstrated by these athletes). Therefore, the effects of EIH are probably less than those imposed by limitations in maximum stroke volume and cardiac output.

6. BLOOD FLOW REDISTRIBUTION

Another potential limiting role for ventilation depends on whether or not there is competition of blood flow between locomotor muscles and respiratory muscles during high intensity exercise when the heart's capacity for increasing cardiac output is limited. When the level of ventilatory requirement is such that severe expiratory flow limitation is realized, the oxygen cost of breathing may exceed 15% of total $\dot{V}O_2$ (1). Theoretically, this work and metabolic cost of breathing incurred by the primary respiratory and stabilizing muscles of the chest wall, plus their demand for perfusion to meet this oxygen cost, may "steal" blood flow from locomotor muscles, thereby limiting their work output. This requires that vasoconstriction occurs in working limb muscle during heavy exercise and that there is a redistribution of blood flow to the chest wall. This would likely depend in part upon the relative strength of local and autonomic controlled reflexes in the different muscle vascular beds (11) and also on the magnitude of the increased respiratory muscle work. The fact that high and progressively increasing levels of ventilation occur throughout prolonged exercise would indicate that total respiratory muscle power output remains unaffected and appropriate to requirements of CO_2 elimination. Accordingly, it seems reasonable to postulate that the respiratory muscles would receive a preferential share of blood flow at the expense of limb locomotor muscles under conditions where total cardiac output was at or near maximal.

There has been limited research to date investigating these possibilities. Secher et al (13) determined that adding arm work to ongoing exercise with the legs which were already exercising at 58-78% $\dot{V}O_{2max}$ resulted in vasoconstriction and reduced blood flow to the legs. Therefore, increased blood flow was made available to the working arms at the expense of the legs. Leg $\dot{V}O_2$ was reduced while mean arterial blood pressure was unchanged. Presumably, vasoconstriction in the legs must have occurred to prevent a fall in blood pressure. Interestingly, this vasoconstriction occurred at submaximal levels when it would be thought that a preferable alternative would have been to increase cardiac output to accommodate the increased demand of the extra muscle mass. More recent work has shown that adding arm work to leg work at 70% $\dot{V}O_{2max}$ caused increased sympathetic excitation of the legs, as evidenced from increased norepinephrine spillover (12). However, vasoconstriction (reduced blood flow or vascular conductance) did not occur in the legs. Perhaps clear, consistent demonstration of vasoconstriction might only occur when muscle mass is added at truly maximal work loads where both cardiac output and the arterio-venous oxygen difference are at maximum levels.

How does this postulate of preferential blood flow redistribution relate to human performance? Even if respiratory muscles did steal blood flow from locomotor muscles during heavy exercise, $\dot{V}O_{2max}$ would not likely be compromised. According to the Fick equation, $\dot{V}O_{2max}$ would only be reduced in this case if the arterio-venous DO_2 was lower in respiratory muscles compared to locomotor muscles. This is highly unlikely. Therefore, the major effect of the proposed redistribution of blood flow to respiratory muscles on exercise performance would be to reduce the exercise performance by either lowering maximal work rate at $\dot{V}O_{2max}$ or a shortening performance time for sustained intense exercise.

7. CONCLUSION

The results and speculations presented in this discussion are consistent with the concept that significant pulmonary limitations to exercise are most likely to occur under very high demand conditions (ie, intense exercise) in the young adult male with a high aerobic capacity, whether via arterial hypoxemia or the generation of excessive ventilatory work. Several possibilities were presented explaining how ventilation might limit human performance. These possibilities were not meant to be inclusive as other potential pulmonary-related limitations to human performance exist. These include the inability of the respiratory muscles to generate pressure to meet ventilatory requirements, respiratory muscle fatigue, and/or the sum of all chemoreceptor and mechanoreceptor inputs to give rise to severe dyspneic sensations. It is likely that interactions occur between these mechanisms as well.

8. ACKNOWLEDGMENT

Support by NHLB1 A0115469.

9. REFERENCES

1. Aaron, E.A., K.C. Seow, B.D. Johnson, and J.A. Dempsey. Oxygen cost of exercise hyperpnea: implications for performance. *J. Appl. Physiol.* **72**:1818-1825, 1992.
2. Clanton T.L., G.F. Dixon, J. Drake, and J.E. Gadek. Effects of swim training on lung volumes and inspiratory muscle conditioning. *J. Appl Physiol.* **62**:39-46, 1985.
3. Dempsey, J.A., P.G. Hanson, and K.S. Henderson. Exercise-induced arterial hypoxaemia in healthy human subjects at sea level. *J. Physiol.* **355**:161-175, 1984.

4. Harms, C.A., and J.M. Stager. Low chemoresponsiveness and inadequate hyperventilation contribute to exercise-induced hypoxemia. *J. Appl Physiol.* **79**:575-580, 1995.
5. Hopkins, S.R., and D.C. McKenzie. Hypoxic ventilatory response and arterial desaturation during heavy work. *J. Appl Physiol.* **67**:1119-1124, 1989.
6. Johnson, B.D., K.W. Saupe, and J.A. Dempsey. Mechanical constraints on exercise hyperpnea in endurance athletes. *J. Appl. Physiol.* **73**:874-886, 1992.
7. Linstedt, S.L., R.G. Thomas, and D.E. Leith. Does peak inspiratory flow contribute to setting $\dot{V}O_{2max}$? *Resp. Physiol.* **95**:109-118, 1994.
8. Martin, B.J., J.V. Weil, K.E. Sparks, R. McCullough, and R.F. Grover. Exercise ventilation correlates positively with ventilatory chemoresponsiveness. *J. Appl. Physiol.* **45**:557-564, 1978.
9. Powers, S.K., S. Grinton, J. Lawler, D. Criswell, and S. Dodd. High intensity exercise training-induced metabolic alterations in respiratory muscles. *Resp. Physiol.* **89**:169-177, 1992.
10. Powers, S.K., J. Lawler, J.A. Dempsey, S. Dodd, and G. Landry. Effects of incomplete pulmonary gas exchange on $\dot{V}O_{2max}$. *J. Appl. Physiol.* **66**:2491-2495, 1989.
11. Rowell, L. Integration of cardiovascular control systems. In: *Handbook of Physiology*, edited by L. Rowell New York: Oxford University Press, 1995.
12. Rowell, L.B., and D.S. O'Leary. Reflex control of the circulation during exercise: chemoreflexes and mechanoreflexes. *J. Appl. Physiol.* **69**:407-418, 1990.
13. Secher, N.H., J.P. Clausen, K. Klausen, I. Noer, and J. Trap-Jensen. Central and regional circulatory effects of adding arm exercise to leg exercise. *Acta Physiol. Scand.* **100**:288-297. 1977.
14. Wagner, P.D. Influence of mixed venous PO_2 on diffusion of O_2 across the pulmonary blood:gas barrier. *Clin. Physiol.* **2**:105-115, 1982.
15. Weibel, E.R., L.B. Marques, M. Constantinopol, R. Doffey, P. Gehr, and C.R. Taylor. The pulmonary gas exchanger. *Resp. Physiol.* **69**:81-100, 1987.

14

BLOOD FLOW REGULATION DURING EXERCISE IN MAN

Niels H. Secher and Bengt Saltin

The Copenhagen Muscle Research Centre
Department of Anaesthesia, Rigshospitalet
University of Copenhagen, Denmark

1. INTRODUCTION

Responses to exercise with one and several muscle groups allows for evaluation of integrative aspects of human physiology. One such area which has received much attention is the cardiovascular system. Exercise is an effective intervention as it not only brings various components of the system to function at its upper limits, but also gives clues to which variables that are primarily regulated. The focus here will be on muscle mass involvement in the exercise and the interplay between oxygen delivery and blood pressure.

2. MAXIMAL OXYGEN UPTAKE: DEPENDENCE UPON MUSCLE MASS

Early studies evaluated limiting factors for maximal oxygen uptake ($\dot{V}O_{2max}$). Taylor *et al.* (32) demonstrated a slightly larger $\dot{V}O_{2max}$ during arm cranking performed together with uphill running than for running alone. This approach was extended by Hermansen (7) to the use of ski poles during uphill treadmill running, which elicited a similar small elevation in $\dot{V}O_{2max}$. These findings were in line with the existing view that peak muscle blood flow was in the order of 50-60 ml 100 g^{-1} min^{-1}. A very large fraction of the muscle mass had to be engaged in the exercise to fully tax the capacity of the heart to deliver a blood flow. In contrast, Åstrand & Saltin (2) found the same $\dot{V}O_{2max}$ for combined arm cranking and cycling as for cycling alone, which was taken to indicate that the heart is limiting $\dot{V}O_{2max}$. The discrepancy between results was evaluated by investigation of subjects with varying level of fitness for arm exercise (25). During exercise with untrained arms, $\dot{V}O_{2max}$ was ~70% of the value obtained during leg exercise; while subjects with trained arms (e.g. rowers) this percentage increased to 90% and it was above that obtained during leg exercise in swimmers. In the former group of subjects there was no or only a minimal effect on $\dot{V}O_{2max}$ when arm exercise was added to leg exercise, while in the subjects with trained arms, $\dot{V}O_{2max}$ increased by 15% or more (24). Indeed, in a canoeist, upper body exercise can elicit almost the same maximal oxygen uptake as when performing regular bicycle exercise (16). The conclusions to be drawn are that the whole muscle mass does not have to be engaged in the exercise to attain $\dot{V}O_{2max}$ and that $\dot{V}O_{2max}$ is influenced significantly by the training status of the involved muscle groups.

It is true that physical training induces marked adaptations in skeletal muscles of humans (19). When these were first described in more detail, they were suggested to be a prerequisite for an enhanced $\dot{V}O_{2max}$ upon training. The capillary proliferation makes it possible to maintain an appropriate mean vascular transit time with the enhancement of the maximal muscle blood flow which occurs after endurance training (17). Whether the enlargement of the mitochondrial volume also plays a role for $\dot{V}O_{2max}$ is debated. What is certain, however, is that enhanced mitochondrial capacity affects the metabolic response with increased reliance on fat metabolism and saving of muscle carbohydrate stores. Whether the capillary proliferation also contributes to the enhanced muscle conductance observed after training is less obvious. However, specific training is related closely to the maximal vascular conductance of the calf (29), and the forearm (26, 28) and $\dot{V}O_{2max}$ to systemic vascular conductance (3).

3. VASOCONSTRICTION IN ACTIVE LIMBS

A more detailed analysis of the cardiovascular physiology in exercise became available with the determination of systemic and regional vascular variables, when several muscle groups were either separately or simultaneously engaged in the exercise. When intense arm cranking is added to leg exercise, blood flow and $\dot{V}O_2$ decrease over the working legs (Fig. 1; ref. 22). Similarly, venous oxygen saturation decreases over the arms, when leg exercise was added to ongoing arm exercise at a constant work rate (Fig. 2). Thus, simultaneous high intensity exercise with both arms and legs reduces blood flow and presumably also $\dot{V}O_2$ for both the arms and the legs. In a subsequent study, no reduction in leg blood flow was observed when arm exercise was combined arm with 2-legged knee-extensor exercise, but $\dot{V}O_2$ was only in the range of 75% $\dot{V}O_{2max}$ for the legs (20). In another study using ordinary cycling and adding arm cranking, several subjects did (but others did not) reduce their leg blood flow, but again the work rate was barely large enough to elicit $\dot{V}O_{2max}$ (13).

Another approach to evaluate whether the pump capacity of the heart is sufficient in humans to supply contracting vasodilated skeletal muscle with an adequate flow is the determination of true peak skeletal muscle blood flow. Above, it was mentioned that for the leg, peak values around 50 ml kg^{-1} min^{-1} were regarded as the upper level. Such values have been obtained with the plethysmograph or ^{133}Xe-washout techniques. As discussed elsewhere (9), these methods markedly underestimate true peak perfusion of skeletal muscle. Andersen & Saltin (1) used the dynamic knee-extensor model and measured the blood flow directly from the same muscle group. They found peak values of 5 to 7 l min^{-1}. Indeed, Richardson et al. (12) observed leg blood flows in the range of 10 l min^{-1} using the same exercise model. As only 2-4 kg of muscle is engaged in the exercise, the perfusion amounts to at least 200-300 ml 100 g^{-1} min^{-1} during maximal exercise. Even higher values have been observed during knee-extensor exercise in hypoxia (14). It is likely that most other skeletal muscles of man can reach similar high flows (18). This means that one kg of skeletal muscle is able to receive a blood flow as large as 2-3 l min^{-1} during intense dynamic exercise. That would require a cardiac output of 60-90 l min^{-1} if all muscles of the

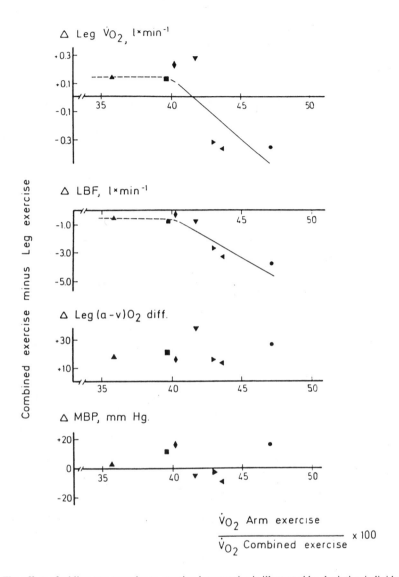

Figure 1. The effect of adding arm exercise to ongoing leg exercise is illustrated by depicting individual changes in leg oxygen uptake (Leg $\dot{V}O_2$), leg blood flow (LBF), regional arterio-venous oxygen difference ((a-v) O_2 diff.), and mean arterial blood pressure (MBP) (22). Reproduced with permission.

body were to be active at a similarly high work rate. However, maximal cardiac output in sedentary humans is 15-25 l min^{-1}, with 42 l min^{-1} being the highest ever recorded.

The decrease in muscle blood flow to an exercising muscle in consequence of the addition of another muscle group to exercise suggests either a limitation of cardiac output and/or a hindrance to flow by vasoconstriction due to an increase in the sympathetic nerve activity. In line with the earlier view that peak muscle blood flow was in the range of 50 ml 100 g^{-1} min^{-1}, allowing for the heart to produce a cardiac output sufficient to fill all

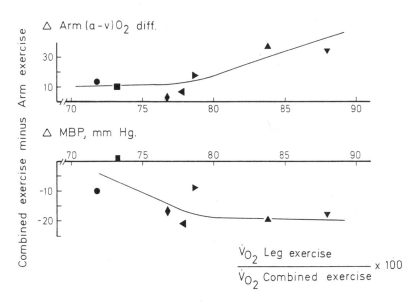

Figure 2. As in figure 1, but leg exercise is added to ongoing arm exercise. For symbols see legend to figure 1 (22). Reproduced with permission.

capillary beds of skeletal muscles when dilated, the hypothesis of "sympatholysis" in an active muscle minimizing a role of sympathetic nerve activity to induce vasoconstriction and decrease blood flow gained support. However, several studies have demonstrated an important role for sympathetic nerve activity in regulating muscle blood flow during exercise in both humans (27, 30) and animals (4, 33). Microneurographic recordings of sympathetic nerve activity cannot be performed in an exercising limb due to movement artifacts and a dominance of motor activity in the nerve. Instead, noradrenaline spillover in a limb has been used as an index of sympathetic nerve activity.

Both Savard *et al.* (20) and Richter *et al.* (13) observed an elevated leg noradrenaline spillover from the leg during combined arm and leg exercise, although, as mentioned above, leg blood flow was not affected on an average. These findings could argue that even enhanced sympathetic nerve activity is unable to modulate muscle flow in humans. Alternatively, the effect of an elevated sympathetic nerve active on muscle blood flow depends on a balance between local vasodilatator influence of metabolites and the intensity of the nerve activity (Figs. 1 and 2). Up to a given level, local vasodilation is not overridden by the sympathetic vasoconstrictor activity. The critical factor is likely to be that peripheral demand for a blood flow exceeds peak cardiac output. Thus, Pawelczyk *et al.* (11) used cardiac beta1-blockade, which resulted in a 3.6 and 1.2 l min^{-1} reduction of cardiac output and leg blood flow, respectively, at the highest work rate. Noradrenaline spillover became markedly elevated only when leg blood flow could not be compensated for by an increased arterio-venous difference for oxygen, i.e. when muscle pH decreased and/or other metabolites accumulated markedly.

In patients with a low cardiac reserve as in chronic heart failure, a reduction in muscle perfusion during exercise is observed already when more than 4 kg of muscle is engaged in the exercise (10). Conversely, when cardiac output is enhanced in such patients already in sinus rhythm by digitalis, leg blood flow is also increased (21). Indeed, in patients with cardiac insufficiency there is also a reduction in the blood flow velocity of the middle cerebral artery (6). Also the normal increase in middle cerebral artery mean flow velocity upon exercise is reduced to approximately half in patients with atrial fibrillation (5).

4. INTEGRATED VIEW

It is a fundamental problem for human vascular control that the blood volume is smaller than the potential size of the vascular bed. At rest the size of the vascular bed is adjusted by sympathetic vasoconstriction. Thus, when sympathetic tone is lost, blood pressure may be affected, as illustrated during regional anaesthesia (23) where vascular tone is lost (15). Equally, during dynamic exercise with large muscle groups, a significant part of the vascular bed represented by the active muscle dilates, and circulatory and blood pressure control are challenged. This is especially so at the onset of exercise (8) and even more so when combined with orthostasis in the transition from the supine to the upright position (34). With continued exertion, blood pressure becomes stable because the muscle pump secures that the central blood volume is elevated and the plasma concentration of the atrial natiuretic peptide is elevated. Evidence suggests that muscle vasodilatation is not allowed to progress unlimited. Reflex vasoconstriction takes place when presumably metabo- (31) and mechanoreceptors (35) in working muscle are activated. Significant reduction of muscle vasodilatation has been demonstrated under two circumstances. One is when muscle blood flow becomes limited due to a disproportionate, high work rate of another muscle group, even when the muscle mass is small. The other circumstance that limits regional blood flow is when exercise is performed with a reduced cardiac output. In both situations, evidence supports sympathetic vasoconstriction in the active muscles.

REFERENCES

1. Andersen, P., and B. Saltin, B. Maximal perfusion of skeletal muscle in man. *J. Physiol. Lond.* **366**:233-249, 1985.
2. Åstrand, P.-O., and B. Saltin. Maximal oxygen uptake and heart rate in various types of muscular activity. *J. Appl. Physiol.* 16:977-981, 1961.
3. Clausen, J.P. Circulatory adjustments to dynamic exercise and effects of physical training in normal subjects and in patients with coronary artery disease. *Prog. Cardiovas. Dis.* **18**:459-495, 1976.
4. Donald, D.E., D.J. Rowlands, and D.A. Ferguson. Similarity of blood flow in the normal and the sympathectomiced dog limb during graded exercise. *Clin. Res.* **26**:185-199, 1970.
5. Gulløv, A.L., F. Pott, B.K. Koefoed, P. Peteren, and N.H. Secher. Transcranial Doppler determined cerebral arterial blood velocity during cycling in arterial fibrillation. XV Nor. Cong. Cardiol., Malmø, Sweden, 1995, p. 37.
6. Hellström, G., G. Magnusson, B. Saltin, and N.G. Wahlgren. Cerebral haemodynamic effects of physical exercise in patients with chronic heart failure (abstract). *Nord. Neurol. Soc.* 1994.
7. Hermansen, L. Oxygen transport during exercise in human subjects. *Acta Physiol. Scand.* Supp. **299**:1-104, 1973.
8. Holmgren, A. Circulatory changes during muscular work in man. *Scand. J. Clin. Lab. Invest.* **8**: suppl. 24., 1956.
9. Kim, C.K., S. Strange, J. Bangsbo, and B. Saltin. Skeletal muscle perfusions in electrically induced dynamic exercise in humans. *Acta Physiol Scand* **153**; 279-287, 1995.
10. Magnusson, G., L. Kaijser, C. Sylven, K.-E. Karlberg, B. Isberg, and B. Saltin. Peak skeletal muscle perfusion is maintained in patients with chronic heart failure. In press 1995.
11. Pawelczyk, J.A., B. Hanel, R.A. Pawelczyk, J. Warberg, and N.H. Secher. Leg vasoconstriction during dynamic exercise with reduced cardiac output. *J. Appl. Physiol.* **73**:1838-1846, 1992.

12. Richardson, R.S., D.C. Poole, D.R. Knight, S.S. Kurdak, M.C. Hogan, B. Grassi, E.C. Johnson, K.F. Kendrick, B.K. Erickson, and P.D. Wagner. High muscle blood flow in man: is maximal O_2 extraction compromised? *J. Appl. Physiol.* **75**:1911-1916, 1993.

13. Richter, E.A., B. Kiens, M. Hargreaves, and M. Kjær. Effects of arm-cranking on leg blood flow and noradrenaline spillover during leg exercise in man. *Acta Physiol. Scand.* **144**:9-14, 1992.

14. Rowell, L.B., B. Saltin, B. Kiens, and N.J. Christensen. Is peak quadriceps blood flow in humans even higher during exercise with hypoxemia. *Am. J. Physiol.* **251**:H1038-H1044, 1986.

15. Rørdam P., H.L. Olesen, J. Sindrup, and N.H. Secher. Effect of epidural anaesthesia on dorsal pedis arterial diameter and blood flow. *Clin Physiol* **15**: 143-149, 1995.

16. Saltin, B. Aerobic and anaerobic work capacity at an altitude of 2,250 meters. In: *Proc. Symp. on Physical Performance at Altitude*, edited by U. Luft, 1967, pp. 97-102.

17. Saltin, B. Malleability of the system in overcoming limitations: functional elements. *J. Exp. Biol.* **115**: 345-354, 1985.

18. Saltin, B. Maximal oxygen uptake; limitation and malleability. In: *International Perspectives in Exercise Physiology*, edited by K. Nazar, R.L. Terjung, H. Kaciuba-Uscilko, and L. Budohoski. Champaign, USA: Human Kinetics, 1990, pp. 26-40.

19. Saltin, B., and P.D. Gollnick. Skeletal muscle adaptability: significance for metabolism and performance. In: *Handbook of Physiology: Skeletal Muscle*, sect. 10, edited by L.D. Peachey, R.H. Adrian, and S.R. Geiger. Bethesda, USA: Amer. Physiol. Soc., 1983, pp 555-631.

20. Savard, G.K., E.A. Richter, S. Strange, B. Kiens, N.J. Christensen, and B. Saltin. Norepinephrine spillover from skeletal muscle during exercise in humans: role of muscle mass. *Am. J. Physiol.* **257**:H1812-H1818, 1989.

21. Schmidt T.A., H. Bundgaard, H.L. Olesen, N.H. Secher, and K. Kjeldsen. Digoxin affects potassium homeostasis during exercise in patients with heart failure. *Cardiovas. Res.* **29**: 506-511, 1995.

22. Secher, N.H., J.P. Clausen, K. Klausen, I. Noer, and J. Trap-Jensen. Central and regional circulatory effects of adding arm exercise to leg exercise. *Acta Physiol. Scand.* **100**:288-297, 1977.

23. Secher, N.H., J. Jacobsen, D.B. Friedman, and S. Matzen. Bradycardia during reversible hypovolaemic shock: associated endocrine changes and clinical implications. *Clin. Exp. Pharm. Physiol.* **19**:733-743, 1992.

24. Secher, N.H., and Oddershede, I. Maximal oxygen uptake during swimming and bicycling. In: *Swimming II*, edited by L. Lewillie and J.P. Clarys. Baltimore, USA: Univ. Park Press, 1975, pp 137-142.

25. Secher, N.H., N. Ruberg-Larsen, R.A. Binkhorst, and F. Bonde-Petersen. Maximal oxygen uptake during arm and combined arm plus leg exercise. *J. Appl. Physiol.* **36**:315-318, 1974.

26. Sinoway L.I., T.I. Musch, J.R. Minotti, and R. Zelis. Enhanced maximal metabolic vasodilatation in the dominant forearms of tennis players. *J. Appl. Physiol.* **61**:673-678, 1986.

27. Sinoway, L., and S. Prophet. Skeletal muscle metaboreceptor stimulation opposes peak metabolic vasodilation in humans. *Cir. Res.* **66**:1576-1586, 1990.

28. Sinoway L.I., J. Shenberger, J.S. Wilson, D. McLaughlin, T. Musch, and R.A. Zelis. 30-day forearm work protocol increases maximal forearm blood flow. *J. Appl. Physiol.* **62**:1063-1067, 1987.

29. Snell, P.G., W.H. Martin, J.C. Burckey, and C.G. Blomqvist. Maximal vascular leg conductance in trained and untrained men. *J. Appl. Physiol.* **62**:606-610, 1987.

30. Strandell, T., and J.T. Shepherd. The effect in humans of increased sympathetic activity on the blood flow to active muscles. *Acta Med. Scand. Supp.* 472:146-167, 1967.

31. Strange, S., N.H. Secher, J.A. Pawelczyk, J. Kappakka, N.J. Christensen, J.H. Mitchell, and B. Saltin. Neural control of cardiovascular responses and of ventilation during dynamic exercise in man. *J. Physiol. (Lond.)* **470**:693-704, 1993.

32. Taylor, H.L., E. Buskirk, and A. Henchel. Maximal oxygen intake as an objective measure of cardiorespiratory performance. *J. Appl. Physiol.* **8**:73-80, 1955.

33. Thompson, L.P., and D.E. Mohrman. Blood flow and oxygen consumption in skeletal muscle during sympathetic stimulation. *Am. J. Physiol.* **245**:H66-H71, 1983.

34. Wieling, W., and J.J. van Lieshout. Circulatory adaptation upon standing. In: *New Trends in Autonomic Nervous System Research*, edited by M. Yoshikawa, M. Uono, H. Tanabe, and S. Ishikawa. Amsterdam, Excerpta Medica, 1991, pp. 200-204.

35. Williamson, J.W., J.H. Mitchell, H.L. Olesen, P. Raven, and N.H. Secher. Reflex increase in blood pressure induced by leg compression in man. *J. Physiol. (Lond.)* **475**: 351-357, 1994.

15

THERMOREGULATION AND FLUID BALANCE AS POSSIBLE LIMITING FACTORS IN PROLONGED EXERCISE IN HUMANS

R. J. Maughan, S. D. R. Galloway, Y. Pitsiladis, S. M. Shirreffs, and
J. B. Leiper

Department of Environmental and Occupational Medicine
University Medical School
Foresterhill, Aberdeen AB9 2ZD, Scotland

1. INTRODUCTION

Fatigue is the inevitable result of prolonged strenuous exercise, but the nature of the fatigue process and the time for which exercise can be sustained will be influenced by many factors. The most important of these is undoubtedly the intensity of the exercise in relation to the cardiovascular and metabolic capacity of the individual, and the most effective way to delay the onset of fatigue and improve performance is by training. The primary cause of fatigue in exercise lasting more than one hour but not more than 4-5 hours is usually considered to be the depletion of the body's carbohydrate reserves: in bicycle exercise (6), but less convincingly in running (9), the glycogen content of the exercising muscles appears to be crucial. Systematic training results in many adaptations to the cardiovascular system and to the muscles, allowing them to increase delivery and use of oxygen and enhancing the extent to which they can use the relatively unlimited fat stores as a fuel and thus spare the rather small amounts of carbohydrate which are stored in the liver and in the muscles. In the same way, in situations where exercise capacity is limited by carbohydrate availability, improvements can be demonstrated to result from feeding carbohydrate before or during exercise.

It is equally clear, however, that exercise tolerance is also affected by the environmental conditions: when temperature and humidity are high, performance is invariably reduced relative to the capacity to perform the same task in more moderate climatic conditions. Figure 1 demonstrates that there is clearly an optimum environmental temperature for the performance of prolonged exercise. This suggests a major role for thermoregulatory limitations to exercise in the heat, although it does not exclude an effect of elevated temperature on other aspects of physiological function, and also raises questions as to how far the same mechanisms can influence performance at lower ambient temperatures.

The Physiology and Pathophysiology of Exercise Tolerance
edited by Steinacker and Ward, Plenum Press, New York, 1996

a. exercise performance is influenced by ambient temperature or by alterations in muscle and/or core temperature;

b. performance is altered by the pre-exercise hydration status; and

c. administration of fluids can increase exercise capacity.

In addition to demonstrating that these effects exist, it is also necessary that a plausible explanation of the mechanism of action should exist. A review of the literature leaves no doubt as to the fact that thermoregulation and fluid balance can influence exercise capacity, but the mechanisms by which they do so are less clear.

2. EXERCISE AND TEMPERATURE REGULATION

Fluid loss during exercise is primarily a consequence of the need to maintain body temperature - or at least the temperature of the body core - within narrow limits. The resting oxygen consumption of the average human is about 250 ml/min, corresponding to a rate of metabolic heat production of about 70 W. Thermoregulation is achieved in most situations by behavioural mechanisms: the amount of clothing worn is adjusted or the ambient temperature is changed so that the rate of heat production or heat gain is balanced by the rate of heat loss. During exercise, the rate of heat production can be increased to many times the resting level. At very high power outputs, muscle temperature will rise markedly without major changes in body core temperature, but exercise duration is inevitably short and it is not clear that core temperature will rise sufficiently to limit performance.

In sustained rhythmic exercise, muscle blood flow is high and the rise in muscle temperature is limited by convective heat removal. In an event such as running on the level, the rate of heat production is determined primarily by running speed and body mass, with individual variations in mechanical and metabolic efficiency being of secondary importance. The elite marathon runner can achieve an oxygen consumption of 5 l/min or more, and highly trained athletes can sustain a power output of about 80% of this level (equivalent to 80 kJ/min or 20 kcal/min) for more than 2 hours (29). When the ambient

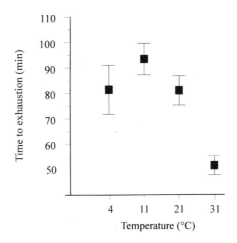

Figure 1. Exercise capacity during cycle exercise at an intensity corresponding to about 70% of $\dot{V}O_2$max in eight male subjects who were moderately trained but not heat acclimated. All trials were significantly different from each other, except for those at 4 and 21°C. Based on data from Ref. 15.

temperature is higher than skin temperature, heat will also be gained from the environment by physical transfer, adding to the heat load on the body. In spite of this, marathon runners normally maintain body temperature within 2-3°C of the resting level, indicating that heat is being lost from the body almost as fast as it is being produced. The highest temperatures in exercising subjects have generally been observed in runners competing in events at distances of 10-15 km. Sutton (45) reported more than 30 cases where rectal temperature exceeded 42°C in competitors in a 14 km road race: in such conditions, heat illness and collapse are not uncommon. These observations indicate that a high absolute rate of heat production places the greatest stress on the thermoregulatory system.

At high ambient temperatures, the only mechanism by which heat can be lost from the body is evaporation, and even at low ambient temperatures high sweat rates are observed in some individuals (26). Evaporation is an extremely effective heat loss mechanism, and evaporation of 1 l of water from the skin surface will remove 2.4 MJ (580 kcal) of heat from the body. For the 2 h 30 min marathon runner with a body mass of 70 kg to balance the rate of metabolic heat production by evaporative loss alone would therefore require sweat to be evaporated from the skin at a rate of about 1.6 l/h: at such high sweat rates, an appreciable fraction of the sweat secreted drips from the skin without evaporating, and a sweat secretion rate of about 2 l/h is likely to be necessary to achieve this rate of evaporative heat loss. This is possible, but would result in the loss of 5 l of body water, corresponding to a loss of more than 7% of body mass for a 70 kg runner. Typical body mass losses in marathon runners range from about 1-6% (0.7-4.2 kg) at low (10°C) ambient temperatures (27) to more than 8% (5.6 kg) in warmer conditions (8). Figure 2 demonstrates that the rate of sweat production is related to running speed, but it is apparent that there is a large inter-individual variation in the sweating response even when the running speed (and hence the power output) is the same.

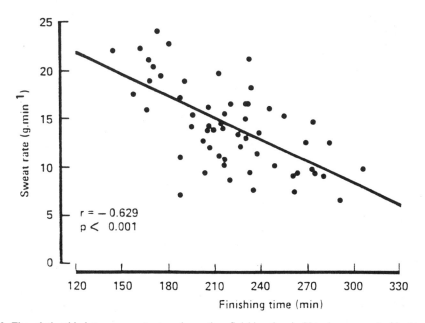

Figure 2. The relationship between sweat rate and marathon finishing time in 59 male runners. Ambient temperature during the event was 8-10°C. Reproduced with permission from Ref. 26.

Some water will also be lost by evaporation from the respiratory tract. During prolonged hard exercise in a hot dry environment, this can amount to a significant water loss, although this is not generally considered to be a major heat loss mechanism in humans. The rise of 2-3°C in body temperature which normally occurs during marathon running means that some of the heat produced is stored, but the effect on heat balance is minimal: a rise in mean body temperature of 3°C for a 70 kg runner would reduce the total requirement for evaporation of sweat by less than 300 ml.

Fluid losses are distributed in varying proportions among the plasma, extracellular water, and intracellular water. The decrease in plasma volume which accompanies dehydration may be of particular importance in influencing work capacity; blood flow to the muscles must be maintained at a high level to supply oxygen and substrates, but a high blood flow to the skin is also necessary to convect heat to the body surface where it can be dissipated (34). When the ambient temperature is high and blood volume has been decreased by sweat loss during prolonged exercise, there may be difficulty in meeting the requirement for a high blood flow to both these tissues. In this situation, skin blood flow is more likely to be compromised, allowing central venous pressure and muscle blood flow to be maintained but reducing heat loss and causing body temperature to rise sharply (39).

3. EFFECTS OF BODY HEATING AND COOLING ON EXERCISE PERFORMANCE

Although it has long been known that warm-up prior to exercise may enhance performance (4), it is equally clear that there is an optimum temperature above which a detrimental effect is observed. The process of warming up also has many effects other than simply on muscle temperature, and these effects may confound the interpretation of these studies. Several investigations have shown that manipulation of the temperature of human limbs - usually by immersion in water at different temperatures, can influence the capacity for exercise performance: because of the obvious practical difficulties, there are rather few similar studies with whole body exercise. In isometric contractions at high forces, muscle blood flow is reduced or absent: the ability to sustain submaximal isometric contractions of the forearm muscles is related to the pre-exercise muscle temperature, with the greatest duration being observed when the muscle temperature is about 27°C, compared with the normal resting value of about 34°C (7).

Bergh and Ekblom (5) studied swimming in cold water or submaximal cycling exercise to alter core (oesophageal, TES) or muscle (m vastus lateralis, TM) temperature prior to a combined arm and leg exercise test: average work time decreased from 6.8 min at normal temperature (TES 37.7°C, TM 38.5°C) to 4.4 min at reduced temperature (TES 35.8°C, TM 36.5°C) and was further reduced at the lowest temperatures studied (TES 34.9°C, TM 35.1°C) to 3.1 min. Elevation of temperature (TES 38.4°C, TM 39.3°C) had no significant effect on exercise time (6.2 min) relative to the normal condition.

In the exercising dog, cooling of the trunk with ice packs has been shown to attenuate the rise in both muscle and rectal temperature, and to extend exercise time by about 45% compared with the mean exercise time without cooling of 57 min (21): this study was carried out at an ambient temperature of 20°C.

4. EFFECTS OF AMBIENT TEMPERATURE ON EXERCISE PERFORMANCE

Many studies have investigated the effects of alterations in ambient temperature on the capacity to perform different types of exercise. In many of these investigations, only two temperature conditions have been compared, and in some of these, exercise trials at low (0-10°C) and high (40°C or higher) temperatures have been compared without a trial under more moderate conditions, but others have compared exercise in the heat with more moderate conditions. The results of these tests have invariably shown that exercise performance is impaired when the environmental temperature is high: indeed, a reduction in exercise tolerance in the heat is a matter of common experience.

The effects of high ambient temperatures on exercise capacity can be extremely large: Suzuki (46) reported that exercise time at a work load of 66% of $\dot{V}O_2$max was reduced from 91 min when the ambient temperature was 0°C to 19 min when the same exercise was performed in the heat (40°C). In an unpublished study in which 6 subjects exercised to exhaustion at 70% $\dot{V}O_2$max on a cycle ergometer, we found that exercise time was reduced from 73 min at an ambient temperature of 2°C to 35 min at a temperature of 33°C. Many other studies have observed that performance of exercise in the heat at temperatures around 30-35°C dry bulb may be substantially reduced compared to a cooler (20-25°C) environment (1,21,25,37,43).

More recently, we have measured the endurance capacity of subjects cycling at an exercise intensity of about 70% of $\dot{V}O_2$max at ambient temperatures of 4, 11, 21 and 31°C (15), and the results are shown in Figure 1. It is clear that there is an optimum temperature for performance of this type of exercise, and that this optimum occurs at about 10°C, with shorter exercise times being achieved at lower and at higher temperatures. Although fatigue at these work intensities is generally considered to result from depletion of the muscle glycogen stores, this is clearly not the case when the ambient temperature is high: the total amount of carbohydrate oxidised was much less at a high (31°C) ambient temperature, because of the short exercise time, and it seems highly improbable that muscle glycogen stores were exhausted when the subjects stopped exercising.

Maw et al. (30) have shown that ambient temperature can influence the subjective perception of effort during exercise: subjects reported a lower level of perceived exertion at an ambient temperature of 8°C or 20°C relative to exercise at the same power output at 40°C. Similar results were obtained by Galloway and Maughan (15), and these subjective responses support the notion of a reduced exercise tolerance in the heat.

5. HYDRATION STATUS AND EXERCISE PERFORMANCE

It is often reported that exercise performance is impaired when an individual is dehydrated by as little as 2% of body mass, and that losses in excess of 5% of body mass can decrease the capacity for work by about 30% (42). The original data on which these figures are based have proved elusive, but there are many reports in the literature of reductions in exercise performance resulting from hypohydration induced prior to exercise, and these studies have been comprehensively reviewed (44).

Several different methods have been used to induce body water loss, including thermal sweating, exercise at high or moderate ambient temperatures, restriction of fluid (and food) intake, and the administration of diuretic agents. The different effects of these various procedures can account for some of the variability in response which has been reported. Where exercise is used to provoke sweat loss, hyperthermia and depletion of muscle and liver glycogen stores will also be possible causes of reductions in subsequent exercise performance, and the results of these studies must be treated with caution.

Nielsen *et al.* (36) showed that prolonged exercise, which resulted in a loss of fluid corresponding to 2.5% of body mass, resulted in a 45% fall in the capacity to perform high intensity exercise lasting about 7 min. More recently, Walsh *et al.* (48) have shown that prevention of dehydration by administration of 1 litre of saline solution during 1 hour of exercise at 70% of $\dot{V}O_2$max resulted in a prolonged exercise time in a subsequent exercise test at 90% of $\dot{V}O_2$max: exercise time was 9.8 min, compared with 6.8 min on the control trial where body mass was reduced by 1.3 kg.

Passive heat exposure in a sauna or similar environment will result in an elevation of body temperature and loss of sweat, but will not affect tissue carbohydrate stores. Nielsen *et al.*. (36) found that sauna exposure (2.5% reduction in body mass, rectal temperature 39°C) reduced performance in a cycling task lasting about 7 min to 65% of the control value. Thermal sweating (6 hours exposure at 46°C) with (4.3% reduction in body mass) or without (1.9% reduction) fluid restriction reduced treadmill walking time by 48% and 22% respectively in a task that could be sustained for about 5-10 min: Ref. 12). In another study, passive heat exposure (56°C) was used to induce dehydration corresponding to 2, 4 or 5% of initial body mass: subsequent performance of a 30s maximum power output test was not affected by dehydration (19).

Saltin (41) found that exercise time in a standard test (about 6 min in the control condition) on the cycle ergometer was reduced by both sauna dehydration (2.8% reduction in body mass: 20% reduction in exercise time), hard work in a mild environment (2.5-4 h at 70% of $\dot{V}O_2$max at an ambient temperature of about 18°C, resulting in a 2.6% reduction in body mass: 25% reduction in exercise time) and by moderate work in the heat (2.5-4 h at 56% of $\dot{V}O_2$max at an ambient temperature of about 37°C, resulting in a 2.8% reduction in body mass: 19% reduction in exercise time). Other studies which have used exercise in the heat to induce sweating have shown similar results (e.g. Ref. 36).

Diuretic administration (80 mg Lasix, 2.5% reduction in body mass) reduced exercise performance by 18% in the study of Nielsen *et al.* (36). Armstrong *et al.* (3) used the same drug (40mg) to reduce body mass by 1.6-2.1%: mean running speed in simulated races over distances of 1,500 m, 5,000 m and 10,000 m was reduced by 3.1%, 6.7% and 6.3% respectively after dehydration.

Attempts have been made to induce a state of relative hyperhydration prior to exercise by administration of glycerol solutions. These have the effect of increasing total body water, but result in an increase in plasma osmolality, and so may be considered to result in a relative hypohydration, even though total body water is increased (16). There have, however, been some recent reports of improvements in the ability to perform prolonged exercise after glycerol administration (32).

There are few reports of the effects of hypohydration on mental performance and cognitive function, but some negative effects have been recorded. Gopinathan *et al.* (17) reported a reduction in a variety of tasks involving arithmetic ability, short term memory and a visual tracking test after dehydration (2-4% of body mass) induced by exercise in the heat and water restriction: performance was not affected by a 1% reduction in body mass. Other authors have reported similar effects (2,22).

6. EFFECTS OF FLUID INGESTION ON PERFORMANCE

The effects of feeding different types and amounts of beverages during exercise have been extensively investigated, using a wide variety of experimental models. Not all of these studies have shown a positive effect of fluid ingestion on performance, but, with the exception of a few investigations where the composition of the drinks administered was such as to result in gastrointestinal disturbances, there are no studies showing that fluid ingestion will have an adverse effect on performance. These studies have been the subject of a number of extensive reviews which have concentrated on the effects of administration of CHO, electrolytes and water on exercise performance, and the results of the individual studies will not be considered in detail here (11,23,27,33).

Many different fluid replacement regimens have been investigated, and in most studies, drinks administered during exercise trials have contained a variety of types and amounts of carbohydrate as well as electrolytes and other components. The problems in resolving the effects of fluid replacement from those of the supply of exogenous substrate are formidable. This is particularly true as the most effective fluid replacement is known to be achieved with drinks that contain small amounts of glucose and sodium (24). However, performance is improved by ingestion of carbohydrate, whether in liquid or solid form (10), and it seems clear that the most effective way to improve exercise performance is to replace both water and substrate. The precise balance between these two objectives that will be most effective in any given situation will depend on a number of factors, including the exercise intensity and duration, the ambient temperature and humidity and the physiological and biochemical characteristics and nutritional status of the individual (27).

There have been rather few systematic investigations of the effects of variations in the composition of ingested fluids or in the amount of fluid ingested on exercise performance. Much of the information is the literature consists of reports of what are effectively tests of candidate solutions for optimising performance: where two or more solutions are compared, they often differ in several respects, making interpretation of the results difficult. There are more coherent data relating to fluid replacement, and it is clear that increasing the amount of carbohydrate contained in ingested fluids will slow gastric emptying, thus limiting the possible rate of fluid replacement, while increasing the amount of substrate delivered to the small intestine (47). The optimum rate of water absorption in the small intestine is achieved with hypotonic (osmolality of about 220-250 mosmol/kg) solutions with relative low glucose concentrations and relatively high sodium concentrations (29).

As mentioned above, there are many studies showing improvements in performance with the administration of plain water, while other studies show that solid carbohydrate feeding can also improve performance. In one investigation, however, where several different treatments were compared, administration of a dilute glucose-electrolyte solution was found to be more effective in improving performance than was a concentrated carbohydrate solution (28).

7. POSSIBLE MECHANISMS OF REDUCED EXERCISE TOLERANCE IN THE HEAT

With exercise at different temperatures, many studies have shown changes in cardio-respiratory, thermoregulatory and metabolic function related to the ambient temperature.

There are, however, inconsistencies in the reported changes with increases in oxidation rates of both carbohydrate and lipid being reported during exercise in hot or neutral environments (14,20) compared with a cool environment, and increased oxygen costs of exercise observed in both cold (18) and hot (13) environments. Overall the reported metabolic responses to heat exposure do not provide strong evidence for substrate depletion as the primary cause of fatigue during exercise in the heat. More systematic studies examining the effects of exposure temperature on the metabolic responses to exercise in the heat are required in order to ascertain the major causes of reduced exercise tolerance.

Hypohydration is the most mechanism proposed for reduced exercise tolerance in the heat. Since the early experiments of Adolph and associates (2), the effects of hypohydration on exercise performance have been studied extensively. However, it is not clear whether the hypovolemia or the hypertonicity associated with hypohydration is responsible for the subsequent elevations in core temperature, the changes in sweat rate and the large cardiovascular drift which occur during exercise in the hypohydrated state. Montain and Coyle (31) have highlighted the inconsistent effects of fluid ingestion on core temperature and sweat rate. Typically, core temperature and heart rate are reduced and peripheral blood flow is maintained following fluid administration compared with no fluid ingestion. This therefore strongly suggests that fluid balance is a key factor determining exercise tolerance in a hot environment.

During moderate to intense exercise in the heat, the combined increased demands for perfusion of the working muscles and of the skin can result in difficulties in meeting both requirements (38). With hypohydration, peripheral blood flow is usually reduced and heart rate increases in an attempt to maintain central vascular pressure. A reduction in peripheral blood flow will reduce the ability to lose heat causing body temperature to rise. The dual demand of the working muscle and peripheral circulation in the heat may also compromise hepatic, splanchnic and muscle blood flow which may ultimately limit exercise capacity (40). This again suggests that fluid balance is a key factor determining exercise capacity. In contrast to this hypothesis Nielsen (35) did not observe a reduction in muscle perfusion during exercise and heat stress and suggested that a high core temperature was the more likely cause of fatigue in the heat. Similarly, Saltin et al. (43) suggested that a high muscle temperature would limit exercise performance in the heat.

It is clear that further investigation of thermoregulation and fluid balance during exercise in hot environments is necessary in order to understand the mechanisms for the reduced exercise tolerance observed in the majority of studies conducted in the heat. It is also apparent that the thermoregulatory response to exercise varies between individuals. It is unlikely, therefore, that this or any other factor will contribute to the subjective sensation of fatigue to the same degree in all individuals or in all situations. Nonetheless, fluid replacement appears to have a large influence on exercise tolerance during heat exposure, although the mechanisms involved and the interactions with thermoregulation remain obscure.

8. REFERENCES

1. Adams, W.C., R.H. Fox, A.J. Fry, and I.C. Macdonald. Thermoregulation during marathon running in cool, moderate, and hot environments. *J. Appl. Physiol.* **38**: 1030-1037, 1975.
2. Adolph, E.D., and associates. *Physiology of Man in the Desert*. New York: Wiley. 1947.
3. Armstrong, L.E., D.L. Costill, and W.J. Fink. Influence of diuretic-induced dehydration on competitive running performance. *Med. Sci. Sports Ex.* **17**: 456-461, 1985.
4. Asmussen, E., and O. Boje. Body temperature and capacity for work. *Acta Physiol. Scand.* **10**: 1-22, 1945.

5. Bergh, U., and B. Ekblom. Physical performance and peak aerobic power at different body temperatures. *J. Appl. Physiol.* **46**: 885-889, 1979.

6. Bergstrom, J., L. Hermansen, E. Hultman, and B. Saltin. Diet, muscle glycogen and physical performance. *Acta Physiol. Scand.* **71**: 140-150, 1967.

7. Clarke, R.S.J., R.J. Hellon, and A.R. Lind. The duration of sustained contractions of the human forearm at different muscle temperatures. *J. Physiol. (Lond.)* **143**: 454-473, 1958.

8. Costill, D.L. Physiology of marathon running. *JAMA* **221**: 1024-1029, 1972.

9. Costill, D.L., W.M. Sherman, W.J. Fink, C. Maresh, M. Witten, and J.M. Miller. The role of dietary carbohydrates in muscle glycogen synthesis after strenuous running. *Am. J. Clin. Nutr.* **34**: 1831, 1981.

10. Coyle, E.F. Timing and method of increased carbohydrate to cope with heavy training, competition and recovery. In: *Foods, Nutrition and Sports Performance*, edited by C. Williams, and J.T. Devlin. London: Spon, 1992, pp. 35-62.

11. Coyle, E.F., and A.R. Coggan. Effectiveness of carbohydrate feeding in delaying fatigue during prolonged exercise. *Sports Med.* **1**: 446-458, 1984.

12. Craig, F.N., and E.G. Cummings. Dehydration and muscular work. *J. Appl. Physiol.* **21**: 470- 674, 1966.

13. Dimri, G.P., M.S. Malhotra, J. Sen Gupta, T. Sampath Kumar, and B.S. Arora. Alterations in aerobic-anaerobic proportions of metabolism during work in heat. *Eur. J. Appl. Physiol.* **45**: 43-50, 1980.

14. Fink, W.J., D.L. Costill, and P.J. Van Handel. Leg muscle metabolism during exercise in the heat and cold. *Eur. J. Appl. Physiol.* **34**: 183-190, 1975.

15. Galloway, S.D.R., and R.J. Maughan. Effects of ambient temperature on the capacity to perform prolonged exercise in man. *J. Physiol. (Lond.)* **489**: 35-36, 1995.

16. Gleeson, M., R.J. Maughan, and P.L. Greenhaff. Comparison of the effects of pre-exercise feeding of glucose, glycerol and placebo on endurance and fuel homeostasis in man. *Eur. J. Appl. Physiol.* **55**: 645-653, 1986.

17. Gopinathan, P.M., G. Pichan, and V.M. Sharma. Role of dehydration in heat stress-induced variations in mental performance. *Arch. Environ. Health* **43**: 15-17, 1988.

18. Hurley, B.F., and E.M. Haymes. The effects of rest and exercise in the cold on substrate mobilization and utilization. *Aviat. Space Environ. Med.* **53**: 1193-1197, 1982.

19. Jacobs, I. The effect of thermal dehydration on performance of the Wingate anaerobic test. *Int. J. Sports Med.* **1**: 21-24, 1980.

20. Jacobs, I., I.I. Romet, and D. Kerrigan-Brown. Muscle glycogen depletion during exercise at 9°C and 21°C. *Eur. J. Appl. Physiol.* **54**: 35-39, 1985.

21. Kozlowski, S., Z. Brzezinska, B. Kruk, H. Kaciuba-Uscilko, J.E. Greenleaf, and K. Nazar. Exercise hyperthermia as a factor limiting physical performance: temperature effect on muscle metabolism. *J. Appl. Physiol.* **59**: 766-773, 1985.

22. Ladell, W.S.S. The effects of water and salt intake upon the performance of men working in hot and humid environments. *J. Physiol. (Lond.)* **127**: 11-46, 1955.

23. Lamb, D.R., and G.R. Brodowicz. Optimal use of fluids of varying formulations to minimize exercise-induced disturbances in homeostasis. *Sports Med.* **3**: 247-274, 1986.

24. Leiper, J.B., and R.J. Maughan. Experimental models for the investigation of water and solute uptake in man: implications for oral rehydration solutions. *Drugs* **36**(Suppl 4): 65-79, 1988.

25. MacDougall, J.D., W.G. Reddan, C.R. Layton, and J.A. Dempsey. Effects of metabolic hyperthermia on performance during heavy prolonged exercise. *J. Appl. Physiol.* **61**: 654-659, 1974.

26. Maughan, R.J. Thermoregulation and fluid balance in marathon competition at low ambient temperature. *Int. J. Sports Med.* **6**: 15-19, 1985.

27. Maughan, R.J. Fluid and electrolyte loss and replacement in exercise. In: *Oxford Textbook of Sports Medicine*, edited by M. Harries, L.J. Micheli, W.D. Stanish and C. Williams. Oxford: Oxford Univ. Press, 1994, pp. 82-93.

28. Maughan, R.J., and J.B. Leiper. Aerobic capacity and fractional utilisation of aerobic capacity in elite and non-elite male and female marathon runners. *Eur. J. Appl. Physiol.* **52**: 80-87, 1983.

29. Maughan R.J., C.E. Fenn, M. Gleeson, and J.B Leiper. Metabolic and circulatory responses to the ingestion of glucose polymer and glucose/electrolyte solutions during exercise in man. *Eur. J. Appl. Physiol.* **56**: 365-362, 1987.

30. Maw, G.J., S.H. Boutcher, and N.A.S. Taylor. Ratings of perceived exertion and affect in hot and cool environments. *Eur. J. Appl. Physiol.* **67**: 174-179, 1993.

31. Montain, S.J., and E.F. Coyle. Influence of graded dehydration on hyperthermia and cardiovascular drift during exercise. *J. Appl. Physiol.* **73**: 1340-1350, 1992.

32. Montner, P., T. Chick, M. Reidesel, M. Timms, D. Stark, and G. Murata. Glycerol hyperhydration and endurance exercise. *Med. Sci Sports Ex.* **24**: S157, 1992.

33. Murray, R. The effects of consuming carbohydrate-electrolyte beverages on gastric emptying and fluid absorption during and following exercise. *Sports Med.* **4**: 322-351, 1987.

34. Nadel, E.R. Circulatory and thermal regulations during exercise. *Fed. Proc.* **39**: 1491-1497, 1980.

35. Nielsen, B., R. Kubica, A. Bonnesen, I.B. Rasmussen, J. Stoklosa, and B. Wilk. Physical work capacity after dehydration and hyperthermia. *Scand. J. Sports Sci.* **3**: 2-10, 1981.

36. Nielsen, B., G. Savard, E.A. Richter, M. Hargreaves, and B. Saltin. Muscle blood flow and muscle metabolism during exercise and heat stress. *J. Appl. Physiol.* **69**: 1040-1046.

37. Nielsen, B. Heat stress and acclimation. *Ergonomics* **37**: 49-58, 1994.

38. Rowell, L.B., G.L. Brengelmann, J.R. Blackmon, R.D. Twiss, and F. Kusumi. Splanchnic blood flow and metabolism in heat-stressed man. *J. Appl. Physiol.* **24**: 475-484, 1968.

39. Rowell, L.B. Human cardiovascular adjustments to exercise and thermal stress. *Physiol. Rev.* **54**: 75-159, 1974.

40. Rowell, L.B. *Human Circulation.* New York: Oxford University Press, 1986.

41. Saltin, B. Aerobic and anaerobic work capacity after dehydration. *J. Appl. Physiol.* **19**: 1114-1118, 1964.

42. Saltin, B., A.P. Gagge, U. Bergh, and J.A.J. Stolwijk. Body temperatures and sweating during exhaustive exercise. *J. Appl. Physiol.* **32**: 635-643, 1972.

43. Saltin, B., and D.L. Costill. Fluid and electrolyte balance during prolonged exercise. In: *Exercise, Nutrition, and Metabolism*, edited by E.S. Horton and R.L Terjung. New York: Macmillan, 1988, pp. 150-158.

44. Sawka, M.N., and K.B. Pandolf. Effects of body water loss on physiological function and exercise performance. In: *Fluid function and Exercise Performance*, edited by C.V. Gisolfi and D.R. Lamb. Carmel, USA: Benchmark, 1990, pp. 1-38.

45. Sutton, J.R. Clinical implications of fluid imbalance. In: *Perspectives in Exercise Science and Sports Medicine, vol 3, Fluid Homeostasis during Exercise*, edited by C.V. Gisolfi and D.R. lamb. Indianapolis, USA: Benchmark, 1990, pp. 425-448.

46. Suzuki, Y. Human physical performance and cardiocirculatory responses to hot environments during submaximal upright cycling. *Ergonomics* **23**: 527-542, 1980.

47. Vist, G.E., and R.J. Maughan. The effect of increasing glucose concentration on the rate of gastric emptying in man. *Med. Sci. Sports Ex.* **26**: 1269-1273, 1994.

48. Walsh, R.M., T.D. Noakes, J.A. Hawley, and S.C. Dennis. Impaired high-intensity cycling performance time at low levels of dehydration. *Int. J. Sports Med.* **15**: 392-298, 1994.

16

ENDOCRINE REGULATION OF METABOLISM DURING EXERCISE

H. Weicker and G. Strobel

Department of Sports Medicine
University of Heidelberg, Germany

1. INTRODUCTION

The efficiency of hormonal stimulation on metabolic regulation not only depends on the plasma concentration of the hormones, their receptors, post-receptor, and second messenger systems, but also on the enzymatic and metabolic adaptation triggered by physical activity. Insulin and catecholamines, as well as the mechanisms, which elicit the metabolic response, will be discussed. The previous assumption of the insulin mediated glucose transport through the cell membrane after insulin receptor interaction and post-receptor activation is challenged by the function of the glucose transporter Glut 4 on the plasma and intracellular membranes. New findings on this subject, such as additive effects of insulin and exercise, will be reported. The exercise induced catecholamine overflow, which can exceed resting values by up to 50 fold, is attenuated by adrenoreceptor down-regulation, post-receptor adaptation, activation of second messenger systems, metabolic and renal clearance rates, as well as by the sulfconjugation of catecholamines. The metabolic regulation of carbohydrate- and lipid metabolism by insulin and catecholamines will be addressed.

2. INSULIN

The biochemical structure of insulin, a 5 Kd peptide, is identified, its chemical and genetechnological synthesis has been completed. The insulin interaction with its receptor on the plasma membrane has been well established. However, the insulin effectiveness on glucose uptake and the post-receptor signaling for intracellular glucose utilization are less well elucidated.

2.1. Insulin Biosynthesis and Secretion

Insulin is stored in the β cells of the pancreas as a prohormone. It is secreted from the β cells together with C peptide by exocytosis. The extent of the secretion is strongly

The Physiology and Pathophysiology of Exercise Tolerance
edited by Steinacker and Ward, Plenum Press, New York, 1996

influenced by the concentration of blood glucose and, to a lesser degree, by some amino acids, such as arginine and alanine (22).

Insulin release is inhibited by stimulation of α adrenoceptors on β cells. This occurs, for instance, during exhaustive exercise, due to elevated catecholamine·levels. The inhibition of the insulin release reduces hepatogenic and muscular glycogenesis and intensifies the glycogenolytic and gluconeogenetic glucose output from the liver and glycogen mobilization in muscle.

Insulin enhances the lipogenesis in adipocytes and is the strongest hormonal inhibitor of lipolysis, due to the activation of the enzyme phosphodiesterase. This enzyme catalyzes the degradation of cAMP to AMP, which prevents the phosphorylation of the hormone sensitive lipases via protein kinase. Moreover, insulin enhances the glucose uptake and the glycerolphosphate formation, which supports lipogenesis in fat cells, and also diminishes the release of free fatty acids by the activation of the endothelial adipocyte proteinlipase. Therefore, the α adrenergic inhibition of the insulin secretion supports the substrate supply during work.

2.2. Insulin Receptor

The insulin receptor is a tetrameric glycoprotein of two α 135 Kd and two β 90 Kd subunits, which has been formed after several cleavages of the insulin receptor precursors in the cell. After the receptors have been incorporated into the plasma membrane, the carbohydrate moiety is covered by sialic acid, which prevents hydrolysis by galactosidase (3,18). The mediation of the insulin effect via insulin receptors is regulated by different mechanisms (Fig.1).

First, after insulin has bound to the α subunits of the receptor, the receptor becomes converted, now specifically activating the tyrosine kinase at the catalytic site of the intracellular β subunits. This activation seems to be responsible for the intracellular glucose utilization, depending on a cascade of further enzymatic steps, in which tyrosine kinase has a key function. However, with regard to glucose uptake, the significance of activated tyrosine kinase is still not yet completely clarified. Besides the specific activation of tyrosine kinase by phosphorylation, threonine and serine phosphorylation sites of the insulin receptor may be phosphorylated by cAMP sensitive protein kinase C after β adrenergic stimulation by catecholamines (3,18). Phosphorylation of the insulin receptor via protein kinase C reduces the insulin affinity of the α subunits during intense exercise, for example, diminishing glycogenesis and lipogenesis and enhancing lipolysis and glycogenolysis.

Both specific tyrosine kinase phosphorylation as well as autophosphorylation via protein kinase C enhance the internalization of the hormone receptor complex and its degradation. Both effects might cause receptor down-regulation, which occurs considerably during hyperinsulinemia (3,18).

2.3. Glucose Transporters (Glut 4 and Glut 1)

Glucose transporters on plasma membranes of skeletal muscle, adipocytes and some other tissues, facilitate the glucose uptake. In skeletal muscle, glucose uptake can also be elicited by exercise without insulin stimulation (1,5,14,15,16,17,19), challenging the previously postulated exceptional function of the insulin mediated glucose uptake. In peripheral tissues, glucose transporters Glut 4 and Glut 1 are most prominent.

The glucose transporters Glut 4 and Glut 1 are 40 Kd proteins with a carbohydrate moiety of 12-15 % (1,4,5,7). The glucose transport through the plasma membrane is trig-

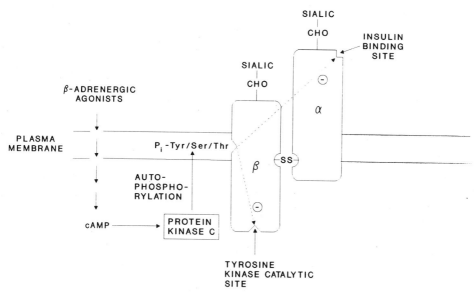

Figure 1. The effect of β-adrenergic agonists mediated by autophosphorylation of threonine-serine phosphorylation sites and following activation of protein kinase C. Autophosphorylation (1) reduces insulin binding to the α-subunit of the insulin receptor and (2) reduces tyrosine phosphorylation at the tyrosine kinase catalytic site, i.e. reduces the effectiveness of insulin. For clarity, only one of the two α and β subunits of the insulin receptor is shown.

gered by glucose itself, eliciting a conformational change of the glucose transporter. The energy supply, necessary for this conformational alteration of the glucose transporters has not yet been identified. Glut 1 transporters are distributed in many fetal and adult tissues, apparently in human red cells and also in skeletal muscle. Glut 4 transporters are primarily distributed on skeletal muscle, adipocytes and heart. After biosynthesis, the Glut 4 transporter is stored in low density microsomes (LDM) or is expressed in the plasma membrane. Both biosynthesis and translocation from LDM to the plasma membrane are additively triggered by insulin and exercise, thus regulating the extent of glucose uptake.

2.3.1. Glucose Transport Into Skeletal Muscles. Glut 1 mainly facilitates the glucose transport at rest. Its membrane density is somewhat higher on red than on white muscle fiber membranes. Glut 1 membrane density cannot be influenced by exercise or physical training. On the contrary, Glut 4 membrane density is enhanced by insulin as well as by exercise and training. The exercise dependent increase in Glut 4 density is more pronounced in red than in white muscle fibres. Both insulin and exercise increase Glut 4 membrane density. However, the mechanisms by which insulin and exercise increase the translocation of Glut 4 from LDM to plasma membrane are different (1,5,7,17,19,Fig.2). Insulin activated tyrosine kinase seems to trigger Glut 4 exocytosis on LDM which supports the Glut 4 translocation to the plasma membrane. At the same time, insulin inhibits the pretranslational biosynthesis of Glut 4, leading to a marked reduction of the Glut 4 pool of LDM. Exercise also increases the Glut 4 recruitment from LDM to the plasma membrane, but unlike insulin, exercise also increases the biosynthesis of Glut 4, leading to a minor decrease in the Glut 4 pool of LDM during exercise (1,5,6,7,12,13).

Glucosetransporter

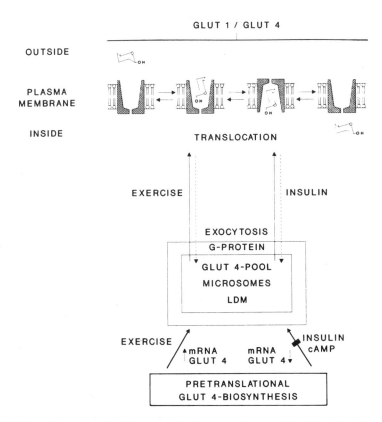

Figure 2. Glucose transport mechanism across the plasma membrane and the effect of insulin and exercise on the regulation of the Glut 4 translocation from low density michrosomes (LDM) to plasma membrane and pretranslational Glut 4 biosynthesis.

2.3.2. Adipose Tissue. The Glut 4 and Glut 1 transport and function of adipocytes is similar to the muscle. However, it is important to mention that insulin but not exercise accelerates the Glut 4 translocation velocity from the LDM compartment to the plasma membrane in fat cells (1,5,11).

2.4. Clinical Implications

The improvement of glucose uptake by exercise, due to the increase of Glut 4 density in plasma membrane, may have therapeutical consequences for both Type I and Type II diabetes mellitus. With the insulin deficiency of Type I, exercise and especially training induces an insulin independent increase of glucose uptake by Glut 4 due to the rise of plasma membrane Glut 4 density. This might partially compensate for the lack of the insulin stimulated glucose uptake. In Type II diabetes, which is mainly accompanied by receptor and post-receptor defects, as well as by insulin resistance; training and exercise also have a benefit on the improvement of Glut 4 function. In this context, it should be mentioned that in rats oral treatment of diabetes mellitus Type II with sulfonylurea stimulates the Glut 4 recruitment as well (10), thus enhancing the translocation and increase of membrane Glut 4 density. This should be kept in mind when Type II diabetics with obesity do not respond to dietary regimens together with physical activation by special training modes which are, at present, the first choice of treatment of Type II diabetes. However, it has to be taken into account that neuromuscular impairment, which often occurs with diabetes mellitus of both types, reduces the capability to exercise and training, thus reducing its benefits of Glut 4 translocation from LDM to plasma membranes (7,11,14,16,17,20).

3. CATECHOLAMINES

Catecholamines are derived from the amino acid tyrosine, which enters the chromaffine cells in the adrenal medulla, and the postganglionary neurons. From tyrosine, L-dopa is formed by the rate-limiting enzyme tyrosine hydroxylase and is then converted to dopamine by decarboxylation. Dopamine is taken up into the storage vesicles, in which dopamine β hydroxylase synthesizes norepinephrine. In the cytosol, norepinephrine is N-methylated to epinephrine and is then taken up again in the storage vesicles. Whether the postganglionary neurons function as dopaminergic, noradrenergic or adrenergic neurons depends on the enzyme equipment. For instance, dopaminergic neurons do not possess dopamine β hydroxylase and are therefore not able to synthesize norepinephrine. Catecholamines may serve as neurotransmitters and as hormones. In humans, the ratio of epinephrine and norepinephrine released into the blood stream by the adrenal medulla amounts to 8:1. In the brain, the neurotransmitter dopamine prevails, followed by norepinephrine and epinephrine. In the peripheral postganglionary sympathetic neurons, norepinephrine is the preferred neurotransmitter. Its release is regulated by presynaptic α2 adrenoceptors (inhibition) and β2 adrenoceptors (activation). Catecholamines, serving as hormones or neurotransmitters, are primarily regulated by the central sympatho-adrenergic stimulation due to afferent or efferent impulses required for metabolic or cardiocirculatory adaptation. During exercise, the central sympatho-adrenergic stimulation depends on the work load and duration of exercise and on the recruited muscle mass, but also on the emotional stress (21,22). The sympatho-adrenergic center in the brain stem stimulates the preganglionary neurons, followed by the excitation of their postganglionary neurons or the adrenal medulla, resulting in the release of the hormones or neurotransmitters. The metabolic or cardiovascular effects of catecholamines are mediated via the subtypes of α and/or β adrenoceptors.

As already mentioned above, the hormonal impact does not only depend on the hormonal blood concentration and on the receptor- and post-receptor-systems, but also on the enzymatic and metabolic adjustment and the neuromuscular excitation during work. To emphasize this point of view, we would like to discuss the well known example of muscle glycogenolysis.

3.1. Glycogenolysis

Epinephrine dependent β2 stimulation, which activates the adenylate cyclase system via Gs protein, results in the formation of the 3,5-cAMP. 3,5-cAMP phosphorylates a protein kinase, activating phosphorylase kinase B, which for its part converts phosphorylase b to a, catalyzing glycogen mobilization. Since phosphorylase a cleaves the glycosyl units phosphorolytically to glucose-1-phosphate, the ATP yield from glycogen compared to blood-borne glucose is higher. Besides this well known pathway, the glycogenolytic rate due to exercise depends primarily on the cytosolic calcium, i.e. on the activation of phospholipase C. Phospholipase C may be activated in skeletal muscle first by noradrenergic stimulation of the α1 adrenoceptor on the T tubular system and also by the voltage dependent calcium channels and the dihydropyridine receptors (22). Activated phospholipase C increases the second messengers, namely inositoltrisphosphate and the membrane-bound diacylglycerol. Inositoltrisphosphate enhances the calcium release from the SR through calcium channels, which are identical with ryanodine receptors (9). These channels are sensitised by voltage sensitive sensors between the triades and the T tubules (22). The increased calcium efflux triggers troponin C, eliciting the myofilament coupling in muscle contraction. The cytosolic surplus of calcium is bound by calmodulin, a calcium binding protein of the cytosol. Ca2+ calmodulin then activates the above mentioned enzyme cascade leading to the conversion of phosphorylase b to a. In addition, diacylglycerol supports the function of the membrane-attached protein kinase allosterically, increasing the glycogen mobilization (22, Fig.3). Thus, the glycogen mobilization will be directly adapted to the muscle contraction and the intensity of the T tubular excitation. It will also be potentiated by the hormonal impact. The cooperation between the β2 adrenergic hormonal stimulation and the additive enzymatic metabolic regulation, which is elicited by the T tubular excitation and SR coupling, indicates that the glycogen liberation already

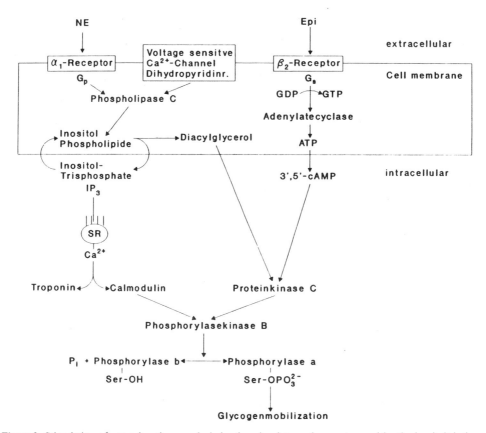

Figure 3. Stimulation of muscular glycogenolysis by the adenylate cyclase system and by the inositoltrisphosphate/diacylglycerol system. Reproduced with permission from Ref. 22.

starts before significant catecholamine increase in blood is obvious. This theoretical consideration supports the practical findings that the substrate mobilization from muscular glycogen depots depends on the intensity and duration of muscle contraction.

3.2. Lipolysis

β2 adrenergic stimulation of lipolysis and fatty acid release in adipocytes is also mediated by hormonal interaction with enzymatic and metabolic adjustment. Activation of the adenylate cyclase via Gs protein, after the binding of the ligands to the β2 adrenoceptors, elicits the protein kinase phosphorylation by cAMP. This activates the hormone sensitive lipases and leads to the hydrolysis of triacylglycerol to fatty acids and glycerol. The β2 adrenergic catecholamine stimulation prevails in human lipolysis, whereas ACTH and glucagon, which especially takes place in animals, are only able to trigger the lipolysis in humans if catecholamine production is impaired (22). α2 adrenoceptor stimulation inhibits the adenylate cyclase activation via Gi protein. Thus, the cAMP dependent phosphorylation of protein kinase C is reduced.

3.3. cAMP/Receptor Density

Desensitised post-receptor regulation resulting in a reduced cAMP formation which might be caused, for instance, by exhaustive long term exercise can be indicated by the decreased ratio of cAMP/adrenoceptor density. The marked increase of catecholamines during long-lasting exhaustive exercise increases the β2 adrenoceptor density of the target cells, but inhibits cAMP production, influenced by the extent of β2 adrenergic stimulation (8). This finding may be explained by an overproportional phosphorylation of protein kinase, which is able to phosphorylate threonine and serine at the intracellular glycoprotein moiety of the β adrenoceptors. The phosphorylation of these amino acids inhibits the Gs function upon the adenylate cyclase activation and consequently decreases cAMP production (8).

3.4. Measurement of the Overall Sympathetic Activity

Physical exercise represents the most powerful stimulus of the sympatho-adrenergic system (2), which is important for the adaptation of metabolism and the cardiovascular system to exercise. Therefore, knowledge about the sympatho-adrenergic activation during various types of exercise and training seems of interest. Due to the short half-lives, the free catecholamines are less appropriate to reflect the overall sympatho-adrenergic activity over a prolonged period of time. Recent studies provide evidence that the single measure of plasma norepinephrine sulfate reflects the overall spillover of norepinephrine during prolonged periods of exercise (Fig.4). First, norepinephrine sulfate is formed from free norepinephrine by the enzyme phenol sulfotransferase. Second, norephinephrine sulfate in plasma is concentrated about 2 to 4 times higher than free norepinephrine. Third, the plasma half-life of norepinephrine sulfate compared to free norepinephrine is longer (hours vs. minutes). Fourth, plasma norepinephrine sulfate accumulates with the duration of exercise and rises with increasing workloads. Fifth, the increase in plasma catecholamines during exercise depends on the total work performed, but not on the pattern of the power output. These findings were established in continuous and discontinuous exercise tests (21). Therefore, measurement of plasma norepinephrine sulfate after exercise reflects the overall spillover of norepinephrine during the whole period of prolonged exercise per-

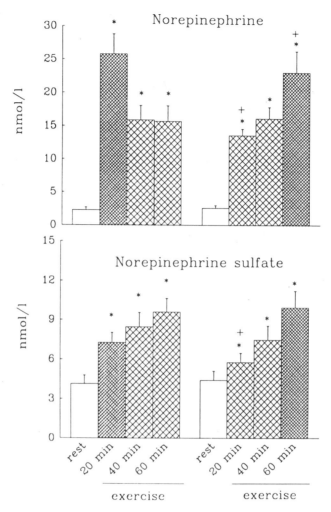

Figure 4. Plasma free and sulfated norepinephrine at rest, as well as 20, 40 and 60 min after beginning of continuous exercise from ten well-trained subjects. Values are means ± standard error. * significantly different from resting values at p < 0.05. + significantly different between corresponding values of both exercise tests at p<0.05. Results indicate that plasma norepinephrine is closely following the pattern of power output. Norepinephrine sulfate seems to accumulate during exercise, reflecting the overall norepinephrine spillover throughout the duration of exercise. (Modified from Ref. 21 with permission).

formance, such as interval training sessions, soccer, handball and other competitive sport events which last longer than one hour.

4. REFERENCES

1. Bonen, A., J. C. McDermott, and M. H. Tan. Glucose transport in skeletal muscle. *Intern. Series On Sport Sciences: Biochemistry of Exercise VII*, **21**: 295-317, 1990.

2. Cryer, P. E. Physiology and pathophysiology of the human sympathoadrenal neuroendocrine system. *New Engl. J. Med.* **303**: 436-443, 1980.

3. Czech, M. P. The nature and regulation of the insulin receptor: structure and function. *Ann. Rev. Physiol.* **47**: 357-381, 1985.

4. Etgen Jr., G. J., A. R. Memon, G. A. Thompson Jr., and J. L. Ivy. Insulin- and contraction-stimulated translocation of GTP binding proteins and Glut 4 protein in skeletal muscle. *J. Biol. Chem.* **268**: 20164-20169, 1993.

5. Farrell, P. A. Exercise effects on regulation of energy metabolism by pancreatic and gut hormones. *Perspectives in Exerc. Science and Sports Med.: Energy Supply in Exercise and Sport* **5**: 383-434, 1992.

6. Flores-Riveros, J. R., K. H. Kästner, K. S. Thompson, and M. D. Lane. Cyclic AMP-induced transcriptional repression of the insulin-responsive glucose transporter (Glut 4) gene: identification of a promoter region required for down-regulation of transcription. *Biochem. Biophys. Res. Commun.* **194**: 1148-1154, 1993.

7. Goodyear, L. J., M. F. Hirshman, P. M. Valyou, and E. S. Horton. Glucose transporter number, function, and subcellular distribution in rat skeletal muscle after exercise training. *Diabet.* **41**: 1091-1099, 1992.

8. Hausdorff, W. P., M. G. Caron, and R. J. Lefkowitz. Turning off the signal: desensitization of beta-adrenergic receptor function. *FASEB J.* **4**: 2881-2889, 1990.

9. Herrmann-Frank, A., and G. Meissner. Regulation of Ca 2+ release from skeletal muscle sarcoplasmatic reticulum. *Muscle Fatigue Mechan. in Exerc. and Train.* **34**: 11-19, 1992.

10. Jakobs, D. B., and Y. C. Jung. Sulfonylurea potentiates insulin-induced recruitment of glucose transport carrier in rat adipocytes. *J.. Biol. Chem.* **260**: 2593-2596, 1985.

11. Kono, T., F. W. Robinson, T. L. Blevins, and O. Ezaki. Evidence that translocation of the glucose transporter acitivity is the major mechanism of insulin action on glucose transport in fat cells. *J. Biol. Chem.* **18**: 10942-10947, 1982.

12. Neufer, P. D., M. H. Shinebarger, and G. L. Dohm. Effect of training and detraining on skeletal glucose transporter (Glut 4) content in rats. *Can. J. Physiol. and Pharmacol.* **70**: 1286-1290, 1992.

13. Piper, R. C., D. E. James, J. W. Slot, C. Puri, and J. C. Lawrence Jr. Glut 4 phosphorylation and inhibition of glucose transport by dibutyryl cAMP. *J. Biol. Chem.* **268**: 16557-16563, 1993.

14. Ploug, T., H. Galbo, and E. A. Richter. Increased muscle glucose uptake during concentrations: no need for insulin. *Am. J. Physiol.* **247**: E726-E731, 1984.

15. Ploug, T., H. Galbo, J. Vinten, M. Jorgensen, and E. A. Richter. Kinetics of glucose transport in rat muscle: effects of insulin and concentrations. *Am. J. Physiol.* **253**: E12-E20, 1987.

16. Richter, E. A., T. Ploug, and H. Galbo. Increased muscle glucose uptake after exercise. *Diabet.* **34**: 1041-1048, 1985.

17. Rodnick, K. J., E. J. Henriksen, D. E. James, and J. O. Holloszy. Exercise training, glucose transporters, and glucose transport in rat skeletal muscles. *Am. J. Physiol.* **262**: C9-C14, 1992.

18. Rosen, O. M. After Insulin Binds. *Science* **237**: 1452-1458, 1987.

19. Slentz, C. A., E. A. Gulve, K. J. Rodnick, E. J. Henriksen, J. H. Youn, and J. O. Holloszy. Glucose transporters and maximal transport are increased in endurance trained rat soleus. *Am. J. Physiol.* **73**: 486-492, 1992.

20. Smith, R. L., and J. C. Lawrence Jr. Insulin action in denervated skeletal muscle. *J.. Biol. Chem.* **260**: 273-278, 1985.

21. Strobel, G., B. Friedmann, J. Jost, and P. Bärtsch. Plasma and platelet catecholamine and catecholamine sulfate response to various exercise tests. *Am. J. Physiol.* **267**: E537-E543, 1994.

22. Weicker, H., G.Strobel. *Sportmedizin:Biochemisch-physiologische Grundlagen und ihre sportartspezifische Bedeutung.* Stuttgart, Germany: Gustav Fischer Verlag, Stuttgart, 1994.

17

THE DYNAMICS OF BLOOD FLOW CHANGES IN LOWER LIMB ARTERIES DURING AND FOLLOWING EXERCISE IN HUMANS

S. T. Hussain,[1] R. E. Smith,[1] A. L. Clark,[1] R. F. M. Wood,[1] S. A. Ward,[3] and B. J. Whipp[2]

[1] Professorial Surgical Unit
St. Bartholomew's Hospital
London, United Kingdom
[2] Department of Physiology
St. George's Hospital Medical School
London, United Kingdom
[3] Division of Human and Exercise Physiology
School of Applied Science
South Bank University
London, United Kingdom

1. INTRODUCTION

In the steady state of dynamic muscular exercise in humans, blood flow to exercising muscles ($\dot{Q}_M$) increases essentially linearly with respect to work rate and oxygen uptake ($\dot{V}O_2$). The profiles of the nonsteady state $\dot{Q}_M$ response however are poorly understood, reflecting in large part the technical difficulties of accurately determining limb blood flow during periods of rapid change. Previous attempts to establish $\dot{Q}_M$ kinetics have been hampered by uncertainties relating to the assumptions of the techniques. For example, the constant indicator infusion approach assumes that the indicator is uniformly dispersed throughout the flow pulse, and none is lost prior to reaching the sampling site. Also, "standard" Doppler estimates of flow from blood velocity assume both that the focus of the beam on to the vessel and the vessel geometry itself remain unchanged throughout the exercise - with no assurance, of course, that they are. The duplex Doppler technique, in contrast, is not prey to these assumptions as it (a) visualizes the vessel throughout the measurement and (b) determines the vessel diameter and therefore can establish blood flow in addition to velocity (3,4) - assuming only that the vessel cross-section is circular.

The aim of the present investigation was therefore to utilize duplex Doppler monitoring of both the common femoral artery (CFA) and its major branches - the superior femoral and the profunda femoris arteries (SFA, PFA) - to establish the dynamic profiles of arterial blood flow to the lower limbs during exercise and recovery.

2. METHODS

Five healthy young subjects performed light- and moderate-intensity square-wave exercise (6-minute bouts) of the quadriceps muscles in a semi-recumbent posture that allowed the flow-sensing probe to remain stable during the exercise. The probe was placed in the inguinal region of the inner thigh. The exercise comprised alternating knee extensions against restraining rubber bands, arranged as stirrups over the dorsal surfaces of the feet. The extensions were from the vertical to an angle of ~75° (with stops provided to set the excursion limits), the cadence of 40/sec being set by a metronome. The return to the initial position was achieved chiefly through the device recoil rather than by contraction of the antagonist muscles: this had the effect of "isolating" the work to the quadriceps muscles. Rotation of the inguinal region was minimized by means of a strap placed over the thigh 5 cm rostral to the upper margin of the patella. The length and recoil characteristics of the bands were selected so as to induce the required increments of metabolic rate (i.e. $\dot{V}O_2$ averaging 1.3 and 1.7 l/min for the light and moderate intensities, respectively). CFA flow was measured by duplex Doppler imaging (Ultramark, Advanced Technology Laboratories, Seattle, USA) (3,4) at rest and then at 15-20 second intervals throughout the exercise and subsequent 6-minute recovery phase. All procedures were approved by the appropriate Institutional Ethics Committee.

As subjects typically experienced difficulties in reliably maintaining the square-wave profile at high work rates, a separate series of experiments was undertaken in which six subjects completed a standard 30-second "Wingate" test on a cycle ergometer (Monark 814e, Varburg, Sweden) (1). Flow in the CFA, SFA and PFA was measured with the subjects recumbent at rest and then at approximately 2, 15, 30 and 60 minutes post-exercise. Monitoring was not undertaken during the exercise bout, as motion artefacts preclude accurate Doppler monitoring during such a maximum-effort test.

3. RESULTS

3.1. Light and Moderate Exercise

In response to the square-wave workrate forcings, CFA flow increased in a curvilinear fashion to reach a new steady state within 2-3 minutes (Fig. 1).

The individual on- and off-transient CFA flow responses were well fit to a mono-exponential function. That is, for the on-transient:

$$\dot{Q}_M(t) = \dot{Q}_M(0) + \Delta\dot{Q}_M(ss) \cdot (1 - \exp(-t/\tau on))$$

where $\dot{Q}_M(t)$ is the $\dot{Q}_M$ response at time t after exercise onset; $\dot{Q}_M(0)$ is the resting or baseline value; $\Delta\dot{Q}_M(ss)$ is the steady-state $\dot{Q}_M$ increment above baseline; and τon is the time constant of the response.

For the off-transient:

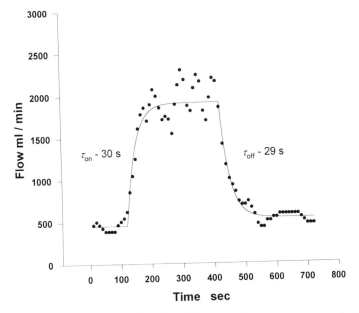

Figure 1. Individual response profile of CFA blood flow to a moderate square-wave workrate forcing.

$$\dot{Q}_M(t) = \dot{Q}_M(0) + \Delta\dot{Q}_M(ss) \cdot \exp(-t/\tau off)$$

where $\dot{Q}_M(t)$ is the $\dot{Q}_M$ response at time t after the cessation of the work; $\dot{Q}_M(0)$ is the recovery baseline value; $\Delta\dot{Q}_M(ss)$ is the steady-state $\dot{Q}_M$ between exercise and recovery; and τoff is the time constant of the response.

The group-mean τ for the on-transient (τon) was not significantly different for the light and moderate intensities of exercise, averaging 0.47 and 0.49 min (or 28.2 and 29.4 s), respectively (Table 1).

The off-transient responses for both intensities were also mono-exponential, with time constants (τoff) of 0.65 min (39 s) and 0.75 min (45 s), respectively. That is, on average (but not in all cases) they were slower than for the corresponding on-transient (Table 1).

3.2. High-Intensity Exercise

As was the case for the lower exercise intensities, the decline in CFA flow back to baseline during the recovery phase could also be characterized by a monoexponential function (Fig. 2). This was also the case for the SFA and PFA responses. However, the recovery was in all cases substantially slower than for light and moderate exercise, with τoff averaging more than 12 min.

Table 1. Group-mean parameter estimates

	Light-Intensity CFA	Moderate-Intensity CFA	High-Intensity CFA	High-Intensity SFA	High-Intensity PFA
Q_M τon(min)	0.47	0.49	-	-	-
Q_M τoff(min)	0.65	0.75	14.8	21.2	12.6

4. DISCUSSION

The control mechanisms for the kinetics of $\dot{V}O_2$ during muscular exercise remains a topic of controversy. The two major competing theories to account for this control process contend, on the one hand, that the kinetics are controlled by enzymatically-controlled changes in intramuscular high-energy phosphate profiles, with tissue blood flow being adequate or even excessive for the current demands and, on the other, that the kinetics are controlled by oxygen delivery to the contractile elements (i.e. control via blood flow limitation). It seems unlikely that this important issue will be conclusively resolved in humans until techniques are available which allow valid and reliable high-density determinations of limb blood flow throughout the transient phase of the exercise.

Previous attempts to establish these flow dynamics have been limited by the broad assumptions inherent in the techniques, such as those based on inferences from indicator-dilution (2,6) and blood velocity (7,8) measurements. The duplex Doppler technique allows limb blood flow to be quantifed directly without many of the possibly-unwarranted assumptions of the other techniques. The major problem with this means of determining

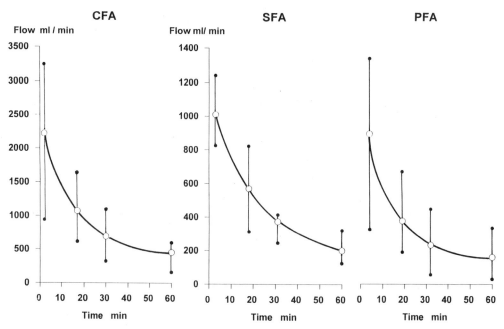

Figure 2. Group-mean responses of CFA, SFA and PFA flow at the off-transient of the Wingate test. Thick curves represent best fits to the group-mean responses (o); thin vertical lines indicate the ranges of individual values (•).

flow, however, remains that of the signal-to-noise ratio, especially when the scanning site itself moves as a result of the exercise. Our technique of knee-extension exercise allows the quadriceps to be contracted (resulting in metabolic rates approaching 2 l/min) with little or no movement of the scanning region. This resulted in profiles of CFA flow which were adequate for kinetic analysis during the exercise (Fig. 1). Naturally, the off-transient kinetics were less noisy. It should be possible, therefore, to obtain higher-resolution profiles of the flow dynamics by superimposing replicates of the transient; i.e. improving the signal-to-noise ratio. This, however, presupposes that the noise on the limb blood flow measurement is white and Gaussian, as it is for $\dot{V}O_2$ (5). There is currently no information available on these signal features or on the reproducibility of the parameter estimates.

As it is currently not technically feasible to measure the blood flow transients validly during high-intensity cycle-ergometry, we restricted our analysis of the Wingate test to the recovery phase. Here the flows were readily determined, with kinetics that were markedly long with respect to those of the lower intensities (Table 1). This prolonged flow recovery phase has recently been shown to be markedly long compared to the recovery profile of $\dot{V}O_2$ (9), but similar to those of exercise-induced metabolites such as lactate, pH and catecholamines, but not potassium (4).

We therefore conclude that duplex Doppler techniques have the potential to provide the required resolution of limb blood flow dynamics to address the issue of whether the dynamics of muscle O_2 utilisation during exercise are controlled by and limited by limb perfusion. The ability to discriminate the flow profiles in vessels distal to the bifurcation of the common femoral artery also provides the potential for such considerations in leg muscles other than the quadriceps femoris. This also opens for consideration the important issue of the distribution of blood flow in different modes of leg exercise.

5. REFERENCES

1. Bar-Or, O., 1987, The Wingate test. An update on methodology, reliability and validity. *Sports Med.* **4**: 381-394.
2. Cerretelli, P., and C. Marconi, 1986, Blood flow in exercising muscles. *Adv. Cardiol.* **35**: 65-78.
3. Gill, R.W., 1985, Measurement of blood flow by ultrasound: accuracy and sources of error. *Ultrasound Med. Biol.* **11**: 625-641.
4. Hussain, S.T., R.E. Smith, S. Medbak, R.F.M. Wood, and B.J. Whipp, 1996 (In press), Haemodynamic and metabolic responses of the lower limb after high intensity exercise in humans. *Exp. Physiol.* **81**: 173-187.
5. Lamarra, N., B.J. Whipp, S.A. Ward, and K. Wasserman, 1987, Effect of inter-breath fluctuations on characterizing exercise gas exchange kinetics. *J. Appl. Physiol.* **62**: 2003-2012.
6. Poole, D.C., W. Schaffartzik, D.R. Knight, T. Derion, B. Kennedy, H.J. Guy, R. Prediletto, and P.D. Wagner, 1991, Contribution of exercising legs to the slow component of oxygen uptake kinetics in humans. *J. Appl. Physiol.* **71**: 1245-1253.
7. Shoemaker, J.K., L. Hodge, and R.L. Hughson, 1994, Cardiorespiratory kinetics and femoral artery blood velocity during dynamic knee extension exercise. *J. Appl. Physiol.* **77**: 2625-2632.
8. Walloe, L., and J. Wesche, 1988, Time course and magnitude of blood flow changes in the human quadriceps muscles during and following rhythmic exercise. *J. Physiol. (Lond.)* **405**: 257-273.
9. Withers, M., T. Hussain, M. Donlon, S.A. Ward, and B.J. Whipp, 1996, Pulmonary O_2 uptake kinetics and femoral artery blood flow following high-intensity exercise in humans. *Proc. Physiol. Soc.*, Feb. meeting, p. C23.

18

HYPOXIA AND ANAEMIA PRODUCE EQUIVALENT EFFECTS ON MAXIMAL OXYGEN CONSUMPTION

Guido Ferretti

Département de Physiologie
Centre Médical Universitaire
1 rue Michel Servet, 1211 GENEVE 4, Switzerland

1. INTRODUCTION

In humans, oxygen flows from inspired air to the mitochondria against a resistance, which is overcome by an overall O_2 pressure gradient. According to the O_2 conductance equation, this resistance is provided by the sum of numerous in-series resistances, so that a progressive drop in O_2 partial pressure from inspired air to the mitochondria takes place. This approach was applied to the study of the O_2 transfer system at maximal exercise (22). In this case, at the steady-state, the flow of O_2 is the same across each resistance and is equal to $\dot{V}o_2$max. Afterwards, a multifactorial model of $\dot{V}o_2$max limitation was developed, assuming that each of the resistances in-series (or their reciprocal, the conductances) provides a sizeable fraction of the overall limitation to $\dot{V}o_2$max. Three major groups of conductances were identified, namely 1) the pulmonary conductances, related to alveolar ventilation and to alveolar-capillary O_2 transfer, 2) the cardiovascular conductance (Gq), due to cardiovascular oxygen transport, and 3) the muscular conductances, related to tissue O_2 diffusion and utilisation. The second turned out to provide 60-to-70% of the overall $\dot{V}o_2$max limitation, at least during exercise with big muscle groups in normoxia (4, 5, 10).

However, Gq includes factors that respond linearly and others that respond non-linearly (see below). The former include, for example, changes in haemoglobin concentration (Hb), as in the case of anaemia or of autologous blood transfusion. The latter vary when the inspired O_2 partial pressure (Pio_2) is modified (hypoxia - hyperoxia). Since in the former case $\dot{V}o_2$max varies linearly with Hb (5), whereas the $\dot{V}o_2$max decrease in hypoxia is curvilinear (2, 9), one could expect that modifying the linear vs the non-linear terms defining Gq may affect $\dot{V}o_2$max in a different manner. This hypothesis, which is the basis for this theoretical paper, is challenged by the results obtained.

The Physiology and Pathophysiology of Exercise Tolerance
edited by Steinacker and Ward, Plenum Press, New York, 1996

129

2. THEORY

The flow of O_2 across Gq is equal to:

$$\dot{V}o_2max = Gq \, (Pao_2 - P\bar{v}o_2) \tag{1}$$

where Pao_2 and $P\bar{v}o_2$ are the O_2 partial pressures in arterial and mixed venous blood, respectively. The conductance term Gq is in turn equal to:

$$Gq = \dot{Q} \, \beta b \, (Pao_2 - P\bar{v}o_2) = 1.34 \, \dot{Q} \, Hb \, \sigma b \tag{2}$$

where $\dot{Q}$ is cardiac output, and the constant 1.34 (in $mlO_2 \, g^{-1}$) the physiological oxygen binding coefficient of haemoglobin. The term βb indicates the oxygen transport coefficient of blood (in $mlO_2 \, Torr^{-1}$) otherwise equal to the average slope of the oxygen dissociation curve:

$$\beta b = (CaO_2 - C\bar{v}o_2)/(Pao_2 - P\bar{v}o_2) \tag{3}$$

where Cao_2 and $C\bar{v}o_2$ indicate oxygen concentrations in arterial and mixed venous blood, respectively. Similarly, the term σb (in $Torr^{-1}$), which can be defined as an oxygen saturation coefficient of blood, is equal to:

$$\sigma b = (Sao_2 - S\bar{v}o_2)/(Pao_2 - P\bar{v}o_2) \tag{4}$$

where Sao_2 and $S\bar{v}o_2$ are the haemoglobin oxygen saturations in arterial and mixed venous blood, respectively. While σb is the non-linear term of Eq. 2, Hb and $\dot{Q}$ represent the linear term of the same equation. The former variable can be modified by changing either the O_2 pressure gradient or the characteristics of the O_2 equilibrium curve (e.g. by means of pathological haemoglobins).

3. ASSUMPTIONS AND CONSTRAINTS

This analysis is an application of the five-site model of $\dot{V}o_2max$ limitation (5). Control values for $\dot{V}o_2max$ in normoxia of 2.9 l min^{-1} and for Hb of 150 g l^{-1} were used (3). An invariant $\dot{Q}$ of 20 l min^{-1}, irrespective of Pio_2, was also assumed. Mitochondrial Po_2 was assumed to be near zero (14). Hill's model of the oxygen equilibrium curve was used. A steady-state for gas exchange was imposed, implying a gas exchange ratio equal to one. The latter assumption allows taking into account the Bohr effect in the computation of sb and bb. No changes in other O_2 conductances than Gq were admitted.

The simulation was carried out along two lines. First, Hb was let to vary, and the ensuing Gq (or Rq) calculated. The resulting $\dot{V}o_2max$ was then computed by means of the same iterative procedure employed in a previous study (10). The constancy of O_2 flow in mixed venous blood (11, 12) was used as a criterion to interrupt the simulation and to retain the resulting $\dot{V}o_2max$ value. Then Hb was fixed at its reference value, while Pio_2, and thus Pao_2, were let to vary. The ensuing Gq and $\dot{V}o_2max$ values were then computed by the same principles as for the previous case. This procedure gave a remarkable reproduction of the curve describing the decrease of $\dot{V}o_2max$ in hypoxia (2, 9).

Hill's model of the oxygen equilibrium curve has been used for its simplicity. Although some of the assumptions it was based on were falsified by further experimental evidence, it still provides a reliable empirical description of the oxygen equilibrium curve, at least in the saturation range from 0.20 to 0.98. Also the constraint that no conductance except Gq was let to vary contributed to the simplification of the simulation. However, the same constraint limits the range of validity of the results. For instance, these do not refer to hypobaric hypoxia (altitude exposure), because the decrease in barometric pressure and thus in gas density which occurs at altitude results in a higher ventilatory conductance to O_2 flow. The other assumptions have been discussed in detail in a previous paper (10).

4. RESULTS AND DISCUSSION

Figure 1 describes the changes in $\dot{V}o_2max$ ensuing from induced changes in Hb. It has been constructed from the equation of the multifactorial model for $\dot{V}o_2max$ limitation which holds when only one resistance to O_2 flow is varied (5). The slope of the line (= 0.57) is equal to the fractional limitation to $\dot{V}o_2max$ imposed by Rq (Fq). Experimental values from different sources in the literature (1, 6, 7, 18, 20, 23, 24) were added to the figure. These appear to be in good agreement with the theoretical data, at least in the physiological range. Nevertheless, Rq, calculated by linear regression from the experimental data, appears to be 0.7, *i.e.* higher than the theoretical value. This may be due to some distortions introduced by the imposed constraints on the theoretical line, which have little influence around the physiological range, but may become particularly important towards the extremes.

This representation could not be used for the hypoxic-hyperoxic simulation, because it requires constant Pio_2. Thus the classical $\dot{V}o_2max$ *vs* Pio_2 curve was used to represent the simulation results (Figure 2). The curve obtained reproduces quite well that described

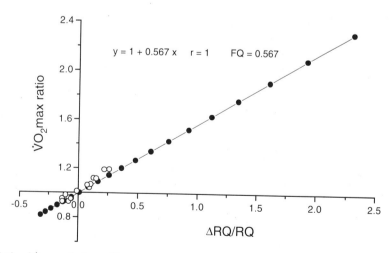

Figure 1. Ratio of $\dot{V}o_2max$ before to $\dot{V}o_2max$ after an acute change in haemoglobin concentration as a function of the induced change in cardiovascular resistance to oxygen flow (Rq). The slope of the line is equal to the fractional limitation to $\dot{V}o_2max$ imposed by Rq (Fq, 10). The filled dots, to which the equation refers, represent the results of the present simulation. The open dots refer to experimental data from various literature sources (1, 6, 7, 18, 20, 23, 24)..

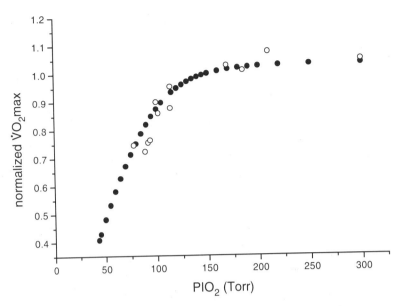

Figure 2. Normalised V̇o₂max (normoxic value set equal to one) as a function of inspired oxygen pressure (Pio_2). The filled dots represent the results of the present simulation. The open dots refer to experimental data from various literature sources (8, 13, 15, 16, 17, 19, 21).

by the experimental data (8, 13, 15, 16, 17, 19, 21). It is noteworthy that the shape of Fig 2 implies a progressive increase of conductance (= $\dot{Q}/\Delta P$ or, in this case, V̇o₂max $/Pio_2$), likely due to changes in Gq imposed by the characteristics of the O₂ equilibrium curve (10).

Both Figs 1 and 2 refer to conditions that lead to changes in Cao_2. These, however, are the result of a linear response in the former case, but of a non-linear response in the latter. This being the case, one could expect that the two considered conditions may not have the same effects on V̇o₂max for a given change in Cao_2. However, this is not so. In fact the relationships between V̇o₂max and Cao_2 in the two tested conditions, which are reported in Fig 3, are described by two identical linear equations. This means that, for any given change in Cao_2, independent of the reason behind it, equivalent variations in V̇o₂max are to take place, once the same starting condition (normocythaemia in normoxia) is maintained. In practical terms, this is to say that anaemia and hypoxia have the same effects on V̇o₂max.

It should be noticed, however, that this does not imply that the fractional limitation to V̇o₂max is the same in the two tested conditions. In fact, Fig 2 does not inform on the fractional limitation to V̇o₂max, as it implies a change in the overall pressure gradient which prevents from computing Fq. It has been argued that Fq may be lower in hypoxia than in normoxia. Consequently, if one is blood-doped in hypoxia, for the same change in Hb, a smaller increase in V̇o₂max is to be expected than in normoxia.

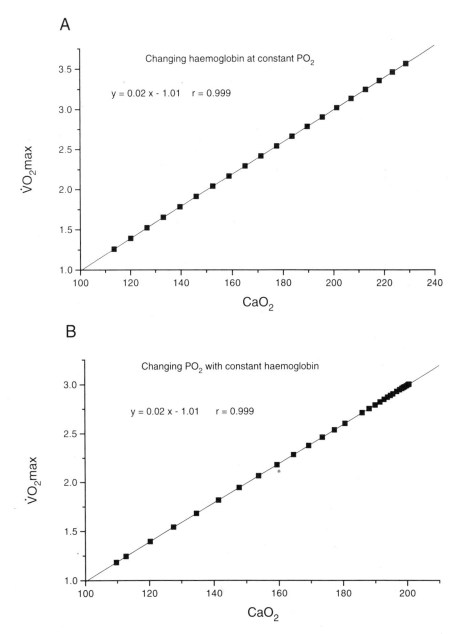

Figure 3. $\dot{V}o_2$max as a function of arterial oxygen concentration (Cao_2) as simulated from changing haemoglobin concentration (anaemia/polycythaemia, A) or from changing inspired oxygen pressure (hypoxia/hyperoxia, B).

5. REFERENCES

1. Buick, F. J., N. Gledhill, A. B. Froese, L. L. Spriet, and E. C. Meyers. Effects of induced erythrocythemia on aerobic work capacity. *J. Appl. Physiol.* **48**: 636–642, 1980.

2. Cerretelli, P. Gas exchange at altitude. In: *Pulmonary gas exchange. II. Organism and environment*, edited by J. B. West. New York: Academic Press, 1980, pp 97–147.

3. Cerretelli, P., and P. E. di Prampero. Gas exchange at exercise. In: *Handbook of Physiology. The Respiratory System. Vol IV, Gas Exchange.* edited by L. E. Farhi and S. M. Tenney, Bethesda, USA:Amer. Physiol. Soc., 1987, pp 555–632.

4. di Prampero, P. E. Metabolic and circulatory limitations to $\dot{V}o_2$max at the whole animal level. *J. Exp. Biol.* 115: 319–331, 1985.

5. di Prampero, P. E., and G. Ferretti. Factors limiting maximal oxygen consumption in humans. *Respir. Physiol.* 80: 113–128, 1990.

6. Ekblom, B., A. N. Goldbarg, and B. Gullbring. Response to exercise after blood loss and reinfusion. *J. Appl. Physiol.* 33: 175–180, 1972.

7. Ekblom, B., G. Wilson, and P. O. Åstrand. Central circulation during exercise after venesection and reinfusion of red blood cells. *J. Appl. Physiol.* 40: 379–383, 1976.

8. Fagraeus, L., J. Karlsson, D. Linnarsson, and B. Saltin. Oxygen uptake during maximal work at lowered and raised ambient air pressure. *Acta Physiol. Scand.* 87: 411–421,1973.

9. Ferretti, G. On maximal oxygen consumption in hypoxic humans. *Experientia* 46: 1188–1194, 1990.

10. Ferretti, G., and P. E. di Prampero. Factors limiting maximal O_2 consumption: effects of acute changes in ventilation. *Respir. Physiol.* 99: 259–271, 1995.

11. Ferretti, G., B. Kayser, .F Schena, D. L. Turner, and H. Hoppeler. Regulation of perfusive O_2 transport during exercise in humans: effects of changes in haemoglobin concentration. *J. Physiol. Lond.* 455: 679–688, 1992.

12. Ferretti, G., B. Kayser, and F. Schena. Effects of hypoxia on cardiovascular oxygen transport in exercising humans (Abstract). *Pflügers Arch.* 424: R21, 1993.

13. Fulco, C. S., P. B. Rock, L. Trad, V. Forte, and A. Cymerman. Maximal cardiorespiratory responses to one- and two-legged cycling during acute and long-term exposure to 4300 m altitude. *Eur. J. Appl. Physiol.* 57: 761–766, 1988.

14. Gayeski, T. E. J., and C. R. Honig. Intracellular PO_2 in long axis of individual fibers in working dog gracilis muscle. *Am. J. Physiol.* 254: H1179-H1186, 1988.

15. Greenleaf, J. E., E. M. Bernauer, W. C. Adams, and L. Juhos. Fluid-electrolyte shift and $\dot{V}o_2$max in man at simulated altitude (2287 m). *J. Appl. Physiol.* 44: 652–658, 1978.

16. Hartley, L. H., J. A. Vogel, and K. Landowne. Central, femoral and brachial circulation during exercise in hypoxia. *J. Appl. Physiol.* 34: 87–90, 1973.

17. Hughes, R. L., M. Clode, R. H. T. Edwards, T. J. Goodwin, and N. L. Jones. Effect of inspired O_2 on cardiopulmonary and metabolic responses to exercise in man. *J. Appl. Physiol.* 24: 336–347, 1968.

18. Kanstrup, I. L., and B. Ekblom. Acute hypervolemia, cardiac performance, and aerobic power during exercise. *J. Appl. Physiol.* 52: 1186–1191, 1982.

19. Lawler, J., S. K. Powers, and D. Thompson. Linear relationship between $\dot{V}o_2$max and $\dot{V}o_2$max decrement during exposure to acute hypoxia. *J. Appl. Physiol.* 64: 1486–1492, 1988..

20. Spriet, L. L., N. Gledhill, A. B. Froese, and D. L. Wilkes. Effect of graded erythrocythemia on cardiovascular and metabolic responses to exercise. *J. Appl. Physiol.* 61: 1942–1948, 1986.

21. Stenberg, J., B. Ekblom, and R. Messin. Hemodynamic response to work at simulated altitude (4000 m). *J. Appl. Physiol.* 21: 1589–1594, 1966.

22. Taylor C. R., and E. R. Weibel. Design of the mammalian respiratory system. I. Problem and strategy. *Respir. Physiol.* 44: 1–10, 1981.

23. Turner, D. L., H. Hoppeler, C. Noti, H. P. Gurtner, H. Gerber, F. Schena, B. Kayser, and G. Ferretti. Limitations to $\dot{V}o_2$max in humans after blood retransfusion. *Respir. Physiol.* 92: 329–341, 1993.

24. Woodson, R. D., R. E. Wills, and C. Lenfant. Effect of acute and established anemia on O_2 transport at rest, submaximal and maximal work. *J. Appl. Physiol.* 44: 36–43, 1978.

19

EFFECT OF DRY AND HUMID HEAT ON PLASMA CATECHOLAMINES DURING PROLONGED LIGHT EXERCISE

Juhani Smolander, Olli Korhonen, Raija Ilmarinen, Kimmo Kuoppasalmi, Matti Härkönen, and Ilmari Pyykkö

Institute of Occupational Health
Helsinki, Finland
University of Helsinki
Finland

1. INTRODUCTION

During dynamic exercise in the heat, plasma noradrenaline (NA) levels are higher than during the same exercise in a thermoneutral environment (2). However, the exponential noradrenaline-heart rate relationship is unaltered in exercise with exogenous heat (2). Heat stress seems to have no added effect on plasma adrenaline (A) concentration (2). The aim of our study was to determine the effect of dry and humid heat on plasma NA and A concentrations during prolonged light exercise, and their relationships to core temperature and hemodynamic variables.

2. METHODS

Eight healthy, physically fit, and unacclimated men volunteered for the study. Their mean (±1 SD) age, height, weight, and maximal oxygen uptake ($\dot{V}O_{2max}$) were 32 ± 3 years, 175 ± 6 cm, 72 ± 6 kg, and 59 ± 5 ml/min/kg, respectively.

The tests included three prolonged treadmill tests at $30\%\dot{V}O_{2max}$ in standard working overalls (clothing insulation value 0.7 clo units) as follows: once in a thermoneutral ($20°C/40\%$), once in a warm humid ($30°C/80\%$), and once in a hot dry ($40°C/20\%$) environment. The tests were done at one week intervals in a random order at the same time of day. The exercise consisted of seven 30 min work periods separated by 5 min pauses for weighing and blood sampling (total duration 4 hours). During the tests, water was offered *ad libitum*, and drinking was encouraged.

Before each test, a Teflon catheter was inserted into an antecubital vein, and the subject then lay in a supine position at least 30 min before the resting blood sample was

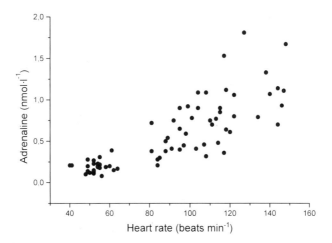

Figure 1. Plasma adrenaline concentration in relation to heart rate at rest and during prolonged light exercise in thermoneutrality, and under heat stress.

taken. During exercise, blood samples for NA and A analysis were taken after 1 h and 3 h of excercise. Plasma NA and A were analyzed by a single-isotope radioenzymatic technique (1). The intra-assay coefficients of variation for A and NA were 12.6% (at the level of 0.25 nmol/l) and 6.4% (2.2 nmol/l), respectively. In addition, rectal temperature (Tre), heart rate (HR), cardiac output (CO) (CO_2-rebreathing technique), stroke volume (SV), and skin thermal conductance (C) were determined. The methods and results of these variables have been presented elsewhere (3).

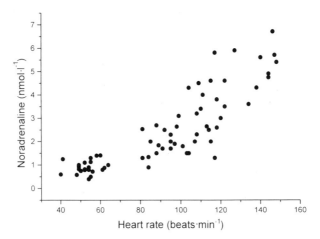

Figure 2. Plasma noradrenaline concentration in relation to heart rate at rest and during prolonged light exercise in thermoneutrality, and under heat stress.

3. RESULTS

At rest NA, Tre, and HR were not significantly different between the three climates. At rest, NA averaged 0.88±0.35, 0.88±0.22, and 0.92±0.24 nmol/l in the thermoneutral, warm humid, and hot dry environment tests, respectively. Plasma A was slightly lower (p<0.05) before the test in the thermoneutral environment (0.17±0.04 nmol/l) as compared to the hot dry environment (0.23±0.09 nmol/l), but similar to the warm humid environment (0.19±0.07 nmol/l).

During exercise, HR, Tre, and C were significantly higher and SV was lower in the warm humid and the hot dry environment in comparison to the thermoneutral environment. Cardiac outputs did not differ significantly between the climates (Table 1). The mean $\dot{V}O_2$ was approximately 1.3 l/min with no difference between the three climatic conditions.

In all climates, NA and A were significantly higher during exercise as compared to rest. As compared to thermoneutrality, heat stress had a significant added effect on NA an A concentrations both after 1 h and 3 h of exercise (Table 1).

Plasma NA correlated significantly with HR (0.90), Tre (0.79), C (0.70), and SV (-0.44). Plasma A correlated significantly with HR (0.81), Tre (0.76), C (0.54), and SV (-0.34).

4. CONCLUSIONS

In our study, very little difference in heat strain was observed between the two hot climates. Dry and humid heat stress during prolonged light exercise had an added effect on NA and A. Heightened skin blood flow under heat stress presumably required greater compensatory vasocontriction, especially in inactive regions, because CO was unaltered. Thus, increased NA levels probably indicated greater sympathetic outflow to splanchnic area, kidneys, inactive muscles and perhaps also to active muscles. Elevated A levels may be related to metabolic factors or psychologic factors accompanying prolonged physical work in the heat.

Table 1. Plasma catecholamines, rectal temperature, and hemodynamic variables after 3 hours of light exercise in a thermoneutral, a warm humid, and a hot dry climate. Mean±SD for eight subjects

Variable	Thermoneutral	Warm humid	Hot dry
Noradrenaline (nmol/l)	1.93 ± 0.46	4.54 ± 1.09[a]	4.34 ± 1.56[a]
Adrenaline (nmol/l)	0.47 ± 0.13	1.00 ± 0.40[a]	0.98 ± 0.39[a]
Rectal temperature (°C)	37.7 ± 0.2	38.3 ± 0.2[a]	38.2 ± 0.3[a]
Cardiac output (l/min)	11.8 ± 1.9	12.8 ± 0.9	12.3 ± 2.4
Heart rate (beats/min)	90.0 ± 6.0	136.0 ± 16.0[a]	125.0 ± 14.0[a]
Stroke volume (ml)	131.0 ± 24.0	100.0 ± 14.5[a]	106.0 ± 22.0[a]
Conductance (Km²/W)	33.6 ± 4.9	112.6 ± 35.6[a]	106.5 ± 26.1[a]

[a] p<0.05 as compared to the thermoneutral environment

REFERENCES

1. Endert, E. Determination of noradrenaline and adrenaline in plasma by a radioenzymatic assay using high pressure liquid chromatography for the separation of the radiochemical products. *Clin. Chim. Acta* **96**: 233–239, 1979.
2. Rowell, L.B., G.L. Brengelmann, and P.R. Freund. Unaltered norepinephrine-heart rate relationship in exercise with exogenous heat, *J. Appl. Physiol.* **62**: 646–650, 1987.
3. Smolander, J., R. Ilmarinen, O. Korhonen, and I. Pyykkö. Circulatory and thermal responses of men with different training status to prolonged physical work in dry and humid heat. *Scand. J Work Environ. Health* **13:** 37–46, 1987.

20

COORDINATION OF BREATHING AND WALKING AT DIFFERENT TREADMILL SPEED AND SLOPE LEVELS AND ITS EFFECTS ON RESPIRATORY RATE AND MINUTE VENTILATION

B. Raßler[1] and J. Kohl[2]

[1] Carl Ludwig Institute of Physiology
University of Leipzig
D - 04103 Leipzig, Germany
[2] Institute of Physiology
University of Zurich
CH - 8057 Zurich
Switzerland

1. INTRODUCTION

During simultaneous motor actions, e.g. in walking, the respiratory rhythm changes due to an entrainment with locomotor rhythm. This entrainment is in accord with the phenomenon of "relative coordination" (5). Since central nervous coordination is characterized by reciprocal interactions (5), an influence of the respiratory rhythm on the walking pattern might also be expected. The biological importance of coordination is still a matter of debate, although it is often stated that such entrainment reflects economy of the involved motor actions (2–4).

The purpose of the present study was to investigate the coordination between breathing and walking in humans with emphasis on the following items:

a. Does the strength of coordination depend on the metabolic load or on the rhythm of locomotion?

b. Is coordination between breathing and leg movement a truly mutual interaction, including influences on the temporal stride pattern exerted by the respiratory rhythm?

c. Does coordination cause a change of the respiratory rate (RR) with functional consequences for ventilation (Ve) and to energetic cost of the motor exercise?

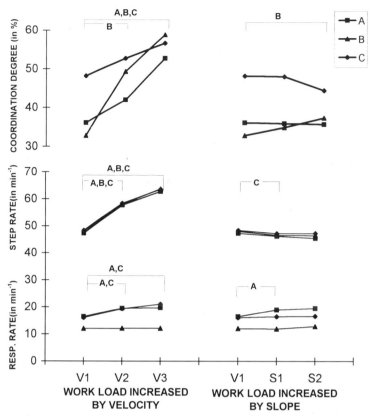

Figure 1. Coordination degree, step rate, and respiratory rate at different levels and kinds of load. Significant differences between load levels are marked by ☐ with A, B, C labelling the appropriate part of the test.

2. METHODS

We examined 18 healthy volunteers walking on a treadmill (Woodway) at three workload levels produced by combinations of different velocities and grades. The experimental protocol consisted of five walking tests: V1 (velocity 1 m/s, slope 0); V2 (velocity 1.5 m/s, slope 0); V3 (velocity 1.8 m/s, slope 0); S1 (velocity 1 m/s, slope 5%); S2 (velocity 1 m/s, slope 10%). Each walking test consisted of three parts with different means of guiding the respiratory rhythm: A - "spontaneous breathing" (with no acoustic pacing of the respiratory rhythm); B - "externally paced breathing" (with metronome acoustic signal to pace the onset of inspiration and expiration, typically at 10 breaths/min); C - "step-related breathing" (using a step-related click signal to trigger the onset of inspiration).

Respiratory variables (air flow, tidal volume, inspiration and expiration times, respiratory rate, and concentrations of respiratory gases) were measured by an ultrasonic pneumotachograph and by a mass spectrometer, respectively, and ventilatory and pulmonary gas exchange variables were derived breath-by-breath using an automated respiratory analysis system; i.e. respiratory rate (RR), inspiration time (TI), expiration time (TE), tidal volume (VT), minute ventilation ($\dot{V}e$), oxygen uptake ($\dot{V}O_2$). Leg movements were recorded by mechano-electrical goniometers.

Coordination implies a constant phase relation between both processes at an integer rate ratio (cf. Fig. 2). The degree of coordination (CD) is expressed as the frequency of the preferred phase relation (in % of the total number of phase relations within one test). The exact procedure of CD determination is described in detail in Ref. 12.

Respiratory changes during coordination are reflected by the differences between non-coordinated and coordinated periods (ΔRR, ΔVT, ΔVe, and ΔVO_2) within one test. The relationship between these parameters has been analyzed by linear correlation and regression. Within each test for which the number of both coordinated and non-coordinated breaths was at least 15 breaths (n = 145, cf. Table 1), we additionally compared respiratory and step parameters in periods of coordination with non-coordinated periods: (RR, TI, TE, VT, Ve, VO_2, and step rate (SR). We tested the significance of differences by the Mann-Whitney U test.

3. RESULTS

3.1. Dependence of Coordination Strength on Metabolic Load and Walking Speed

Coordination strength improves distinctly with increasing walking speed and step rate. The respiratory rate does not influence the degree of coordination. Metabolic loads

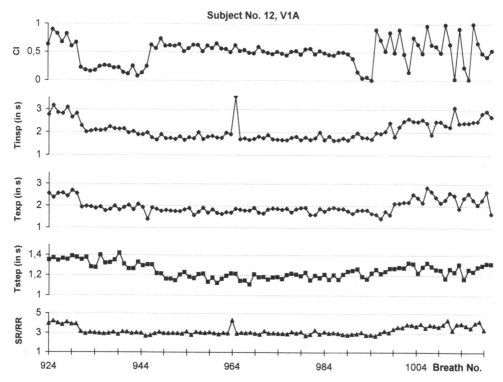

Figure 2. Sequential presentation of cross-intervals (CI), inspiratory and expiratory times (Tinsp, Texp), step duration (Tstep), and rate ratio (SR/RR) of subject 12, V1A.

Table 1. Percentage of analogous and inverse changes of TI and TE (at least one parameter being significant) and percentage of significant RR and SR differences between coordinated and non-coordinated periods. n = number of coordination periods (only for periods lasting at least 15 breaths). Asterisks mark significant differences to part A (chi-squared test; *:p<0.05, **:p<0.01, two-tailed probability)

	n	Analogous T_I/T_E changes	Inverse T_I/T_E changes	significant RR differences	significant SR differences
A:	42	57.1 %	4.8 %	54.8 %	33.3 %
B:	32	12.5 % **	12.5 %	12.5 % **	37.5 %
C:	71	14.1 % **	42.3 % **	12.7 % **	53.5 % *

imposed by increasing the treadmill slope have only a slight effect on the coordination degree (Fig. 1). This effect of slope becomes obvious in cases of constant respiratory and step rates (part B, comparing V1 with S1 and S2).

3.2. Mutual Interaction of Breathing and Walking Rhythms

In non-paced breathing (part A), we found the temporal pattern of respiration to change in all coordinated periods. The variability of the respiratory cycle time - expressed by the standard deviation of RR - decreases by about 35%. In about 55% of coordinated periods, significant changes of the mean TI and TE occurred as well. These alterations mostly occur in the same direction (Table 1, left hand side), i.e. both inspiration and expiration are prolonged or both are shortened and, hence, result in significant RR shifts (Table 1, right hand side). We term these "analogous" changes.

In most coordination periods, we also found the step rate to become more regular, but clearly to a lesser extent than the respiratory rate. The standard deviation of step rate decreases by about 22%, and the percentage of significant changes of SR during coordination is only 33%. Figure 2 shows a characteristic example of the temporal respiratory and step pattern modulation during coordination.

Pacing the respiratory rhythm (parts B and C) modifies the respiratory rate and changes the SR/RR ratio towards an integer ratio. The respiratory rhythm becomes less flexible to coordinative motor influences. Alterations of TI are often accompanied by inverse changes of TE (cf. Table 1). Phase-coupling is, in more cases than in part A, achieved by modifying SR. The percentage of significant SR differences is distinctly higher than in non-paced breathing. On the contrary, the number of significant RR changes is reduced.

3.3. Influence of Coordination and RR Modulation on VT, V̇E, and V̇O$_2$

Alterations of TI and TE at the onset of coordination are associated with changes of tidal volume. The absolute amounts of ΔVT and ΔRR are about 12% of the values during non-coordinated periods in non-paced breathing. Pacing the respiratory rhythm allows only smaller variations of ΔVT and ΔRR (about 7.5%). In about 60% of all coordination periods, the VT modifications correlate negatively with those of RR, the correlation coefficient between ΔVT and ΔRR being -0.73, p<0.001. Therefore, V̇e modulations are (with about 6%) clearly smaller than those of RR and VT. Significant changes of V̇e occur in about 40% of coordinated periods lasting 15 or more breaths, the number of increases and decreases being similar (21% and 17%). In non-paced breathing (part A), $\Delta\dot{V}e$ correlates more tightly with ΔRR than with ΔVT (Fig. 3). In paced breathing (parts B and C), we found $\Delta\dot{V}e$ to be much more dependent on ΔVT (r = 0.61; p < 0.001) than on ΔRR (r = 0.03).

Although step-related pacing of the respiratory rhythm (part C) improves coordination, we found no significant differences of $\dot{V}O_2$ between parts C and A. Changes of $\dot{V}O_2$ occur in about 20% of coordinated periods lasting 15 or more breaths. With the exception of Part B, significant increases predominate by far (Table 2). The absolute amount of $\dot{V}O_2$ change during coordination ($\Delta\dot{V}O_2$) is under all conditions about 5%. Changes of $\dot{V}O_2$ during coordination correlate significantly with changes of minute ventilation ($\Delta\dot{V}e$) (r = 0.60; p < 0.001). In only 30% of all coordination periods does $\Delta VO2$ respond inversely relative to $\Delta\dot{V}e$.

4. DISCUSSION

4.1. Dependence of Coordination Strength on Metabolic Load and Walking Speed

Recent studies in human running (2, 6) showed that coordination becomes closer at higher work loads achieved by increased treadmill speed. In decerebrate cats, Kawahara *et al.* (8) found coordination to improve with increasing stepping frequency. Results of Loring *et al.* (10) indicate that entrainment between breathing and walking occurred more frequently at higher walking speeds and at 10% treadmill inclination. Therefore, the question arises whether it is the higher work load or the increased stride rate that affects the

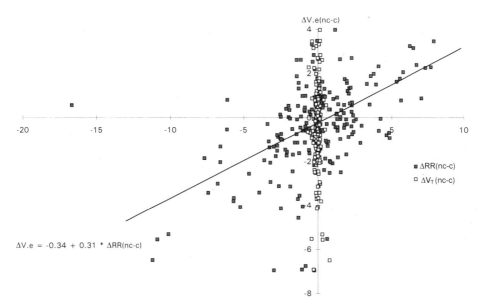

Figure 3. Linear regression of the differences between non-coordinated and coordinated breaths in part A: $\Delta\dot{V}e$ vs. ΔRR (full squares), and $\Delta\dot{V}e$ vs. ΔVT (open squares), total n = 207 tests.

Table 2. Percentage of significant VO2 changes in periods of coordination and portions of significant increases and significant decreases (for number of coordination periods in each part, see Table 1)

In periods of coordination	spontaneous respiration(part A)	respiratory rate paced by metronome (part B)	respiratory rate paced by step rhythm (part C)
significant V̇.O2 changes	16.7 %	25.0 %	19.7 %
- significant increases	14.3 %	12.5 %	16.9 %
- significant decreases	2.4 %	12.5 %	2.8 %

degree of coordination. Our results provide evidence that the treadmill speed and, hence, the stride rate has a much stronger effect on the coordination degree than an increased work load achieved by walking uphill at constant treadmill speed. Neither increased respiratory rate associated with increased work load nor the SR/RR ratio have a substantial effect on coordination degree.

4.2. Mutual Interaction of Breathing and Walking Rhythm

During coordination between breathing and other voluntary movements, the respiratory rhythm usually subordinates to the movement rhythm (1, 11). Influences of respiration on the additional movement occur as well (14), but to a lesser extent (13). In coordination, the extent of each rhythm's frequency variation depends on the relationship between their ability to defend their intrinsic frequency against coordinative influences and the strength of the attraction (5). The results of the within-test comparisons revealed that, although both respiratory and stepping rhythms become more regular during coordination and - in some cases - their rates change significantly, the alterations of the respiratory rhythm predominate over those of the walking rhythm. We assume that voluntary attention paid to walking, larger energy expenditure of the walking movement and a stronger sensory feedback (caused, for example, by touching the ground or by the treadmill speed) are factors limiting variations of the walking rhythm. This assumption is supported by our result that higher work load *per se* can enhance coordination independent of walking and respiratory rates, and agrees with findings reported in literature. Coordination of pairwise limb movements with breathing is more stable than in movements of a single limb (13). Fictive movements entrain the respiratory rhythm to a lesser extent than real movements (9). Since the respiratory rhythm is less constrained than the stepping rhythm, it can adjust its rate in a wider range to other movements' rates.

4.3. Influence of Coordination and RR Modulation on VT , V̇E, and V̇O$_2$

In order to meet metabolic demands, alveolar ventilation has to be stabilized against variations of respiratory rate and tidal volume. Since these two parameters are subject to numerous non-homeostatic influences, such as trunk position and movements, psycho-affective influences, communicative functions (speaking, laughing, singing etc.) and also central coordination with concomitant motor activities, any variation of one of these parameters requires a compensatory modulation of the other one. Human subjects are able to match VT with imposed changes in RR and *vice versa* based on judging both respiratory rate and tidal volume (7). In the present experiments, we found coordinative changes of respiratory rate being accompanied with changes of tidal volume. The number of significant V̇e changes in periods of coordination and the positive correlation between ΔRR and ΔV̇e demonstrate the compensation by tidal volume to be incomplete. According to v. Holst's theory, coordination causes increase or decrease of the attracted rhythm's rate (that means the respiratory rate) dependent on the velocities of the concomitant rhythms and on their actual phase relation in order to strive after stable rate ratio and phase relations(1). Since ΔV̇O2 is positively correlated with ΔV̇e and hence, in periods of coordination, is often higher than in non-coordinated periods, an economy of energy is unlikely to be an essential effect of coordination during walking.

5. CONCLUSIONS

Coordination depends more on the temporal pattern of the dominant rhythm than on metabolic load. Coordination between breathing and walking is a real reciprocal interaction the influence of walking on the respiratory rhythm being stronger than vice versa.. Coordination can cause modifications of respiratory rate that are associated with alterations of VT. Since VT modifications do not fully compensate for changes of respiratory rate, significant changes of V̇e and V̇O2 can occur during coordination with no clear preference of increase or decrease. Consequently, energy economy is not considered to be the main biological importance of coordination during walking.

6. REFERENCES

1. Bechbache, R.R., and J. Duffin. The entrainment of breathing frequency by exercise rhythm. *J. Physiol. (London)* **272**: 553–561, 1976.
2. Bernasconi, P., and J. Kohl. Analysis of co-ordination between breathing and exercise rhythms in man. *J. Physiol. (London)* **471**: 693–706, 1993.
3. Garlando, F., J. Kohl, E.A. Koller, and P. Pietsch. Effect of coupling the breathing- and cycling rhythms on oxygen uptake during bicycle ergometry. *Eur. J. Appl. Physiol.* **65**: 570–578, 1985.
4. Hildebrandt, G., and F.-J. Daumann. Die Koordination von Puls- und Atemrhythmus bei Arbeit. *Int. Z. angew. Physiol. Arb.-physiol.* **21**: 27–48, 1965.
5. Holst, E.v. Die relative Koordination als Phänomen und als Methode zentralnervöser Funktionsanalyse. *Erg. Physiol.* **42**: 228–306, 1939.
6. Jasinskas, C.L., B.A. Wilson, and J. Hoare. Entrainment of breathing rate to movement frequency during work at two intensities. *Respir. Physiol.* **42**: 199–209, 1980.
7. Katz-Salamon, M. Assessment of ventilation and respiratory rate by healthy subjects. *Acta Physiol. Scand.* **120**: 53–60, 1984.
8. Kawahara, K., S. Kumagai, Y. Nakazono, and Y. Miyamoto. Coupling between respiratory and stepping rhythms during locomotion in decerebrate cats. *J. Appl. Physiol.* **67**: 110–115, 1989.
9. Kawahara, K., Y. Nakazono, Y. Yamauchi, and Y. Miyamoto. Coupling between respiratory and locomotor rhythms during fictive locomotion in decerebrate cats. *Neurosci. Lett.* **103**: 326–332, 1989.
10. Loring, S.H., J. Mead, and T.B. Waggener. Determinants of breathing frequency during walking. *Respir. Physiol.* **82**: 177–188, 1990.
11. Perségol, L., M. Jordan, and D. Viala. Evidence for the entrainment of breathing by locomotor pattern in human. *J. Physiol. (Paris)* **85**: 38–43, 1991.
12. Raßler, B., and J. Kohl. Analysis of coordination between breathing and walking rhythms in humans. Unpublished observations.
13. Raßler, B., S. Waurick, and D. Ebert. Einfluß zentralnervöser Koordination im Sinne v. Holsts auf die Steuerung von Atem- und Extremitätenmotorik des Menschen. *Biol. Cybern.* **63**: 457–462, 1990.
14. Viala, D. Evidence for direct reciprocal interactions between the central rhythm generators for spinal "respiratory" and locomotor activities in the rabbit. *Exp. Brain Res.* **63**: 225–232, 1986.

21

INTRACORONARY BLOCKADE OF NITRIC OXIDE SYNTHETASE LIMITS CORONARY VASODILATION DURING SUBMAXIMAL EXERCISE

Patricia A. Gwirtz and Sung-Jung Kim

University of North Texas Health Science Center at Fort Worth
Department of Integrative Physiology
3500 Camp Bowie Boulevard
Fort Worth, Texas 76107–2699

1. INTRODUCTION

The vascular endothelium is important in modulating vasomotor tone (1,3). Among the important vasoactive substances synthesized by endothelial cells is endothelium-derived relaxing factor or nitric oxide (NO), which is synthesized from the amino acid L-arginine and catalyzed by nitric oxide synthase (16,33) in response to many stimuli including flow-induced shear stress at the interface between blood and the endothelial cell surface (7,8). This flow-dependent control seems to be important in the coronary circulation due to the extremely pulsatile nature of flow patterns in this vascular bed.

During exercise, cardiac work and myocardial oxygen demand of the heart increases several-fold due to sympathetic stimulation. Consequently, coronary vasodilation occurs primarily due to local metabolic mechanisms (5). NO may be released from the coronary endothelium due to the elevation in shear stress caused by increases in coronary blood flow and to neurohumoral stimulation (2,10). The present study evaluated the role of NO on regulation of the coronary blood flow during exercise, using the selective intracoronary infusion of the NO synthase inhibitor N^G-monomethyl-L-arginine (L-NMMA) in chronically instrumented, conscious dogs.

2. METHODS

Experiments were performed on seven mongrel dogs (20–30 kg) of either sex. Dogs were chronically instrumented to measure aortic pressure (AoP), left ventricular systolic and end-diastolic pressures (LVSP, LVEDP), maximum rate of rise of left ventricular systolic pressure (+dP/dt$_{max}$), circumflex coronary blood flow (CBF), and heart rate (HR). A circumflex artery catheter was used for selective injection of drugs into the circumflex coronary artery and collection of arterial blood samples. A catheter was inserted into the coronary sinus for myocardial venous blood sampling.

Data was collected while the dog was resting quietly or running on a treadmill. Each animal was subjected twice to a standardized submaximal exercise regimen (9). The first exercise test was a control test. On the next day, L-NMMA (35 mg over 20 min) was infused into the left circumflex artery and the submaximal exercise test repeated. While this dose of L-NMMA did not affect the CBF response to sodium nitroprusside, blockade of NO synthase significantly attenuated the CBF response to acetylcholine (46% at 20 μg/min, p < 0.05). Simultaneous aortic and coronary sinus blood samples were withdrawn at rest and during exercise for measurement of myocardial oxygen extraction and myocardial oxygen consumption (MV̇O$_2$) to determine whether changes in vascular tone by NO during exercise was caused by changes in MV̇O$_2$.

Data are reported as mean ± standard error of the mean (SEM). For all comparisons, data within and among the protocols were analyzed for statistical significance using two way analysis of variance (ANOVA). If the ANOVA detected significant differences within factor means, then these differences were identified with the Student paired t-test for Factor A (with or without L-NMMA) and with the Student-Newman-Keul procedure for Factor B (exercise workload). Statistical significance was accepted at p<0.05.

3. RESULTS

Data from all dogs are summarized in Table 1. In the absence of L-NMMA (control condition), CBF increased as the intensity of exercise increased. At rest and at a mild exercise workload (4.8 kph/0 % incline), CBF was not different between control and L-NMMA conditions. In contrast, the magnitude of increase in CBF was significantly attenuated by intracoronary L-NMMA at workloads greater than 6.4 kph/0 % incline (p<0.05).

For both control and L-NMMA conditions, graded exercise produced progressive increases in MAP, HR, LVSP and LVEDP, and +dP/dt$_{max}$. Note that at rest and during exercise, MAP, HR, LVSP, and LVEDP were not different between control and L-NMMA conditions at any respective level of exercise, indicating that intracoronary infusion of L-NMMA did not affect systemic hemodynamic parameters. However, intracoronary infusion of L-NMMA did attenuate the increase in +dP/dt$_{max}$ during moderate to strenuous levels of exercise (i.e., workloads greater than 4.8 kph/4 % incline; p<0.05).

At rest, MV̇O$_2$ and oxygen extraction were not affected by intracoronary administration of L-NMMA. However, as the intensity of exercise increased for both conditions, arterial-venous oxygen content difference, oxygen extraction, and MV̇O$_2$ were increased significantly (p < 0.05), while coronary sinus oxygen content was decreased (p < 0.05). After L-NMMA blockade, at workloads higher than 6.4 kph/8 % incline, MV̇O$_2$ was significantly reduced compared to respective control values (p < 0.05), but oxygen extraction was not changed.

4. DISCUSSION

The most significant findings of this study are: a) in the conscious resting state and at mild levels of exercise, NO does not appear to contribute to coronary vasodilation; and

Table 1. Myocardial Response to Exercise After Intracoronary L-NMMA (35mg)

	Control							After L-NMMA						
	Rest	3/0	4/0	4/4	4/8	4/12	4/16	Rest	3/0	4/0	4/4	4/8	4/12	4/16
LVSP (mmHg)	120±5	137±6*	139±7*	141±6*	145±6*	150±6*	156±6*	123±5	145±5*	148±5*	149±6*	153±5*	157±7*	161±7*
LVEDP (mmHg)	1±1	4±1	4±2	5±1	6±1	6±1	7±1	1±1	5±1	5±1	6±1	6±1	7±1	7±1
$+dP/dt_{max}$ (mmHg/sec)	1745±147	2423±194†	2677±183*	2813±184*	3003±197*	3308±199*	3740±258*	1768±143*	2145±262*	2185±261*	2350±289†	2510±298†	2720±321†	3080±364†
HR (bpm)	88±6	131±7*	160±4*	170±6*	180±5*	195±7*	216±9*	83±5	129±1*	153±6*	164±6*	177±5*	190±7*	205±9*
Mean AoP (mmHg)	97±5	106±5*	106±5*	107±5*	108±5*	111±5*	115±4*	97±3	112±4*	113±4*	113±5*	114±5*	116±5*	119±5*
CBF (ml/min)	30±3	46±6*	49±6*	52±6*	54±7*	59±8*	63±9*	27±3	43±6*	44±6†	46±6†	47±6†	49±7†	53±7†
$M\dot{V}O_2$ (ml/min)	3.02±0.70	5.11±1.39*			7.03±1.88*		9.38±2.73*	2.55±0.47	5.05±1.45*			6.27±1.67†		9.40±2.38†
O_2 Extraction (%)	60±3	66±2*			74±3*		76±4*	62±1	69±4*			73±4*		77±3*

Values are mean ± SEM, n=7 Abbreviations: LVSP, left ventricular systolic pressure; LVEDP, left ventricular end-diastolic pressure; $+dP/dt_{max}$, maximum rate of use of left ventricular pressure; HR, heart rate; AOP, aortic pressure; CBF, coronary blood flow; $M\dot{V}O_2$, myocardial oxygen consumption.
*p < 0.05 vs respective rest values
†p < 0.05 vs respective rest values and respective control values

b) NO-mediated vasodilation appears to contribute to increases in coronary blood flow during moderate to heavy exercise, since blockade of NO synthesis with L-NMMA attenuates increases in coronary blood flow during exercise. These data indicate that endothelium-dependent flow-induced dilation, in addition to the metabolic vasodilation, may be an important stimulus to maintain myocardial oxygen supply during exercise.

The results of the current study demonstrated that the increase in CBF and dP/dt_{max} during exercise were less after NO blockade and as a result, $M\dot{V}O_2$ was reduced. There are two possible explanations regarding the reduction in dP/dt_{max} and $M\dot{V}O_2$ during exercise after blockade of NO production by L-NMMA. First, a reduction in exercise coronary hyperemia led to the reduction in myocardial contractile function and $M\dot{V}O_2$ as a result of the Gregg phenomenon (5). This reduction of $M\dot{V}O_2$ could have been caused by reduced coronary blood flow because myocardial oxygen extraction did not increase to maintain $M\dot{V}O_2$ at given workload. A second possible explanation of these results is that L-NMMA itself and/or the reduction in NO synthesis directly caused a decrease in myocardial contractility, which led to the decrease in $M\dot{V}O_2$, (1,4,11) although there is little experimental evidence for this possibility. As a result, coronary flow will be lower during exercise.

In summary, our data indicate that at rest and at low levels of exercise, coronary hyperemia is more likely due to local metabolic factors and increases in perfusion pressure rather than NO release. However, during moderate to heavy exercise, NO-mediated vasodilation, in addition to the metabolic-mediated vasodilation, appears to be important in maintaining myocardial oxygen supply during exercise. After NO blockade, the reduction in coronary blood flow during exercise decreases myocardial contractility and myocardial oxygen consumption. This indicates that endothelium-dependent flow-induced dilation, in addition to the metabolic vasodilation, is probably an important stimulus to maintain myocardial oxygen supply during exercise.

5. ACKNOWLEDGMENTS

This research was support by NIH grant #HL-34172 and HL-29232 and a grant from the Texas Affiliate of the American Heart Association.

6. REFERENCES

1. Amrani, J.O., N.J. Allen, S.E. Harding, J. Jayakumar, J.R. Pepper, S. Monaca, and M.H. Yacoub. Role of basal release of nitric oxide on coronary flow and mechanical performance of the isolated rat heart. *J. Physiol.* **456**:681–687, 1992.
2. Berdeaux, A, B. Ghaleh, J.L. Dubois-Rande, B. Vigue, C.D.L. Rochelle, L. Hittinger, and J.F. Giudicelli. Role of vascular endothelium in exercise-induced dilation of large epicardial coronary arteries in conscious dogs. *Circ.* **89**:2799–2808, 1994.
3. Brenner B.M., J.L. Troy, and B.J. Ballermann. Endothelium-dependent vascular responses: mediators and mechanisms. *J. Clin. Invest.* **84**:1373–1378, 1989.
4. Chu, A., D.E. Chamvers, C.-C. Lin, W.D. Kuehl, R.M.J. Palmer, S. Moncada, and F.R. Cobb. Effect of inhibition of nitric oxide formation on basal vasomotion and endothelium-dependent responses of the coronary arteries in awake dogs. *J. Clin. Invest.* **87**:1964–1968, 1991.
5. Feigl, E.O. Coronary physiology. *Physiol. Rev.* **63**:1–205, 1983.
6. Pohl U., J. Holtz, and E. Bassenge. Crucial role of endothelium in the vasodilator responses to increased flow *in vivo*. *Hypertension* **8**:27–44, 1986.
7. Rubanyi R.M., J.C. Romero, and R.M. Vanhoutte. Flow-induced release of endothelium-derived relaxing factor. *Am. J. Physiol.* **250**:H1145–1149, 1986.

8. Shen, W, M. Lundborg, J. Wang, J.M. Stewart, X. Xu, M. Ochoa, and T.H. Hintze. Role of EDRF in regulation of regional blood flow and vascular resistance at rest and during exercise in conscious dogs. *J. Appl. Physiol.* *77*:165–172, 1994.

9. Tipton, C. M., R. A. Carey, W. C. Easten, and H.H.Erickson. A submaximal test for dogs: evaluation of effects of training, detraining and cage refinement. *J. Appl. Physiol.* *37*:271–275, 1974.

10. Vanhoutte, P.M. Endothelium and control of vascular function. *Hypertension* *13*:658–667, 1989.

11. Weyrich, A. S., X. Ma, M. Buerke, T. Murohara, V. E. Armstead, A.M. Lefer, J. M. Nicolas, A. P. Thomas, D. J. Lefer, and J. Vinten-Johansen. Physiological concentrations of nitric oxide do not elicit an acute negative inotropic effect in unstimulated cardiac muscle. *Circ. Res.* *75*:692–700, 1994.

22

ACTIVATION OF THROMBOCYTES AND THE HEMOSTATIC SYSTEM DURING MAXIMAL CYCLE EXERCISE IN HEALTHY TRIATHLETES

Martin Möckel,[1] Natalie-Viviane Ulrich,[1] Lothar Röcker,[2] Andreas Ruf,[3]
Frank Klefisch,[1] Oliver Danne,[1] Jörn Vollert,[1] Reinhold Müller,[4]
Thomas Bodemann,[5] Heinrich Patscheke,[3] Hermann Eichstädt,[6]
Thomas Störk,[7] and Ulrich Frei[1]

[1] Department of Internal Medicine
University Hospital Virchow-Klinikum
Humboldt University, Berlin, Germany
[2] Department of Physiology
Free University Berlin, Germany
[3] Institute for Medical Laboratory Diagnostics
Klinikum Karlsruhe, Germany
[4] Institute for Medical Statistics
Free University Berlin, Germany
[5] Rehabilitation Clinic for Cardiovascular Diseases Nikolassee
Berlin, Germany
[6] Department of Cardiovascular Imaging
University Hospital Virchow-Klinikum
Humboldt University, Berlin, Germany
[7] Department of Cardiology/Angiology
Karl Olga Hospital, Stuttgart, Germany

1. INTRODUCTION

Physical exercise is widely regarded to be important for health. However, heavy physical exertion has been shown to be associated with a higher incidence of myocardial infarction (3, 13, 30). It has also been shown that physical exercise activates fibrinolysis, coagulation and thrombocytes (8–10, 14–24, 26–28). Methodological problems, such as *in vitro* activation of platelets, limited the value of earlier studies. Newly developed analytical tests for molecular markers of the hemostatic system (20) and flow-cytometric techniques for the detection of surface antigens on platelets (25) have enabled more detailed insights into the changes of the hemostatic system during exercise.

For example, Kestin et al. (11) found a flow-cytometric detected activation of platelets in sedentary volunteers, although not in trained subjects. Also, Wang et al. (29) showed that platelet activation may be dependent on the intensity of exercise. Furthermore, Röcker and colleagues (19, 20, 23) described an activation of the hemostatic system with pronounced changes of the fibrinolysis following exercise of different intensities.

The aim of the present study was to examine the influence of exercise intensity on the activation of thrombocytes and the hemostatic system in healthy endurance trained athletes, by documenting the time course of changes in the hemostatic system following exercise.

2. METHODS

2.1. Subjects

The clinical characteristics of the 17 participants (amateur triathletes) are shown in Table 1. 15 athletes (13 men, 2 women) finished the whole protocol. None of them were smokers. No anabolic drugs were taken. 11 volunteers took magnesium tablets, 5 took vitamin E, 12 took other vitamins and 5 had recently taken medical drugs (i. e. antiallergic drugs).

2.2. Protocol

After a 30 minute rest period in the supine position, an exercise test was performed on a cycle ergometer. An aerobic period of 30 minutes came first in which the load was adjusted to yield a heart rate of around 120 min^{-1}. The mean work rate in this period was 125 Watts. An anaerobic period followed with a stepwise increase of the load by 40 Watts every 4 minutes until the volunteers reached fatigue. The mean peak load was 295 Watts. Plasma lactate concentrations were consistently little changed from resting levels in the aerobic phase, but were systematically elevated for the anaerobic phase (Table 2).

Table 1. Clinical characteristics of the 17 participants (15 men, 2 women)

	Median (25% / 75%-percentiles)	Range
Age [years]	25 (21.5/29.5)	19 - 44
Training [years]	5 (2.5/6)	2 - 10
Heart rate [min^{-1}]	72 (58/76)	38 - 100
RRsys [mmHg]	140 (132.5/145)	120 - 165
RRdia [mmHg]	95 (90/100)	75 - 110
Waerob [W]	125 (125/150)	75 - 175
Wmax [W]	295 (230/310)	155 - 350
LV mass index [$g \bullet m^{-2}$]	137.6 (110.2/160.9)	84.1 - 236
FS [%]	31.2 (23.7/39.9)	10 - 50

RRsys, systolic blood pressure; RRdia, diastolic blood pressure; Waerob, workload under aerobic conditions; W_{max}, maximal workload; LV, left ventricular; FS, fractional shortening.

Table 2. Activation of thrombocytes, concentrations of hemostatic markers and hematological variables in 15 participants (13 men, 2 women) during exercise

	REST	AER	ANA	30 min	60 min	90 min
P-selectin [%]	2.3 (2.0/2.4)	2.6 (2.4/3.2)	3.2 (2.9/6.2)	2.3 (2.3/2.9)	2.3 (2.0/2.5)	2.4# (2.1/2.6)
PLT [$10^3 \cdot \mu l^{-1}$]	234 (205/288)	263 (242/310)	284 (257/325)	188 (165/239)	188 (161/236)	178# (146/220)
TAT [$\mu g \cdot l^{-1}$]	1.9 (1.2/3.3)	4.9 (1.8/15.2)	53.3 (33.5/60.0)	39.6 (28.0/60.0)	54.1 (26.5/60.0)	50.0# (34.4/60.0)
F1, F2 [$nmol \cdot l^{-1}$]	0.4 (0.25/0.48)	0.55 (0.37/0.74)	1.41 (0.8/1.78)	0.83 (0.67/1.32)	1.17 (0.59/1.74)	1.0# (0.75/2.04)
PAP [$\mu g \cdot l^{-1}$]	441 (369/765)	948 (768/1629)	4286 (2782/5000)	896 (779/1948)	1017 (567/1353)	780# (699/1281)
FM [$\mu g \cdot ml^{-1}$]	0.43 (0.17/0.64)	0.51 (0.09/0.93)	0.96 (0.7/1.8)	--	--	--
RPP [$10^3 \cdot min^{-1} \cdot mmHg$]	9.7 (7.5/11.8)	21.8 (19.7/24.7)	38.0 (34.2/41.4)	8.6 (7.8/9.6)	8.0 (7.1/9.5)	7.8# (7.0/8.6)
Lactate [$mmol \cdot l^{-1}$]	1.1 (1.0/1.8)	1.8 (1.1/2.0)	7.7 (6.3/10.1)	--	--	--
WBC [$10^3 \cdot \mu l^{-1}$]	4.8 (4.3/6.1)	6.1 (5.2/8.6)	8.7 (7.3/11.2)	5.5 (4.0/7.5)	5.5 (4.2/7.4)	6.2# (5.6/8.3)
HCT [%]	42.6 (40.0/44.2)	44.6 (40.8/46.1)	46.1 (43.7/49.2)	42.6 (39.1/43.6)	40.8 (38.2/42.9)	42.1# (38.7/43.6)

#, $p < 0.001$, Friedman ANOVA; PLT, platelets; WBC, white blood cell count; HCT, hematocrit; for additional abbreviations see text. Values in parentheses are 25% and 75% percentile values.

2.3. Analytical Procedures

Blood samples were drawn at rest and immediately after aerobic and anaerobic exercise. Blood for flow-cytometric (ORTHO, Cytoron-absolute) detection of p-selectin (CD 62, GMP-140) expression was immediately diluted in a special stabilization buffer (Cyfix; 25). *In vitro* activation of platelets was assessed by microscopy of the discoidity of the thrombocytes (25).

The concentrations of thrombocytes (PLT) and white blood cells (WBC), the hematocrit (HCT) and lactate concentration were determined using standard analytical procedures. Concentrations of prothrombin fragments (F1, F2), thrombin-antithrombin complex (TAT) and plasmin-antiplasmin complex (PAP) were measured by enzyme-immuno-assays (EIA; Behringwerke Marburg). Fibrin monomers (FM), to provide a marker of fibrin-generation, were measured by a separate EIA (Boehringer Mannheim).

To rule out heart disease and to determine the degree of physiologic myocardial hypertrophy, all participants underwent an echocardiographic examination. To characterize the left ventricular (LV) systolic function, the fractional shortening was calculated. The LV mass was calculated using a formula by Devereux (6). Heart rate (HR) and systolic blood pressure (SBP) were obtained simultaneously, and the rate pressure product (RPP = HR•SBP) derived.

3. RESULTS AND DISCUSSION

Strenuous exercise, as in the present study, leads to an activation of both the coagulation and the fibrinolytic systems (Table 2, Figs 1 and 2).

In contrast to earlier studies (19, 20, 23), fibrinolysis (as judged by the PAP/TAT ratio) was not pronounced compared to the changes in coagulation variables. This is caused by an extreme increase of TAT after anaerobic exercise with a second peak level after 60 min. In more-highly trained triathletes under less anaerobic conditions (17, 18), the PAP/TAT ratio was increased after exercise. In other studies with exercise in the upright position, no pronounced increases of coagulation markers were found (4). In the follow-up period of 90 min with complete cardiocirculatory recovery (RPP lower than baseline values), TAT remained markedly elevated. As the PAP/TAT ratio further decreased in the post-exercise period, this may be a critical period of increased hemostatic risk for cardiovascular events in some predisposed individuals.

Platelets are activated less by aerobic and markedly more by anaerobic exercise (Fig. 1). These findings are in contrast to the results of Kestin et al. (11), who found platelet activation in sedentary people only. Other studies using different methods confirm our

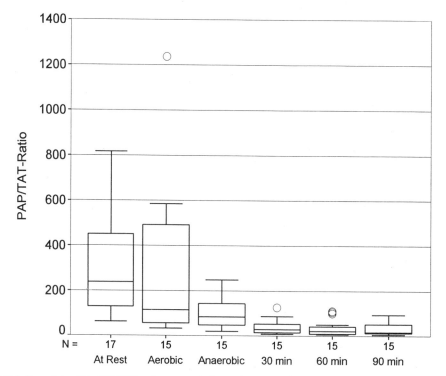

Figure 1. Box-plots of the PAP/TAT ratio at rest, during and post-exercise. Extreme values are displayed as o if they exceed the vertical box length by 1.5%. PAP, plasmin-antiplasmin complex; TAT, thrombin-antithrombin complex.

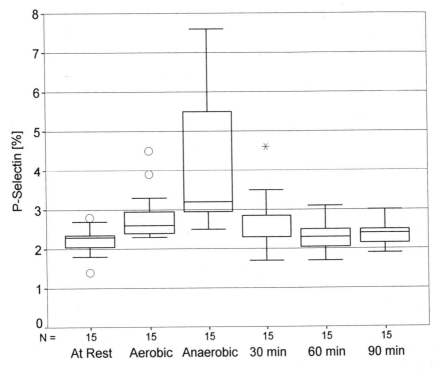

Figure 2. Box-plots of p-selectin levels at rest, during and post-exercise. Extreme values are displayed as o and *
if they exceed the vertical box length by 1.5 and 3.0%, respectively.

findings (29). It is striking that not only median increases of p-selectin expression occur
with maximal exercise, but additionally the variance increases substantially with maximal
exercise. This means that some athletes had a markedly higher increase of platelet activation than others. In these athletes with higher p-selectin values, no additional special characteristics such as age, body mass, left ventricular hypertrophy, other hemostatic variables
or fitness level could be found.

It is yet unclear in which way activated platelets are cleared from the circulation.
Micheleson (12) recently described that platelets possibly lose surface p-selectin but remain functional in the circulation. The possible mechanisms of activation of the hemostatic system include increased shear stress (5, 7, 20), mechanical forces (28) and direct
catecholamine effects (1, 2).

4. CONCLUSIONS

Thrombocytes and plasma coagulation have been shown to be activated less by
moderate and more by maximal physical exercise in endurance-trained athletes. The extent of the activation of thrombocytes varies widely with, for example, the fibrinolytic response being less pronounced for exercise performed in the supine position. Thus,
activation of the hemostatic sytem may therefore contribute to the cardiovascular risk of
heavy physical exertion in a subset of predisposed athletes.

5. REFERENCES

1. Biggs, R., R. G. Macfarlane, and J. Pilling. Observations on fibrinolysis. Experimental activity produced by exercise or adrenaline. *Lancet* I: 402–405, 1947.

2. Chandler, W. L., R. C. Veith, G. W. Fellingham, W. C. Levy, R. S. Schwarz, M. D. Cerquiera, S. E. Kahn, V. G. Larson, K. C. Cain, J. C. Beard, I. B. Abrass, and J. R. Stratton. Fibrinolytic response during exercise and epinephrine infusion in the same subjects. *JACC* 19: 1412–1420, 1992.

3. Curfman, G. D. Is exercise beneficial - or hazardous - to your heart? *N. Engl. J. Med.* 329:1730–1731, 1993.

4. Dag, B., G. Gleerup, A. M. Bak, I. Hindberg, J. Mehlsen, and K. Winther. Effect of supine exercise on platelet aggregation and fibrinolytic activity. *Clin. Physiol.* 14: 181–186, 1994.

5. Davies, P. F., A. Remuzzi, E. J. Gordon, C. F. Dewey, and M. A. Gimbrone. Turbulent fluid shear stress induces vascular endothelial cell turnover in vitro. *Proc. Natl. Acad. Sci.* 83: 2114–2117, 1986.

6. Devereux, R. B., D. R. Alonso, E. M. Lutas, G. J. Gottlieb, E. Campo, I. Sachs and N. Reichek. Echocardiographic assessment of left ventricular hypertrophy: comparision to necropsy findings. *Am. J. Cardiol.* 57: 450–458, 1986.

7. Diamond, S. L., S. G. Eskin, and L. V. McIntire. Fluid flow stimulates tissue plasminogen activator. Secrection by cultured human endothelial cells. *Science* 243: 1483–1485, 1989.

8. Dufaux, B., U. Order, and H. Liesen. Effect of a short maximal physical exercise on coagulation, fibrinolysis, and complement system. *Int. J. Sports Med.* 12: 38–42, 1991.

9. Gough, S. C. L., S. Whitworth, P. J. S. Rice, and P. J. Grant. The effect of exercise and heart rate on fibrinolytic activity. *Blood Coagulation and Fibrinolysis* 3: 179–182, 1992.

10. Herren, T., P. Bärtsch, A. Haeberli, and W. Straub. Increased thrombin-antithrombin III comlexes after 1h of physical exercise. *J. Appl. Physiol.* 73: 2499–2504, 1992.

11. Kestin, A. S., P. A. Ellis, M. R. Barnard, A. Erichetti, B. A. Rosner, and A. D. Michelson. Effect of strenuous exercise on platelet activation state and reactivity. *Circulation* 88:1502–11, 1993.

12. Michelson, A. D., M. R. Barnard, H. B. Hechtman, H. MacGregor, R. J. Connolly, and R. C. Valeri. Circulating degranulated platelets rapidly lose surface p-selectin but continue to circulate and function. *Thromb. Haemostas.* 73: 1000, 1995.

13. Mittleman, M.A., M. Maclure, G. H. Tofler, J. B. Sherwood, R. J. Goldberg, and J. E. Muller. Triggering of acute myocardial infarction by heavy physical exertion. *N. Engl. J. Med.* 329: 1677–1683, 1993.

14. Möckel, M., T. Störk, L. Röcker, A. Ruf, N.-V. Ulrich, O. Danne, R. Müller, H. Patscheke, T. Bodemann, H. Eichstädt, and U. Frei. Activation of thrombocytes and plasmatic coagulation during maximal cycle exercise in healthy triathletes. *Eur. J. Appl. Physiol.* 69 (Suppl. 3): S18, 1994.

15. Möckel, M., T. Störk, N.-V. Ulrich, L. Röcker, A. Ruf, F. Klefisch, O. Danne, R. Müller, T. Bodemann, and U. Frei. Activation of thrombocytes and the hemostatic system during maximal cycle exercise in healthy triathletes. *Ann. Hematol.* 70 (Suppl I): A69, 1995.

16. Möckel, M., T. Störk, N.-V. Ulrich, L. Röcker, A. Ruf, F. Klefisch, O. Danne, R. Müller, T. Bodemann, and U. Frei. Activation of thrombocytes and the hemostatic system during maximal cycle exercise in healthy triathletes. *Thromb. Haemostas.* 73: 1000, 1995.

17. Möckel, M., T. Störk, N.-V. Ulrich, L. Röcker, A. Ruf, H. Riess, F. Klefisch, O. Danne, R. Müller and U. Frei. Activation of thrombocytes and the hemostatic system during triathlon competition. *Thromb. Haemostas.* 73: 1001, 1995.

18. Möckel, M., T. Störk, N.-V. Ulrich, L. Röcker, A. Ruf, J. Vollert, H. Riess, R. Gareis, O. Danne, R. Müller, and U. Frei. Activation of the haemostatic system during triathlon competition. *Eur. Heart J.* 16 (suppl): 345, 1995.

19. Molz, A. B., B. Heyduck, H. Lill, E. Spanuth, and L. Röcker. The effect of different exercise intensities on the fibrinolytic system. *Eur. J. Appl. Physiol.* 67: 298–304, 1993.

20. Röcker, L. Einfluß körperlicher Leistung auf das Hämostasesystem. Erfassung durch sensitive biochemische Marker. *Dtsch. Med. Wschr.* 118: 348–354, 1993.

21. Röcker, L., W. K. Drygas, and B. Heyduck. Blood platelet activation and increase in thrombin activity following a marathon race. *Eur. J. Appl. Physiol.* 55: 374–380, 1986.

22. Röcker, L., and J. W. Franz. Effect of chronic ß-adrenergic blockade on exercise-induced leukocytosis. *Klin. Wschr.* 64: 270–273, 1986.

23. Röcker, L., M. Taenzer, W. K. Drygas, H. Lill, B. Heyduck, and H.-U. Altenkirch. Effect of prolonged physical exercise on the fibrinolytic system. *Eur. J. Appl. Physiol.* 60: 478–481, 1990.

24. Röcker, L., and M. Möckel. Influence of acute physical exercise on the haemostatic system. *Eur. J. Clin. Chem. Clin. Biochem.* 33: A208, 1995.

25. Ruf, A., and H. Patscheke. Flow cytometric detection of activated platelets: comparison of determining shape change, fibrinogen binding and p-selectin expression. *Sem. Thromb. Hemostas.* **21**: 146–151, 1995.

26. Speiser, W., W. Langer, A. Pschaick, E. Selmayr, B. Ibe, E. Nowacki, and G. Müller-Berghaus. Increased blood fibrinolytic activity after physical exercise: Comparative study in individuals with different sporting activities and in patients after myocardial infarction. *Thrombosis research* **51**: 543–555, 1988.

27. Streiff, M., and W. R. Bell. Exercise and hemostasis in humans. *Sem. Hematol.* **31**: 155–165, 1994.

28. Takashima, N., and T. Higashi. Change in fibrinolytic activity as a parameter for assessing local mechanical stimulation during physical exercise. *Eur. J. Appl. Physiol.* **68**: 445–449, 1994.

29. Wang, J., C. J. Jen, H. Kung, L.-J. Lin, T.-R. Hsiue, and H. Chen. Different effects of strenuous exercise and moderate exercise on platelet function in men. *Circulation* **90**: 2877–2885, 1994.

30. Willich, S. N., M. Lewis, H. Löwel, H.-R. Arntz, F. Schubert, and R. Schröder. Physical exertion as a trigger of acute myocardial infarction. *N. Eng. J. Med.* **329**: 1684–1690, 1993.

23

ARTERIAL O$_2$ DESATURATION DURING SUPINE EXERCISE IN HIGHLY TRAINED CYCLISTS

P. K. Pedersen,[1] H. Mandøe,[2] K. Jensen,[3] C. Andersen,[2] and K. Madsen[1]

[1] Department Sports Science
Odense University
Odense M, Denmark
[2] Department Anaesthesia
Odense University Hospital
Odense C, Denmark
[3] Team Danmark Test Center
Odense M, Denmark

1. INTRODUCTION

Performance of intense dynamic exercise in highly trained athletes may be associated with a drop in arterial O$_2$ saturation (%SaO$_2$), also called exercise-induced hypoxemia (EIH) (4, 9). The causes for this gas exchange imperfection are unclear, but ventilation-perfusion mismatch is considered an important factor (5, 6). Blood flow distribution in the lung is greatly influenced by body position, and ventilation-perfusion distribution, reportedly, is more homogenous in the supine (Sup) than in the upright (Up) position (1). In order to examine the influence of body posture on %SaO$_2$ in exercise, we exercised a group of elite road cyclists progressively to maximum in Sup and Up, the hypothesis being that an expected desaturation in upright cycling would disappear or be reduced during supine work as a result of attenuation of uneven ventilation-perfusion distribution.

2. MATERIAL AND METHODS

2.1. Subjects

Eight male athletes, age 19–27 yrs, volunteered as subjects in this study which was approved by the regional Ethics committee. All subjects were non-smokers with no known pulmonary or other diseases or disorders.

2.2. Protocol and Techniques

Two progressive cycle ergometer tests to exhaustion, separated by 60–75 min, were performed in random order, one in Up and the other in Sup. After 10 min of rest and 4 min of work at 160 W and 240 W, respectively, a progressive exercise protocol was initiated where 3 min of work at estimated $\dot{V}O_2$max was followed by 40 W increments every min until exhaustion (80 rev per min). The Douglas bag technique was used for measurements of ventilation ($\dot{V}E$), O_2 uptake ($\dot{V}O_2$), and CO_2 output ($\dot{V}CO_2$). Blood was drawn from an arterial catheter and analysed for O_2 tension (paO_2), CO_2 tension ($paCO_2$), %SaO_2 pH, and lactate concentration (La). Alveolar pO_2 (pAO_2) was estimated from the alveolar gas equation. Alveolar ventilation ($\dot{V}A$) was calculated from $\dot{V}CO_2$ and $paCO_2$.

2.3. Statistics

Data are presented as mean ± SEM. Analysis of variance for repeated comparisons was employed as the main statistical treatment. ANOVA yields two main effects, work intensity and work position, and an interaction term. Paired t-tests were employed for comparison between individual data sets (rest and maximal values) once a significant F-ratio had been found; significance level at $P \ll 0.05$.

3. RESULTS

$\dot{V}O_2$max for upright exercise averaged 75 ml $O_2 \cdot$ min^{-1} per kg body weight (range 70–84) thus supporting the high fitness level of the subjects. The change in working posture from Up to Sup significantly reduced maximal working capacity. Maximal power output decreased by 14% from 505±14 to 436±11 W ($P \ll 0.001$), $\dot{V}O_2$max by 10% from 5.39±0.14 to 4.81±0.12 1 $O_2 \cdot$ min^{-1} ($P \ll 0.001$), $\dot{V}E$max by about 15% from 186.4±4.9 to 159.2±4.9 l.min^{-1} and $\dot{V}A$max by 21% from 125.3 to 99.2 1 . min^{-1} ($P \ll 0.001$). The maximum blood lactate concentration decreased from 10.0±0.7 to 7.8±0.5 mmol.l^{-1} ($P \ll 0.05$).

Individual values for $\dot{V}A$ and La as a function of relative work load (% of activity-specific $\dot{V}O_2$max) during progressive supine and upright exercise are depicted in Fig. 1. $\dot{V}A$ values were generally similar in Up and Sup from rest to near-maximal exercise; however, the Sup value was clearly lower ($P \ll 0.0001$) at maximum as reflected in a significant work intensity-by-posture interaction term. The La interaction term was also statistically significant. If anything, La was slightly higher in heavy submaximal and early maximal exercise but this was reversed at maximum.

Data for blood gases and pH are presented in Table 1. paO_2 dropped gradually from 100 Torr at rest to about 80 Torr at maximum ($P \ll 0.0001$) with no difference between Up and Sup. %SaO_2 dropped ($P \ll 0.0001$) from 98.1±0.2% at rest to 95.2±0.4% at exhaustion in Up, and from 98.3±0.2% to 94.4±0.5% in Sup (Up vs Sup ns). Mean pAO_2 values ranged 100–105 Torr at rest and in submaximal exercise, and increased to 115±1 Torr (Sup) and 118±1 Torr (Up) at exhaustion ($P \ll 0.01$). Arterial pCO_2 was virtually constant during the initial work loads but decreased at work loads above 70–80% of $\dot{V}O2$max. The interaction term was significant, indicating a lower $paCO_2$ at exhaustion in upright exercise than in supine (33.8 vs. 36.2 Torr; $P \ll 0.01$). Average arterial pH-values were between 7.40–7.42 from rest to 80–90% $\dot{V}O_2$max, and then declined to about 7.31 at maximum with no significant difference between Up and Sup. Hemoglobin concentration increased from 8.5 mmol.l^{-1} at rest to 9.4 mmol.l^{-1} at maximum exercise ($P \ll 0.0001$).

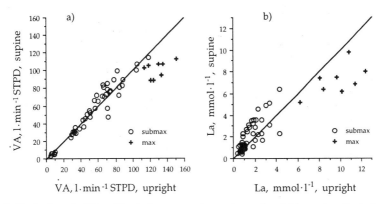

Figure 1. Individual values for V̇A and La during progressive supine and upright exercise. Open circles are from rest and submaximal exercise whereas the crosses are maximum values. Inserted is line of identity (y=x).

4. DISCUSSION

The study confirmed the presence of EIH in highly trained athletes during maximal dynamic exercise. The decreases in $\%SaO_2$ from about 98% to 95% and in paO_2 from around 100 Torr to 80 Torr were clearly significant and all subjects had lower $\%SaO_2$ and paO_2 at exhaustion than at rest. The decrements were, however, smaller than those reported by Dempsey et al. (4) but in accordance with data by Powers et al (9) and Inbar et al. (7), and also similar to our own earlier data (8).

The hypothesis that changing from Up to Sup would diminish EIH during intense exercise was not confirmed. Arterial blood desaturated to the same extent in Sup and in Up, and the drop in paO_2 was also similar in the two situations. A study in pulmonary patients on the effect of upright versus supine position exercise was also unsuccessful in

Table 1. Blood gases, alveolar pO_2 and acid-base status during progressive cycle ergometer exercise in upright (Up) and supine (Sup) position (mean ± SEM; n=8)

| | | Rest | Warm-up | | Maximal exercise | | |
			light	heavy	min 3	min 5	maximum
paO2	Up	99±3	94±1	88±2	84±1	80±2	82±2
Torr	Sup	100±2	90±1	85±1	82±2	79±2	80±5
%SaO₂	Up	98.1±0.2	97.8±0.1	97.3±0.1	96.5±0.2	96.4±0.4	95.2±0.4
	Sup	98.3±0.2	97.6±0.1	97.2±0.2	96.6±0.3	95.9±0.4	94.4±0.5
pAO₂	Up	105±2	101±1	103±1	107±1	109±1	118±1
Torr	Sup	106±2	100±1	105±1	109±1	113±1	115±1
paCO₂	Up	38.8±0.8	40.8±0.6	40.3±0.7	38.8±0.9	37.7±0.8	33.8±0.5
Torr	Sup	39.4±1.1	41.0±0.9	39.7±1.1	37.9±0.9	36.1±0.7	36.2±0.8
pH	Up	7.42±0.007	7.41±0.003	7.41±0.005	7.41±0.007	7.40±0.009	7.31±0.015
	Sup	7.41±0.006	7.41±0.004	7.41±0.006	7.40±0.006	7.40±0.009	7.32±0.013
Base excess	Up	1.0±0.3	1.6±0.4	1.5±0.4	0.4±0.6	-0.9±0.8	-8.5±0.9
mmol . l-1	Sup	1.0±0.4	1.6±0.3	1.1±0.4	-0.7±0.6	-2.2±0.8	-6.3±0.9

paO2 = arterial pO_2 %SaO₂ = % arterial O_2 saturation pAO₂ = alveolar pO_2 paCO₂ = arterial pCO_2

showing a beneficial effect on arterial oxygen saturation on shifting to the supine posture (2). Dempsey et al. (3) examined the effect of varying body posture on pulmonary gas exchange at lower work intensities, i.e. $\dot{V}O_2$ up to about 2.5 l . min^{-1}. At rest, the alveolo-arterial pO2 difference was narrowed in Sup, but during exercise no clear effect was evidenced on this or other oxygen variables, thus supporting the main conclusions of the present study.

Contributing to the failure in eliminating exercise-induced arterial O_2 desaturation may have been a reduced maximal ventilatory capacity in supine position. $\dot{V}$Emax declined by 15% in comparison with Up, and $\dot{V}$Amax by 21%. These decreases are larger than the percentage drop in $\dot{V}O_2$max and indicate an insufficient hyperventilatory response in Sup. It is of interest that the $\dot{V}A$ values to a large extent were similar in Sup and Up for gas exchange levels up to about 90% of $\dot{V}O_2$max. The difference was only seen during the very highest work rates. In accordance with this, the largest drop in %SaO$_2$ was seen from min 5 to exhaustion, i.e. in the final stage of the maximum exercise test. Supportive of this is also the behaviour of paCO$_2$. Whereas the Sup values were virtually constant during the last two observations in the maximum exercise, Up showed a marked drop from 37.7 to 33.8 Torr. The potential role of hyperventilation in the maintenance of arterial O_2 saturation during maximal exercise has been examined by Inbar et al. (7). In experiments with well-trained individuals breathing helium to increase ventilatory function, these authors observed a reduced exercise-induced desaturation in comparison with normoxia.

5. CONCLUSION

Performance of progressive cycle ergometer exercise to maximum in highly trained road cyclists was accompanied by significant decreases in %SaO$_2$ and paO$_2$. This exercise-induced hypoxemia was not abolished or reduced by a change in working posture from upright to supine. An expected favourable influence of gravity towards a more homogenous ventilation-perfusion distribution in Sup may have been opposed or neutralized by a reduced ventilatory capacity.

6. ACKNOWLEDGMENT

Supported by Team Danmark Sports Research Council.

7. REFERENCES

1. Amis, T. C., H. A. Jones, and J. M. B. Hughes. Effect of posture on inter-regional distribution of pulmonary perfusion and $\dot{V}A/\dot{Q}$ ratios in man. *Resp. Physiol.* **56**: 169–182, 1984.
2. Bell, C. W., S. F. Drehsen, I. Kass, and L. W. Burgher. A comparison of vertical and horizontal position exercise on arterial oxygen saturation in pulmonary patients. *Med. Sci. Sports Exerc.* **14**: 132, 1982.
3. Dempsey, J. A., N. Gledhill, W. G. Reddan, H. V. Forster, P. G. Hanson, and A. D. Claremont. Pulmonary adaptation to exercise: effects of exercise type and duration, chronic hypoxia and physical training. *Ann. New York Acad. Sci.* **301**: 243–261, 1977.
4. Dempsey, J. A., P. G. Hansson, and K. S. Henderson. Exercise-induced arterial hypoxemia in healthy human subjects at sea level. *J. Physiol. (Lond.)* **355**: 161–178, 1984
5. Hammond, M. D., G. E. Gale, S. Kapitan, A. Ries, and P. D. Wagner. Pulmonary gas exchange in humans during exercise at sea level. *J. Appl. Physiol.* **60**: 1590–1598, 1985

6. Hopkins, S. R., D. C. McKenzie, R. B. Schoene, R. W. Glenny, and H. T. Robertson. Pulmonary gas exchange during exercise in athletes. I. Ventilation-perfusion mismatch and diffusion limitation. *J. Appl. Physiol.* **77**: 912–917, 1994.

7. Inbar, O., Y. Weinstein, A. Kowalski, S. Epstein, and A. Rotstein. Effects of increased ventilation and improved pulmonary gas exchange on maximal oxygen uptake and power output. *Scand. J. Med. Sci. Sports* **3**: 81–88, 1993.

8. Pedersen, P. K., K. Madsen, C. Andersen, N. H. Secher, and K. Jensen. Arterial oxygen desaturation during dynamic exercise. *Med. Sci. Sports Exerc.* Suppl. **24**: S69, 1992.

9. Powers, S. K, D. Martin, M. Cicale, N. Collop, D. Huang, and D. Criswell. Exercise-induced hypoxemia in athletes: role of inadequate hyperventilation. *Eur. J. Appl. Physiol.* **65**: 37–42, 1992.

PART 4. PATHOPHYSIOLOGY OF EXERCISE INTOLERANCE

24

EXERCISE IN CHRONIC RESPIRATORY DISEASE

C. G. Gallagher and D. D. Marciniuk

Division of Respiratory Medicine
Royal University Hospital
Saskatoon, Saskatchewan S7N OW9
Canada

1. INTRODUCTION

Exercise intolerance is the most distressing consequence of their disease for the majority of patients with chronic respiratory disease. We will briefly review the major abnormalities during exercise in patients with chronic obstructive pulmonary disease (COPD) but will also refer to other chronic diseases. The interested reader is referred to more detailed reviews (4, 13).

The major responses to exercise in patients with COPD are listed in Table 1. Table 2 contrasts usual patterns of exercise response in COPD to those of five other common cardiorespiratory disorders: interstitial lung disease, congestive heart failure, mitral valve disease, pulmonary vascular disease and unfitness.

2. METABOLIC RATE

COPD patients have a small but significant increase in resting oxygen uptake ($\dot{V}_{O2}$) compared to normal subjects. In the past it was frequently assumed that COPD patients had an elevated $\dot{V}_{O2}$ at a given work rate during exercise. However, the available data indicates that there is no significant difference in $\dot{V}_{O2}$ at a given work rate, when compared to normal humans (4). The rise in $\dot{V}_{O2}$ in response to an increase in work rate is slower in COPD patients than in normal humans, causing a delay in attainment of steady state conditions (17). The mechanisms underlying this are unknown but altered skeletal muscle metabolism or cardiovascular abnormalities are most likely (14).

The Physiology and Pathophysiology of Exercise Tolerance
edited by Steinacker and Ward, Plenum Press, New York, 1996

Table 1. Typical responses to exercise testing in patients with COPD*

At Maximal Exercise:
    Low peak $\dot{V}_{O2}$ and peak work rate
    Low $\dot{V}_E$
    High $\dot{V}_E$/MVV
    Low V_T
    Normal or slightly reduced V_T/VC ratio
    Low heart rate (usually)
    Low oxygen pulse
    Reduced metabolic acidosis (usually)
At Submaximal Exercise:
    Normal or near normal $\dot{V}_{O2}$
    High $\dot{V}_E$
    High F, low VT
    High arterial PCO_2, low arterial PO_2
    Normal cardiac output (usually)
    High heart rate, low stroke volume
    Increased metabolic acidosis
    (at least in some patients)

* Data are shown in comparison to normal subjects of the same sex, age and body size.
See text for details. Reproduced with permission from reference 4.

3. VENTILATION

The ventilatory response to exercise in COPD is summarized in Figure 1. During submaximal exercise, COPD patients have increased dead space ventilation and (usually) reduced alveolar ventilation (7). Total ventilation ($\dot{V}_E$) is usually increased during submaximal exercise. The highest level of ventilation (i.e. maximum ventilatory capacity, MVC) that a subject can produce is limited by the maximum inspiratory and expiratory flow volume curves, that is by the highest inspiratory and expiratory flow rates that can be generated (16). COPD patients have reduced maximum expiratory flow rates at all lung volumes, with lesser reductions in maximum inspiratory flow rates (7). Therefore MVC is reduced in COPD. The combination of increased $\dot{V}_E$ and reduced MVC means that the

Table 2. Common patterns of exercise response

	COPD	ILD	CHF	MVD	PVD	UNF
Maximal $\dot{V}_{O2}$	↓	↓	↓	↓	↓	↓
HR at submaximal work	↑	↑ or N	↑	↑	↑	↑
Maximal HR	↓ rarely N	↓	↓ or N	N	↓	N
VE at submaximal work	↑ or N	↑	↑	↑	↑	↑ above LT
Peak $\dot{V}_E$/maximal voluntary ventilation	↑↑	↑	N	N	N	N/↓
Peak VT/vital capacity	N or ↓	N	N	N	?N	N
Arterial O_2 desaturation	+ or -	+	-	-	+	-

COPD = Chronic Obstructive Pulmonary Disease , ILD = Interstitial Lung Disease, CHF = Congestive Heart Failure, MVD = Mitral Valve Disease, PVD = Pulmonary Valve Disease, UNF = Unfitness, LT = Lactate Threshold.
↑, N, ↓ refer to an increase, no change, or decrease compared with normal response; + or - refer to the presence of absence of arterial O_2 desaturation with exercise.
Reproduced with permission from reference 7.

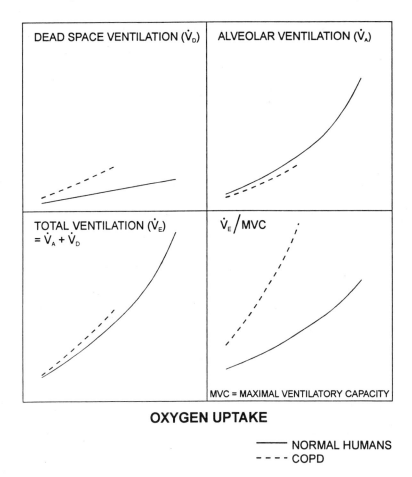

OXYGEN UPTAKE

——— NORMAL HUMANS
- - - - COPD

Figure 1. Schematic comparison of typical ventilatory changes during progressive exercise in COPD patients and in normal humans. See text for details. Reproduced with permission from Reference 4.

ventilatory demands of exercise are a much greater proportion of maximum ventilatory capacity than in normal humans, i.e. the $\dot{V}_E$/MVC ratio is increased at a given work rate during exercise (Figure 1). The reduced MVC constrains the ability of COPD patients to increase ventilation and this is a major cause of exercise limitation in these patients (4).

4. BREATHING PATTERN

Healthy humans initially increase $\dot{V}_E$ during mild exercise by increasing both tidal volume (V_T) and respiratory frequency (f). At higher work rates, V_T usually remains constant at 50–60% of vital capacity and further increases in $\dot{V}_E$ are due to increases in f alone (6, 15). The same general pattern usually occurs in COPD patients. However, V_T is less and f is greater than in healthy humans at similar levels of $\dot{V}_E$, and peak V_T is less (5, 8, 9). Such a tachypneic breathing pattern is not unique to COPD and is seen in the vast majority of patients with different respiratory or cardiac conditions (7, 12). A study (9) of patients with COPD, restrictive lung disease, bronchial asthma and heart disease found a strong linear correlation (R = 0.827, P < 0.0001) between peak exercise V_T and vital capacity when all patients were considered together (Figure 2). The ratio of peak exercise V_T to vital capacity in COPD is usually similar to that of normal humans and patients with other cardiorespiratory diseases (5, 9) though some patients with severe COPD may have a slightly smaller ratio (8). Therefore the altered breathing pattern in COPD and in other patients with respiratory disease is probably largely due to their altered respiratory mechanics (9).

5. RESPIRATORY MUSCLE FUNCTION

As reviewed elsewhere, patients with COPD have to generate much greater inspiratory muscle pressures than normal humans at similar work rates (4). This is because of their increased end expiratory lung volume (dynamic hyperinflation), increased inspiratory flow resistance, increased minute ventilation, decreased lung compliance (frequency dependent), and chest wall distortion. In addition, inspiratory muscle strength is decreased in COPD and this is further decreased because of dynamic hyperinflation during exercise. Therefore the load (pressure generated as a proportion of pressure generating capacity) on inspiratory muscles is much higher in COPD than in normal humans (4). This increased load is the major reason for the increased breathlessness of COPD patients during exercise. It has been suggested that inspiratory muscle fatigue may develop during exercise in COPD (18) but there is no convincing evidence of this at this time (4).

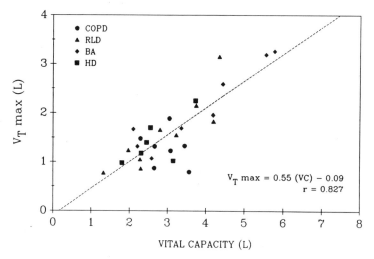

Figure 2. Relation between peak exercise tidal volume (V_T max) and vital capacity (VC) for patients with chronic obstructive pulmonary disease, restrictive lung disease, heart disease, and bronchial asthma. Different symbols are used for the four patient groups. There was a significant linear relation between V_T max and VC (see regression line): V_T max = 0.55 VC - 0.09 litres (R = 0.827, P<0.0001). Reproduced with permission from Reference 9.

6. ARTERIAL BLOOD GASES

As shown in Figure 1, COPD patients usually have a reduced alveolar ventilation during exercise. Therefore arterial PCO_2 is usually higher than in normal humans and, in many patients, arterial PCO_2 may rise during exercise (7). Many patients with severe COPD have arterial oxygen desaturation during exercise. This is partly related to hypoventilation and the effect of a fall in mixed venous PO_2 on low ventilation - perfusion lung units and shunt (19). Some patients with COPD may increase or keep arterial PO_2 the same during exercise. It is not possible to predict, from resting measurements, which COPD patient will desaturate during exercise.

7. CARDIAC FUNCTION

Cardiac output is frequently normal at a given $\dot{V}_{O2}$ during submaximal exercise (7). However, heart rate is usually increased and stroke volume decreased, when compared to normal subjects. Therefore oxygen pulse is reduced. Peak heart rate is usually less than normal, though it may be normal in some patients with mild disease (2, 7). Patients with COPD have increased pulmonary artery pressures during exercise because of their increased pulmonary vascular resistance (10, 11). Most patients with moderate or severe COPD have evidence of right ventricular dysfunction during exercise (7).

8. METABOLIC ACIDOSIS

It was once assumed that COPD patients stop exercise at work rates lower than those at which they would develop significant metabolic acidosis. However, review of studies over the last 20 years shows that this is not the case (7). Many COPD patients have significant metabolic acidosis at end-exercise. In addition, many COPD patients develop metabolic acidosis at lower work rates than normal humans, i.e. their "lactate threshold" is reduced (3, 7). This further increases the ventilatory demands of exercise in these patients. Exercise training improves exercise performance in COPD patients partly by reducing metabolic acidosis at a given work rate (3).

9. LIMB MUSCLES

Traditionally, little attention has been paid to limb muscle function in COPD. Recent studies provide clear evidence of limb muscle dysfunction in this population. Limb muscle strength is frequently reduced and there is evidence of abnormal muscle metabolism during exercise (1, 20). This is discussed further elsewhere (14).

10. EXERCISE LIMITATION

It is therefore clear that COPD patients have abnormal ventilatory, blood gas, cardiovascular, acid-base and limb muscle function during exercise. To what extent do these (or other) abnormalities contribute to the impaired exercise tolerance? Ventilatory/respiratory muscle function probably contributes to exercise intolerance in many, but not all, patients (4). Hypoxemia is probably a significant factor in those patients with oxygen desaturation during exercise. At this time, there are insufficient data regarding the role of limb muscle and cardiovascular function in exercise limitation in COPD. However, we believe that limb muscle dysfunction is probably a significant factor in many patients (14). These issues merit further study.

10. REFERENCES

1. Allard, C., N. L. Jones, and K. J. Killian. Static peripheral skeletal muscle strength and exercise capacity in patients with chronic airflow limitation. *Am. Rev. Respir. Dis.* **139**: A90, 1989.
2. Babb, T. G., R. Viggiano, B. Hurley, B. Staats, and J. R. Rodarte. Effect of mild-to-moderate airflow limitation on exercise capacity. *J. Appl. Physiol.* **70(1)**: 223–230, 1991.
3. Casaburi, R., A. Patessio, F. Ioli , S. Zanaboni, C. F. Donner, and K. Wasserman. Reductions in exercise lactic acidosis and ventilation as a result of exercise training in patients with obstructive lung disease. *Am. Rev. Respir. Dis.* **143**: 9–18, 1991.
4. Gallagher, C. G. Exercise limitation and clinical exercise testing in chronic obstructive pulmonary disease. *Clin. in Chest Med.* **15**: 305–326, 1994.
5. Gallagher, C. G., and M. Younes. Breathing pattern during and after maximal exercise in patients with chronic obstructive lung disease, interstitial lung disease, and cardiac disease, and in normal subjects. *Am. Rev. Respir. Dis.* **133**: 581–586, 1986.
6. Gallagher C. G., E. Brown, and M. K. Younes. Breathing pattern during maximal exercise and during submaximal exercise with hypercapnia. *J. Appl. Physiol.* **63**: 238–244, 1987.
7. Gallagher, C. G. Exercise and chronic obstructive pulmonary disease. *Med. Clin. N. Am.* **74**: 619–641, 1990.
8. Gimenez, M., E. Servera, R. Candina, T. Mohan Kumar, and J. B. Bonnassis. Hypercapnia during maximal exercise in patients with chronic airflow obstruction. *Bull. Eur. Physiopathol. Respir.* **20**: 113–119, 1984.
9. Gowda K., T. Zintel, C. McParland, R. Orchard, and C. G. Gallagher. Diagnostic value of maximal exercise tidal volume. *Chest* **98**: 1351–54, 1990.
10. Light R. W., H. M. Mintz, G. S. Linden, and S. E. Brown. Hemodynamics of patients with severe chronic obstructive pulmonary disease during progressive upright exercise. *Am. Rev. Respir. Dis.* **130**: 391–395, 1984.
11. Mahler D. A, B. N. Brent, J. Loke, B. L. Zaret, and R. A. Matthay. Right ventricular performance and central circulatory hemodynamics during upright exercise in patients with chronic obstructive pulmonary disease. *Am. Rev. Respir. Dis.* **130**: 722–729, 1984.
12. Marciniuk D., R. Watts, and C. G. Gallagher. Dead space loading and exercise limitation in patients with interstitial lung disease. *Chest* **105**: 183–89, 1994.
13. Marciniuk D. D., and C. G. Gallagher. Clinical exercise testing in interstitial lung disease. *Clin. in Chest Med.* **15**: 287–303, 1994.
14. Marciniuk D. D., and C. G. Gallagher. Clinical exercise testing in chronic airflow limitation. *Med. Clin. N. Am.*, In Press.
15. McParland C., B. Krishnan, J. Lobo, and C. G. Gallagher. Effect of physical training on breathing pattern during progressive exercise. *Resp. Physiol.* **90**: 311–323, 1992.
16. McParland C., J. Mink, and C. G. Gallagher. Respiratory adaptations to dead space loading during maximal incremental exercise. *J. Appl. Physiol.* **70**: 55–62, 1991.
17. Nery L. E., K. Wasserman, J. D. Andrews, D. J. Huntsman, J. E. Hansen, and B. J. Whipp. Ventilatory and gas exchange kinetics during exercise in chronic airways obstruction. *J. Appl. Physiol.* **53(6)**: 1594–1602, 1982.
18. Pardy R. L., R. N. Rivington, P. J. Despas, and P. T. Macklem. The effects of inspiratory muscle training on exercise performance in chronic airflow limitation. *Am. Rev. Respir. Dis.* **123**: 426–433, 1981.
19. Wagner P. D., D. R. Dantzker, R. Dueck, J. L. Clausen, and J. B. West. Ventilation-perfusion inequality in chronic obstructive pulmonary disease. *J. Clin. Invest.* **59**: 203–216, 1977.
20. Wuyam B., J. F. Payen, P. Levy, H. Bensaidane, H. Reutenauer, J. F. Le Bas, and A. L. Benabid. Metabolism and aerobic capacity of skeletal muscle in chronic respiratory failure related to chronic obstructive pulmonary disease. *Eur. Respir. J.* **5**: 157–162, 1992.

25

EXERCISE PHYSIOLOGY AND THE IMMUNE SYSTEM

Bente Klarlund Pedersen and Thomas Rohde

The Copenhagen Muscle Research Centre
Department of Infectious Diseases M 7721
Rigshospitalet, Copenhagen N, Denmark

1. INTRODUCTION

It has been suggested that the immunological responses to exercise are a subset of the physical stress reactions that characterize surgery, thermal and traumatic injury, haemorrhagic shock and acute myocardial infarction (1). Therefore, acute exercise can be viewed as a prototype for studying the effects of physical factors on the immune system (2).

2. EXERCISE LEUKOCYTOSIS

The basal levels of circulating leukocytes are rapidly increased by physical activity (3). The leukocytosis is due to increased concentrations of neutrophils, monocytes and lymphocytes. The neutrophil concentration increases during exercise and it continues to increase following exercise. During exercise natural killer (NK), B and T cells are recruited to the blood, which is reflected in an elevated total lymphocyte count. The composition of T cells is altered; thus the CD4/CD8 ratio decreases, because the CD8 count increases more than the CD4 count. Following severe exercise the lymphocyte concentration decreases below its baseline value, and the duration of this suppression depends on the intensity and duration of the exercise stress (1).

3. NATURAL KILLER CELLS

NK cells are thought to play an important role in the first line of defense against tumour and virus-infected target cells. The NK cell activity is determined as the percentage lysis of ^{51}Cr-labelled tumor cells, e.g. K562 target cells. NK cells are highly sensitive to physical exercise and are recruited immediately to the blood during acute exercise. The modulation of the NK cell activity in response to exercise has been investigated extensively (2). During exercise the absolute concentration and the relative fraction of blood mononuclear cells (BMNC) expressing characteristic NK cell markers is markedly enhanced. Simultaneously, the NK cell activity (the cytotoxic activity of the NK cells) increases. Following intense exercise of at least one hour, the NK cell activity is suppressed.

4. ANTIBODY PRODUCTION

The secretory immune system of mucosal tissues such as the upper respiratory tract (URT) is considered to be the first barrier to colonization by pathogenic microorganisms causing URT infections. In mucosal secretions immunoglobulin A (IgA) is the major class of immunoglobulins, and the level of IgA in mucosal fluids correlate more closely with resistance to URT infection than serum antibodies. Suppressed levels of salivary IgA has been reported in cross-country skiers after a race and after 2 hours of intense ergometer cycling (4).

5. LYMPHOCYTE PROLIFERATION

The ability of lymphocytes to proliferate following stimulation with concanavalin A (ConA) and phytohaemagglutinin (PHA) decreases, whereas the proliferative response to interleukin-2 (IL-2) or lipopolysaccharide (LPS) or poke weed mitogen (PWM) increases (2). The low PHA response of cells isolated during bicycle exercise was shown to be due to a decreased contribution to the BMNC-pool of the CD4+ cell subfraction rather than to a changed proliferative response per CD4+ T cell, and the increased ability of cells to proliferate following stimulation with IL-2 was caused by an increased fraction of CD16+ cells, and not by an increased expression of IL-2 receptors. Thus the changes in proliferative response during exercise is not a result of altered function or activation of the individual lymphocyte.

6. CYTOKINES

Cytokines serve as messengers of the immune system, but unlike endocrine hormones (the messengers of the endocrine system) which exert their effects over large distances, the cytokines generally act locally. Physical exercise that includes eccentric muscle contractions increases the production of some cytokines. Increased interleukin-1ß (IL-1ß) in muscle tissue was found for up to 5 days following eccentric exercise. The IL-1 activity in plasma increased following eccentric exercise in untrained subjects. Following long distance running, increased concentrations of interleukin-6 (IL-6) in plasma were found, and others have found increased plasma-tumor necrosis factor-α (TNF-α).

7. ROLE OF STRESS HORMONES

Physical stress increases the concentrations of a number of stress hormones in the blood, including catecholamines, growth hormone, beta-endorphins, adrenocorticotrophic hormone (ACTH) and cortisol, whereas the concentration of insulin declines. Studies where adrenaline, noradrenaline, growth hormone and insulin were selectively infused to obtain plasma concentrations identical with those obtained during exercise showed that catecholamines and growth hormone potentially could mediate exercise-induced immuno-modulation (2). These observations were supported by studies in which pharmacologic blockade of stress hormone receptors was performed.

8. ROLE OF HYPERTHERMIA AND HYPOXIA

Severe and prolonged exercise induces elevated core temperature and, in some instances, arterial hypoxemia. *In vitro* hyperthermia and hypoxemia have numerous effects on the immune system (1). The selective effect of *in vivo* hyperthermia and hypoxia on the human immune system have been investigated. Both hyperthermia and hypoxia induced recruitment of the lymphocytes to the blood, especially the CD16+ NK cells, and the neutrophils. When exercise was performed during hypoxic conditions, a significantly larger increase in the concentration of NK cells was seen compared to exercise during normoxic conditions (1).

9. CONCLUSION

During exercise, lymphoid cells, especially NK cells, are recruited to the blood and if muscle damage occurs the cytokine level is enhanced. Following intense exercise of long duration the concentration of lymphocytes in the blood is suppressed and the function of NK and B cells are inhibited. The mechanisms underlying exercise-induced immuno-modulation are probably multi-factorial and include adrenaline and growth hormone, and possibly also exercise-induced hyperthermia and hypoxia. Furthermore, a lack of glutamine has been suggested to contribute to impaired lymphocyte function after muscular activity. During the time of immunodepression, often referred to as "the open window", the host may be more susceptible to microorganisms bypassing the first line of defense. However, moderate exercise, which induces increases in the concentration of immuno-competent cells in the blood without causing post-exercise immunosuppression, may improve host defense to infections.

10. REFERENCES

1. Pedersen, B. K., M. Kappel, M. Klokker, H. B. Nielsen, N. H. Secher. The immune system during exposure to extreme physiologic conditions. *Int J Sports Med* **15**:116, 1994.
2. Hoffman-Goetz, L. and B. K. Pedersen. Exercise and the immune system: a model of the stress response? *Immunology Today* **15**:382, 1994.
3. McCarthy, D. A. and M. M. Dale. The leucocytosis of exercise: a review and model. *Sports Med* **5**:282, 1988.
4. Mackinnon, L. T. and S. Hooper. Mucosal (secretory) immune system responses to exercise of varying intensity and during overtraining. *Int J Sports Med* **15**:S179, 1994.

26

EXERCISE TOLERANCE AND IMPAIRMENT OF SYMPATHETIC NERVOUS SYSTEM ACTIVITY

Manfred J. Lehmann and Uwe Gastmann

Department of Sports and Performance Medicine
Medical University Hospital Ulm, D-89075 Ulm
Germany

1. INTRODUCTION

The activity of the peripheral sympathetic nervous system can be evaluated based on its influence on the organism as a whole; in terms, for example, of the blood pressure and heart rate responses. Because of the complex regulation of these variables, however, such an approach remains problematic. The evaluation of sympathetic activity can be more reliably based on an analysis of plasma and urinary catecholamine profiles (17–19,31,35). This approach is justified on the basis of correlations that have been demonstrated (i) between the heart rate responses to sympathetic stimulation and the release of endogenous catecholamines into the coronary sinus of the dog (40), and (ii) between sympathetic activity in sympathetic fibers of peripheral nerves, recorded with microelectrodes, and the venous noradrenaline concentration of the respective extremity (38).

There are, of course, differences in noradrenaline concentration in samples of arterial, venous and coronary venous blood drawn simultaneously, but positive correlations have been demonstrated between such samples drawn from various vascular sites (23), especially during exercise (4). The reporting of noradrenaline levels in capillary or arterial blood is considered to be most appropriate (1,19,35), since the concentration in venous blood may be changed considerably following its passage through organs, both by additional release and by elimination. The plasma noradrenaline concentration is determined by release and spillover on the one hand and by plasma clearance on the other.

Since the plasma half-life of noradrenaline of 2–3 min in healthy individuals at rest does not change appreciably during exercise (7,12,23), changes in plasma catecholamine levels in healthy individuals during exercise can be primarily explained by changes in release. Based upon studies using incremental exercise (7,8,12,17,18,30,37), the sympathetic nervous system becomes important for cardiovascular and metabolic function at about halfway to maximum work rates, corresponding to plasma catecholamine levels of 5–6 nmol/l (17,18,35). However, as the sensitivity of the organism to catecholamines increases during physical exercise (3) and the influence of antagonists such as the vagus decreases (5), physiological importance can be assumed even at lower exercise intensities.

2. PHYSICAL TRAINING AND SYMPATHETIC ACTIVITY

Endurance training typically results in an initial increase in vagal activity (28), followed or in parallel with a reduction in sympathetic drive (8,30,39). This has been termed the training-dependent "vegetative" change or adaptation of the organism. It is not, however, induced by resistance training (17). This adaptation is expressed during submaximal exercise in an "economy" of cardiovascular activity, a reduction in myocardial oxygen requirement (9,33) and an increased cardiac capacity. The physiological cardiac hypertrophy and regulative dilatation - the so-called Sports Heart - has to be seen as an significant additional mechanism in high-performance endurance athletes through its effect of further improving cardiac capacity (32). The possibility can be considered that the endurance training-related change in vegetative tone might depend partly on a training-dependent change in afferent neuronal feedback from the musculature (16) - "peripheral command" - as well as a change in "central command" (13). The necessary training volume amounts to about 30-60 min endurance training three times per week, at least 3–4 weeks at an intensity of 60-70% $\dot{V}0_{2max}$ (30,39). Since the plasma clearance shows no training dependency in healthy individuals and slight changes only during exercise (12), the reduction in plasma catecholamine levels reflects primarily a reduction in release, going along with an increase in the sensitivity to catecholamines, reflected by an increase in beta-adreneoreceptor density (2,3,18) that is partly lost again during training cycles with an inadequately high training volume (11).

3. PARASYMPATHETIC TYPE OF OVERTRAINING AND SYMPATHETIC ACTIVITY

The hypothesis of a neuro-vegetative dysfunction in the parasympathetic type of overtraining syndrome (10) is supported by the finding of a 40-70% decrease in the nocturnal basal excretion of urinary free catecholamines (24,25) as confirmed by Naessens *et al.* (29). The reduction in basal catecholamine excretion was in the range observed in patients with autonomic dysfunction (20,21). Catecholamine excretion increased again in affected athletes during the recovery period (25). Basal catecholamine excretion may reflect the intrinsic sympathetic activity, since activating mechanisms of the neuro-vegetative axis, such as (i) "central command" (13), (ii) afferent nervous feedback (16), and (iii) metabolic or nonmetabolic error signals (13) are reduced to a minimum during sleep. Therefore, there is some evidence that a reduction of intrinsic sympathetic activity in the parasympathetic type of overtraining syndrome can be one factor of a complex mechanism which causes "central fatigue" in affected athletes. Besides an overtraining-related decrease in intrinsic sympathetic activity, an increased noradrenaline plasma response has been observed at given submaximal work loads (26) and a decreased maximal exercise response in advanced stages (14). A decrease in senstivity to catecholamines was also observed during training cycles with an inadequately high training volume (11), as is also suspected during overtraining (24).

4. AUTONOMIC DYSFUNCTION AND SYMPATHETIC ACTIVITY

The importance of the sympathetic nervous system for adaptation of the organism to stress, such as orthostasis and physical exertion, becomes particularly clear in patients suffering from a primary insufficiency of the sympathetic nervous system, such as primary orthostatic hypotension (Shy-Drager Syndrome, Bradbury-Eggleston Syndrome) (19,20,36). In Bradbury-Eggleston Syndrome, or in Shy-Drager Syndrome if there is concomitant central-nervous-system involvement, only marginal concentrations of sympathetic neurotransmitters and hormones are released and circulate in the blood in advanced stages (19,20). Twenty four hour renal elimination of catecholamines also decreases to 20-30% of healthy control values as an indication of a reduced total catecholamine metabolism (19,20). In the advanced stage, patients can hardly right themselves without immediate circulatory collapse, there being no significant increase in heart rate (19,20,36). No adequate adjustment of cardiovascular function to exercise-related demands is possible. Even during physical exertion in the supine position, there is a decrease in pressure without an adequate increase in heart rate (19,20). Physical inactivity leads to an additional loss of aerobic work capacity of the skeletal musculature reflected by an inadequate increase in lactate concentration during mild exertion (22,27). An increase in stress tolerance was possible when noradrenaline has been administered subcutaneously in these patients via a programmable micro-dosing pump (22).

5. NEUROCARDIOGENIC SYNCOPE AND SYMPATHETIC ACTIVITY

Beside Shy-Drager and Bradbury-Eggleston Syndromes, in patients with neurocardiogenic syncope (which can be established by tilt-table testing), syncope can be effectively prevented in about 50% of patients by ß-blocker therapy (15); in the responders, syncope was preceded by a higher adrenaline response and sinus tachycardia, pointing to a significant role of endogenous catecholamines in the pathogenesis of neurocardiogenic syncope; that is, exercise or orthostasis tolerance can also be compromized in cases of relative sympathetic overstimulation.

6. PARAPLEGICS VS. TETRAPLEGICS AND SYMPATHETIC ACTIVITY

Because of the interruption of efferent sympathetic fibers with spinal cord transection above T1, no increase in noradrenaline and adrenaline plasma levels and also inadequate blood pressure and heart rate responses have been observed in tetraplegics during wheelchair exercise compared to paraplegics (transection below T1); that is, exercise tolerance in tetraplegics is additionally limited by a secondary dysfunction of the sympathetic nervous system (34), besides the transection-related loss in active musculature.

7. REFERENCES

1. Baumgartner, H., C.J. Wiedermann, H. Hörtnagel, and V. Mühlberger. Plasma catecholamines in arterial and capillary blood. *Naunyn Schmiedebergs Arch. Pharmacol.* **328**:461–463, 1985.

2. Bieger, W., R. Zittel, H. Zappe, and H. Weicker. Einfluß körperlicher Aktivität auf Katecholamin-Rezeptoraktivität. *Dtsch. Z. Sportmed.* **33**:249–252, 1982.

3. Brodde, O.E., A. Daul, and N. O'Hara. ß-adrenoceptor changes in human lymphocytes induced by dynamic exercise. *Naunyn Schmiedebergs Arch. Pharmacol.* **325**:190-192, 1984.

4. Dominiak, P., W. Schulz, W. Delius, G. Kober, and H. Grobecker. Catecholamines in patients with coronary heart disease. In: *Catecholamines and the Heart*, edited by W. Delius, E. Gerlach, H. Grobecker, and W. Kübler. Berlin: Springer, 1981, pp. 223–233.

5. Ekblom, B., A. Kilbom, and A. Soltysiak. Physical training, bradycardia and autonomic nervous system. *Scand. J. Clin. Lab. Invest.* **32**:252–254, 1973.

6. Esler, M., G. Jennings, P. Korner, J. Willet, F. Dudley, G. Hasking, W. Anderson, and G. Lamberg. Assessment of human sympathetic nervous system activity from measurement of noradrenaline turnover. *Hypertension* **11**:3-20, 1988.

7. Hagberg, J.M., R.C. Hickson, J.A. McLane, A.A. Ehsani,and W.W. Winder. Disappearance of norepinephrine from the circulation following strenuous exercise. *J. Appl. Physiol.* **47**:1311–1314, 1979.

8. Hartley, L.H., J.W. Mason, P.R. Hogan, I.G. Jones, T.A. Kotchen, E.H. Mougey, and F.E. Wherry. Multiple hormonal responses to graded exercise in relation to physical training. *J. Appl. Physiol.* **33**:602–606, 1972.

9. Heiss, H.W., J. Barmeyer, K. Wink, G. Huber, J. Breiter, and J. Keul. Durchblutung und Substratumsatz des gesunden menschlichen Herzens in Abhängkeit vom Trainingszustand. *Verh. Dtsch. Ges. Kreislaufforsch.* **41**:247–251, 1975.

10. Israel, S. Zur Problematik des Übertrainings aus internistischer und leistungsphysiologischer Sicht. *Medizin und Sport* **16**:1–12, 1976.

11. Jost, J., M. Weiss, and H. Weicke. Unterschiedliche Regulation des adrenergen Rezeptorsystems in verschiedenen Trainingsphasen von Schwimmern und Langstreckenläufern. In: *Sport, Rettung oder Risiko für die Gesundheit*, edited by D. Böning, K.M. Braumann, M.W. Busse, N. Maassen, and W. Schmid. Köln, Germany: Deutscher Ärzteverlag, 1989, pp.141–145.

12. Kjaer, M. Epinephrine and some other hormonal responses to exercise in man: With special reference to physical training. *Int. J. Sports Med.* **10**:2–15, 1989.

13. Kjaer, M., N.H. Secher, F.W. Bach, S. Sheikh, and H. Galbo. Hormonal and metabolic responses to exercise in humans. Effect of sensory nervous blockade. *Am. J. Physiol.* **257**:95–100, 1989.

14. Kindermann, W. Das Übertraining, Ausdruck einer vegetativen Fehlsteuerung. *Dtsch. Z. Sportmed.* **37**:138–145, 1986.

15. Klingenheben, T., M. Schöpperl, M. Lehmann, and H. Hohnloser. Changes in heart rate and endogenous catecholamines during upright tilt test in patients with neurocardiogenic syncope. *68th Scientific Sessions of American Heart Association*, Abstract, 1995 (In press).

16. Kniffki, K.D., S. Mense, and R.F. Schmidt. Muscle receptors with fine afferent fibers which may evoke circulatory reflexes. *Circ. Res.* **48** (Suppl I):25–31, 1981.

17. Lehmann, M., and J. Keul. Free plasma catecholamines, heart rates, lactate levels and oxygen uptake in competition weight lifters, cyclists and untrained control subjects. *Int. J. Sports Med.* **7**:18–21, 1986.

18. Lehmann, M., H.H. Dickhuth, P. Schmid, H. Porzig, and J. Keul. Plasma catecholamines, ß-adrenergic receptors, and isoproterenol sensitivity in endurance trained and non-endurance trained volunteers. *Eur. J. Appl. Physiol.* **52**:362–369, 1984.

19. Lehmann, M., and J. Keul. Capillary-venous differences of free plasma catecholamines at rest and during graded exercise. *Eur. J. Appl. Physiol.* **54**:502–505, 1985.

20. Lehmann, M., U. Gastmann, R. Tauber, C. Weiler, R. Pilot, F.H Hirsch, W. Auch-Schwelk, and J. Keul. Katecholaminverhalten, Adrenorezeptorendichte an intakten Zellen und Katecholamin-Empfindlichkeit bei einer Patientin mit primärer orthostatischer Hypotonie. *Klin .Wochenschr* . **64**:1249–1254, 1986.

21. Lehmann, M., F.H. Hirsch, W. Auch-Schwelk, J. Alnor, A. Ochs, U. Gastmann, and J. Keul. Primäre orthostatische Hypotonie. Ein Fallbericht. *Z. Kardiol.* **75**:117–121, 1986.

22. Lehmann, M., K.G. Petersen, and A.N. Kalaf. Sympathetic autonomic dysfunction. Programmed subcutaneous noradrenalin administration via microdosing pump. *Klin. Wochenschr.* **69**:872–879, 1991.

23. Lehmann, M., G. Hasenfuß, L. Samek, and U. Gastmann. Catecholamine metabolism in heart failure patients and healthy control subjects. *Drug Res.* **40**:1310-1318, 1990.

24. Lehmann, M., C. Foster, and J. Keul. Overtraining in endurance athletes. A brief review. *Med. Sci. Sports Exerc.* **25**:854–862, 1993.

25. Lehmann, M., W. Schnee, R. Scheu, W. Stockhausen, and N. Bachl. Decreased nocturnal catecholamine excretion. Parameter for an overtraining syndrome in athletes? *Int. J. Sports Med.* **13**:236–242, 1992.

26. Lehmann, M., P. Baumgartl, C. Wieseneck, A. Seidel, H. Baumann, S. Fischer, U. Spöri, G. Gendrisch, R. Kaminski, and J. Keul. Training - overtraining: influence of a defined increase in training volume vs. train-

ing intensity on performance, catecholamines and some metabolic parameters in experienced middle- and long-distance runners. *Eur. J. Appl. Physiol.* **64**:169–177, 1992.

27. Lehmann, M. The impact of exercise metablism, catecholamine levels and individual fitness. In: *Rate Adaptive Cardiac Pacing And Implantable Defibrillators*, edited by E. Alt and H. Blömer. Berlin: . Springer, 1990, pp. 41–51.

28. Meyer, K., M. Lehmann, G. Sünder, J. Keul, and H. Weidemann. Interval vs continuous exercise training after coronary bypass surgery. *Clin. Cardiol.* **13**:851–861, 1990.

29. Naessens, G., J. Lefevre, M. Driessens. Practical and clinical relevance of urinary basal noradrenaline excretion measurements in the follow up of training processes in semi-professional soccer players. *Clin. J. Sports Med.*, 1996 (in press).

30. Péronnet F, J. Cléroux, H Perrault, D. Cousineau, J. de Champlain, and R. Nadeau. Plasma norepinephrine response to exercise before and after training in humans. *J. Appl. Physiol.* **51**:812–815, 1981.

31. DaPrada M, and G. Zürcher. Simultaneous radioenzymatic determination of plasma and tissue adrenaline, noradrenaline and dopamine within the fentomole range. *Life Sci.* **19**:1161–1174, 1976.

32. Reindell H. Das Sportherz, geschichtliche Entwicklung und neue Aspekte. In: *Kardiologie Im Sport*, edited by R. Rost and F. Webering. Köln: Deutscher Ärzte-Verlag, 1987, pp. 1O9–115.

33. Sarnoff S.J., E. Braunwald, C.H. Welch Jr, R.B. Case, W.N. Stainsby, and R. Marcruz. Hemodynamic determination of oxygen consumption of the heart with special reference to the tension-time-index. *Am. J. Physiol.* **12**:148–158, 1958.

34. Schmid, A., M. Lehmann, M. Huonker, J.M. Barturen, J. Keul, and F. Stahl. Plasma catecholamines, cardiocirculatory and metabolic parameterss in tetraplegics and paraplegics at rest and during graded exercise. *Eur. J. Appl. Physiol.* **69** (Suppl):27, 1994.

35. Silverberg, A.B., S.D. Shah, M.W. Haymond, and P.E. Cryer. Norepinephrine. Hormone and neurotransmitter in man. *Am. J. Physiol.* **234**:252–256, 1978.

36. Sobel, B.E., and R. Roberts. Hypotension and syncope. In: *Heart Disease*, edited by E. Braunwald. Philadelphia: Saunders, 1984, pp. 928–939.

37. Trap-Jensen, J., N.J. Christensen, J.P. Clausen, B. Rasmussen, and K. Klausen. Arterial noradrenaline and circulatory adjustment to strenuous exercise with trained and non-trained muscle groups. In: *Physical Fitness*, edited by V. Seliger. Prague: Karlova University Press Prague, 1973, pp. 414–418.

38. Wallin, B.G. Relationship between sympathetic outflow to muscles, heart rate and plasma norepinephrine in man. In: *Catecholamines And The Heart*, edited by W. Delius, E. Gerlach, H. Grobecker, and W. Kübler. Berlin: Springer Berlin, 1981, pp. 11–17.

39. Winder, W.W., R.C. Hickson, J.M. Hagberg, A.A. Ehsani, and J.A. McLane. Training-induced changes in hormonal and metabolic response to submaximal exercise. *J. Appl. Physiol.* **46**:766–771.

40. Yamaguchi, N., J. deChamplain, and R. Nadeau. Correlation between the response of the heart to sympathetic stimulation and the release of endogenous catecholamines into the coronary sinus of the dog. *Circ. Res.* **36**:662–668.

27

NONINVASIVE METHODS TO INVESTIGATE BLOOD SUPPLY TO THE LOWER EXTREMITIES IN PATIENTS WITH PERIPHERAL ARTERIAL OCCLUSIVE DISEASE DURING EXERCISE

Y. Liu, J. M. Steinacker, A. Opitz-Gress, M. Clausen, and M. Stauch

Abteilung Sport- und Leistungsmedizin
University of Ulm, D-89070 Ulm
Germany

1. INTRODUCTION

Peripheral arterial occlusive disease (PAOD) leads to exercise intolerance because of an insufficient blood supply to the lower extremities. As in relatively well-compensated PAOD patients, the insufficient blood supply becomes evident only during exercise, an exercise test has to be applied for its detection and estimation. Although there are several methods which can be used to investigate the peripheral blood supply, the issue remains as to how best quantitate the peripheral blood supply during exercise and, furthermore, to assess its functional reserve.

Among several methods which can be used for estimating peripheral blood supply, the systolic Doppler ankle pressure (DAP) has been widely applied (14). Not only is DAP important for diagnosing PAOD, but also the "Doppler Index" - i.e. DAP versus the systemic systolic blood pressure - is useful for classifying grades of PAOD and for assessing the functional reserve of the peripheral blood supply (7). However, the usefulness of the DAP measurement is limited in some PAOD patients with microvascular angiopathy and it may be invalid and inconvenient to implement during exercise. Venous occlusive plethysmography is another method which can quantitatively measure the blood flow to the lower extremities in PAOD patients (13). However, measurement is impossible during exercise.

Thus, studies on methods which can be used to estimate the peripheral blood supply during exercise are needed. In recent years, we have attempted to estimate the peripheral blood supply to the lower extremities in PAOD patients during exercise using two noninvasive methods - transcutaneous oxygen tension and whole-body Thallium imaging. These are discussed in detail in the following account (4,5).

2. TRANSCUTANEOUS OXYGEN TENSION (TCPO₂)

Originally, tcPO₂ was used to monitor the changes of arterial PO₂ (6). Subsequently, a relationship between tcPO₂ and the blood supply to the site of placement of the tcPO₂-electrode was demonstrated (11). Based on this relationship, tcPO₂ has therefore been taken as an index of the blood supply.

We investigated 20 patients with PAOD stage II (intermittent claudication) using DAP and tcPO₂ during a treadmill test, cycle-ergometry and rowing-ergometry. The tcPO₂ was measured with a Clark-type electrode (heated to 45°C) at two sites for each subject: the peripheral tcPO₂ was measured on the calf where a potential exercise-induced ischemia might be expected in PAOD patients and, as a control, the central tcPO₂ was measured on the chest where a potential ischemia may presumably be excluded.

In comparison to controls, the PAOD patients had a distinctly reduced performance with lower-body exercise (treadmill test and cycle-ergometry), while for upper-body exercise (rowing-ergometry with a fixed seat) there was no significant difference of performance between the two groups. We found that during the upper-body exercise, DAP and tcPO₂ remained almost unchanged, whereas during lower-body exercise they both clearly fell (Figure 1). Since during upper-body exercise, the lower extremities are not (or only slightly) stressed and therefore their peripheral blood supply should not be obviously affected, DAP and tcPO₂ stayed relatively constant. During lower-body exercise, however, the peripheral blood supply to the lower extremities is stressed and the functional reserve becomes exhausted. This implies that the peripheral blood supply to the lower extremities can be reflected in DAP together with tcPO₂. Furthermore, we found a hyperbolic relationship between DAP and tcPO₂ during lower-body exercise (Figure 2). This relationship suggests that the higher the DAP, the greater the tcPO₂. In the lower range of DAP, a

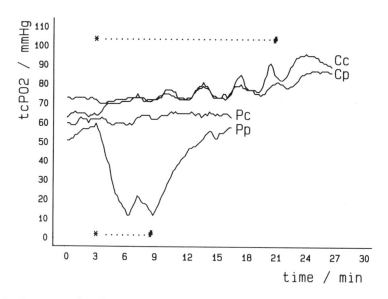

Figure 1. Continuous recording of transcutaneous oxygen tension (tcPO₂) during multiple-stage cycle-ergometry. * and #: start and end of the exercise, respectively. Cc and Cp: central and peripheral tcPO₂ of a control subject, respectively. Pc and Pp: central and peripheral tcPO₂ of a PAOD patient. Reproduced from Ref. 4 with permission.

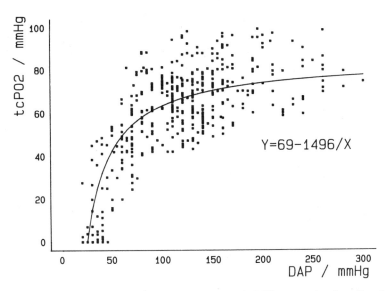

Figure 2. The relationship between transcutaneous oxygen tension (tcPO2) measured at the calf, and systolic Doppler ankle pressure (DAP) during an exercise test. Reproduced from Ref. 4 with permission.

small change of the DAP was accompanied by a large change of $tcPO_2$, which implies that in a critical ischemia the change of $tcPO_2$ is more sensitive than that of DAP.

3. 201-THALLIUM IMAGING

As early as the 1950s, a direct correlation between tracer distribution and fractional blood flow of cardiac output has been reported (8). Strauss *et al.* described, for the first time, that the regional distribution of ^{201}Tl reflected the fractional distribution of the cardiac output (12). Since then, several studies employing ^{201}Tl imaging have been performed in patients with PAOD (3,10). However, most of these studies dealt with resting blood flow. The relationship between regional ^{201}Tl uptake and the blood supply of the lower extremities in patients with PAOD during exercise therefore needs to be established.

We studied the regional blood supply of the calf (RBS, %) in PAOD patients during exercise using ^{201}Tl whole-body imaging, and compared these results with those obtained from $tcPO_2$ measurements. In the $tcPO_2$ study mentioned above, we found that the velocity of $tcPO_2$ fall (VtcPO$_2$↓) during cycle-ergometry was inversely related to DAP. In the present study, VtcPO$_2$↓ was therefore calculated as:

$$VtcPO_2\!\!\downarrow = (tcPO_2 rest - tcPO_2 mi)/2PHT \qquad (1)$$

where: $tcPO_2 rest$ and $tcPO_2 mi$ represent the resting $tcPO_2$ and the minimum $tcPO_2$ during exercise, respectively; and PHT = pressure half time from $tcPO_2 rest$ to $tcPO_2 mi$. 33 patients with PAOD and 10 subjects without PAOD (i.e. controls) were enrolled in this study. A stepwise incremental cycle-ergometer protocol was performed to the limit of tolerance. In the last minute of exercise, 2 mCi of 201Tl were injected intravenously, and whole-body images were taken simultaneously from anterior and posterior projections with a dual-head camera system immediately (stress image) and 4 hours (redistribution image) following exercise. The RBS(%) was calculated from the regional geometric mean counts of the calf divided by the total counts of the whole body. In agreement with the results of the study mentioned above, the exercise performance of the PAOD group during cycle-ergometry was reduced (72 vs 125 W) and the peripheral $tcPO_2$ fell dramatically (51 to 19 mm Hg). The RBS of PAOD patients was clearly reduced both in the stress and redistribution images, compared to controls; there was a significant increase of RBS in the redistribution image (Table 1). Furthermore, nonlinear regression analysis showed a hyperbolic relationship between stress RBS and the velocity of $tcPO_2$ fall during the exercise test (Figure 3). Our results suggest that (i) the ^{201}Tl whole-body images are comparable to the $tcPO_2$ measurement in differentiating patients with PAOD from subjects without PAOD during exercise and (ii) the regional ^{201}Tl uptake reflects regional blood supply in PAOD patients, giving information about the functional reserve. The hyperbolic relationship between peripheral $tcPO_2$ and RBS derived from the ^{201}Tl whole-body imaging implies that the lower the RBS, the faster will be the $tcPO_2$ decrease in a critical ischemia.

4. DISCUSSION

To date, the gold-standard examination for PAOD has remained arterial angiography. This method can directly locate the arterial lesions and the grades of vascular damage can be judged. However, with this method it is difficult to assess the collateral circulation and to estimate the functional reserve (1). The walking distance in a treadmill test therefore has been adopted as a standard means of assessing the functional reserve (9). However, it is recognized that this index is subjective and of limited value in patients with difficulties in walking or maintaining balance. In contrast, DAP is a well-established method for diagnosing PAOD and is used to assess the functional reserve of the peripheral blood supply. Since this is a discontinuous method and is invalid in patients with angiopathy (incompressible vessels), its usefulness is limited during exercise.

PAOD patients can be differentiated from subjects without PAOD by means of $tcPO_2$ measurements. In comparison to the walking distance approach, $tcPO_2$ provides an objective index and, unlike DAP, it is a continuous measurement and is especially convenient for exercise studies. However, if the blood supply is not critically reduced, $tcPO_2$ essentially stays within the normal range because of the hyperbolic relationship between

Table 1. Regional blood supply (%) determined by whole-body ^{201}Tl images in controls and PAOD patients immediately after (stress) and 4 hours (redistribution) following exercise (5)

		n	stress	redistribution	P
calf	control	10	3.1 ± 0.5	3.4 ± 0.5	NS
	PAOD	33	1.5 ± 0.7	2.8 ± 0.7	.01
	P		.01	.05	
thigh	control	10	18.0 ± 2.2	17.3 ± 2.0	NS
	PAOD	33	10.4 ± 3.0	11.4 ± 2.3	NS
	P		.01	.01	
leg	control	10	21.1 ±2.2	20.7 ± 2.1	NS
	PAOD	33	12.0 ±3.4	14.3 ± 2.6	.05
	P		.01	.01	

Values were taken from the right side of controls and the mainly involved side of PAOD patients (mean ± SD). Reproduced from Ref. 5, with permission.

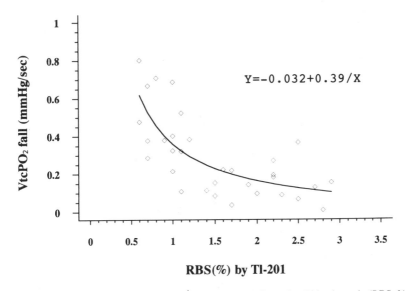

Figure 3. Relationship between $tcPO_2$ fall ($VtcPO_2\downarrow$, mmHg/s) and the regional blood supply (RBS, %) in PAOD patients during a cycle-ergometry ($VtcPO2\downarrow = -0.032 + 0.39/RBS$, $r^2 = 0.54$, $P<0.05$). Reproduced from Ref. 5 with permission.

$tcPO_2$ and blood supply (4,6,11). Therefore, the absolute value of $tcPO_2$ is considered clinically not to be as valuable as the $tcPO_2$ *changes*, and $tcPO_2$ alone seems to be inadequate to provide quantitative information about the blood supply, although in a critical ischemia the $tcPO_2$ change is even more sensitive than that of DAP.

[201]Tl whole-body images have been used to assess resting leg perfusion (2). However, only a few studies have addressed the blood supply during exercise. We have described here our findings on the calf blood supply during cycle-ergometry, using a dual-head camera system which allowed calculation of the geometric mean counts from the simultaneous anterior and posterior images (5). The regional blood flow can be theoretically calculated by multiplying cardiac output by the percentage of the regional ^{201}Tl uptake and thus blood supply can be derived quantitatively. Certainly, the ^{201}Tl whole-body imaging is quite expensive and complicated to undertake, which may limit its routine clinical use.

In conclusion, the noninvasive methods - $tcPO_2$ and ^{201}Tl whole-body imaging - can be used clinically to investigate peripheral blood supply in PAOD patients during exercise. These approaches are therefore especially useful in assessing the functional reserve of peripheral blood supply.

5. REFERENCES

1. Barnes, R.W. Noninvasive diagnostic techniques in peripheral vascular disease. *Am. Heart J.* **97**: 241–258, 1979.
2. Burt, R.W., F.M. Mullinix, D.S. Schauwecker, and B.D. Richmond. Leg perfusion evaluated by delayed administration of Thallium-201. *Radiology.* **151**: 219–224, 1984.
3. Hamanaka, D., T. Odori, H. Maeda, Y. Ishii, K. Hayakawa, and K. Torizuka. A quantitative assessment of scintigraphy of the legs using [201]Tl. *Eur. J. Nucl. Med.* **9**: 12–16, 1984.

4. Liu, Y., J.M. Steinacker, and M. Stauch. Transcutaneous oxygen tension and Doppler ankle pressure during upper and lower body exercise in patients with peripheral arterial occlusive disease. *Angiology.* **46**: 689–698, 1995.

5. Liu, Y., J.M. Steinacker, A. Opitz-Gress, M. Clausen, and M. Stauch. Comparison of whole-body thallium imaging with transcutaneous PO_2 in studying regional blood supply in patients with peripheral arterial occlusive disease. *Angiology.* In press.

6. Lübbers, D.W. Theoretical basis of the transcutaneous blood gas measurements. *Crit. Care. Med.* **9**: 721–733, 1981.

7. Ouriel, K., and C.K. Zarins. Doppler ankle pressure: an evaluation of three methods of expression. *Arch. Surg.* **117**: 1297–1300, 1982.

8. Sapirstein, L.A. Regional blood flow by fractional distribution of indicators. *Am. J. Physiol.* **193**: 161–168, 1958.

9. Sehested, J., S. Bille, and P. Hauge. Assessment of a standard exercise test in peripheral arterial disease. *J. Cardiovasc. Surg.* **28**: 520–523, 1987.

10. Seto, H., M. Kageyama, R. Futatsuya, M. Shimizu, N. Watanabe, Y Wu, and M. Kakishita.et al. Regional distribution of ^{201}Tl during one-leg exercise: comparison with leg blood flow by plethysmography. *Nucl. Med. Comm.* **14**: 810–813, 1993.

11. Steinacker, J.M., and W. Spittelmeister. Dependence of transcutaneous O_2 partial pressure on cutaneous blood flow. *J. Appl. Physiol.* **64**: 21–25, 1988.

12. Strauss, H.W., K. Harrison, and B. Pitt. Thallium-201: Non-invasive determination of the regional distribution of cardiac output. *J. Nucl. Med.* **18**: 1167–1170, 1977.

13. Whitney, R.J. The measurement of volume changes in human limbs. *J. Physiol.* **121**: 1–27, 1953.

14. Zierler, R.E. Hemodynamic considerations in evaluation of arterial disease by Doppler ultrasound. *Clin. Diagn. Ultrasound.* **26**: 13–24, 1990.

28

DETERMINING THE LACTATE THRESHOLD IN PATIENTS WITH SEVERE CHRONIC OBSTRUCTIVE PULMONARY DISEASE

T. L. Griffiths,[2] S. E. Gregory,[3] S. A. Ward,[4] K. B. Saunders,[1] and B. J. Whipp[1]

[1] Division of Physiological Medicine and Department of Physiology
St George's Hospital Medical School
London, SW17 0RE, United Kingdom
[2] Section of Respiratory Medicine
University of Wales College of Medicine
Llandough Hospital, Penarth, CF64 2XX, United Kingdom
[3] Department of Physiotherapy
London Chest Hospital, London, E2 9JX, United Kingdom
[4] School of Applied Science, South Bank University
London, SE1 0AA, United Kingdom

1. INTRODUCTION

The degree of the metabolic stress associated with muscular exercise depends on the intensity of the work being performed. Consequently, the gas exchange, acid/base and ventilatory requirements will also be intensity dependent. Low intensity exercise can be performed utilising ATP formed aerobically through oxidative phosphorylation but supplemented from creatine phosphate stores utilisation. When heavy exercise is undertaken, anaerobic glycolysis further supplements the high energy phosphate transformation with the consequent production of lactic acid - which at cellular pH is dissociated virtually entirely into lactate ions (La^-) and protons (H^+). The consequent H^+ is predominantly buffered by bicarbonate ions (HCO_3^-): the resulting carbonic acid yields extra carbon dioxide to be excreted by the lungs in addition to that produced from cellular respiration. The level of oxygen uptake at which lactate increases has been termed the Anaerobic or Lactate Threshold (LT).

The LT is usually determined during an exercise test in which the work rate is increased progressively to the limit of tolerance. Various methods can be used to discriminate the LT: (a) arterial or arterialised-venous blood may be sampled and analysed to determine the $[La^-]$ / [pyruvate] ratio, the $[La^-]$ and the standard $[HCO_3^-]$. When plotted against $\dot{V}O_2$, the $[La^-]$ / [pyruvate] ratio remains relatively stable despite a small gradual rise in [La⁻] or fall in [HCO₃⁻]. This is followed by a subsequent steepening of the relationship at the LT reflecting increased net production of lactate and hydrogen ions. Log transformation of [La⁻] and [HCO₃⁻] versus $\dot{V}O_2$ plots commonly improves the linearity of the relationship before and after this point (1).

(b) Non-invasive estimates of the LT can be achieved by studying the ventilatory and gas exchange effects of the relative increase in CO_2 output occasioned by buffering of H^+ (6). In normal subjects, a systematic rise in the ventilatory equivalent for oxygen ($\dot{V}E/\dot{V}O_2$) and end tidal PO_2 is seen at the LT without a decrease in the end tidal PCO_2, so ruling out non-specific hyperventilation. The detection of the excess carbon dioxide output in comparison to oxygen uptake - the V-slope method - has also been proposed to provide a valid index of the LT (2). With this relationship, $\dot{V}CO_2$ is plotted against $\dot{V}O_2$ and is usually found to present two linear components which intersect at the LT (if the initial non-linear segment is disregarded). This method together with blood [HCO₃⁻] and plasma [La⁻] measurement has been applied in patients with COPD(4,5).

Determination of the LT serves to partition two important domains of work intensity: one which an individual is able to sustain for a prolonged period (below LT) and another in which the exercise is curtailed due to inability to meet the cardiopulmonary and gas exchange requirements for the task (above LT).

The LT may therefore be used to: 1) evaluate a subject's capacity for wholly aerobic work; 2) differentiate between various mechanisms that lead to effort intolerance; 3) design training regimens for normal subjects and rehabilitation programmes for patients with respiratory or cardiac disease; 4) assess the physiological effect of therapeutic and rehabilitative interventions and 5) assess patients undergoing organ transplantation.

In patients with severe pulmonary disease such as COPD, a number of factors might affect the ability to determine the LT from ventilatory and gas exchange variables during an incremental exercise test. These patients can have a severely reduced exercise capacity and hence peak $\dot{V}O_2$. Consequently, the small change in $\dot{V}O_2$ from rest to peak exercise might make discrimination difficult and, as these patients tend to be severely physically deconditioned, the LT would also be expected to occur at a low absolute metabolic rate. In severely limited patients with COPD, this can occur at the lowest work load in the exercise test protocol. In addition, the ventilatory and gas exchange kinetics at the onset of exercise are known to be slow in these patients (3) and abnormalities of lung mechanics and pulmonary vasculature might also alter the relationship between events occurring in the muscle and resultant changes in pulmonary gas exchange. Finally, the demand for increased ventilation may not be capable of being met at high work rates because of airflow limitation.

2. PURPOSE OF THE INVESTIGATION

We wished to ascertain whether the LT can be validly determined from pulmonary gas exchange variables in patients with *severe* COPD.

3. METHODS

3.1. Subjects

The subjects investigated were 15 (7 female) patients with the clinical features of COPD showing a <20% improvement in FEV_1 after inhalation of 2.5 mg of nebulised salbutamol. Mean (±SD) age was 68 (±8) years; FEV_1, 860 (±352) mL; FEV_1% predicted, 34 (±9) %.

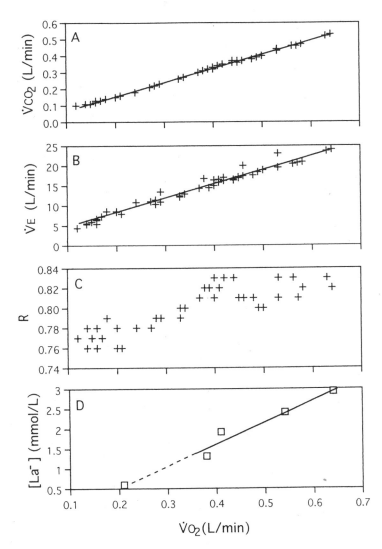

Figure 1. Plots of $\dot{V}CO_{2(A)}$, $\dot{V}E$ (B), R (C) and [La⁻] (D) against $\dot{V}O_2$ in a subject performing a 5 watt/min cycle ergometer incremental exercise test preceded by 3 minutes of "unloaded" cycling.

3.2. Exercise Testing

The subjects performed incremental exercise to the limit of tolerance on a standard electrically braked cycle ergometer (Bosch, ERG 551). We estimated the "unloaded" power output for cycling this ergometer to be 20 Watts at 60 rpm.

In a preliminary study in which a 3 minute "unloaded" pedalling period preceded the onset of work incrementation, it became apparent that a LT could not be discerned in some of our subjects: there being no break point in either the V-slope or the plot of ventilation against $\dot{V}O_2$ (Figure 1, A and B). From these plots it was not clear whether this was because of LT occurring at a very low $\dot{V}O_2$ or whether LT was actually not reached due to severe effort intolerance. A clue to discriminating between these alternatives was provided by the fact that R, the respiratory exchange ratio, rose progressively from an early stage (rather than the usual fall seen at the start of exercise) suggesting early onset of excess CO_2 output (Figure 1, C). Finally, measurements of [La⁻] supported this notion of ongoing net La⁻ production from early in exercise. Thus it appeared that LT had been reached at a $\dot{V}O_2$ associated with "unloaded" pedalling at 60 rpm. Consequently, it was necessary to modify the standard incremental protocol.

After a period at rest to allow acclimatisation to the experimental apparatus, the subjects performed one minute each of "unloaded" cycling at 20, 40 and 60 rpm. This was followed by one minute increments of 5 Watts cycling at 60 rpm to tolerance. A metronome was used to help the subjects maintain regular cycling at the lowest frequency. We found that subjects were able to perform these low rates of pedalling surprisingly well. The protocol produced an acceptably linear ramp of $\dot{V}O_2$ as shown in Figure 2.

Ventilation and pulmonary gas exchange were monitored using a mixing chamber system (Quinton, Q-plex 1). The LT was determined by inspection using the V-slope plot together with plots of $\dot{V}E$, R, $\dot{V}E/\dot{V}O_2$, mixed expired O_2 concentration, $\dot{V}E/\dot{V}CO_2$ and mixed expired CO_2 concentration against $\dot{V}O_2$. The profile in the first minute of exercise

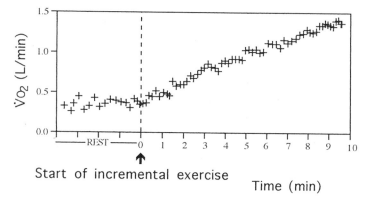

Figure 2. Plot of $\dot{V}O_2$ against work rate in a patient (FEV$_1$ = 0.83 L; 29% predicted) undergoing the modified incremental exercise protocol .

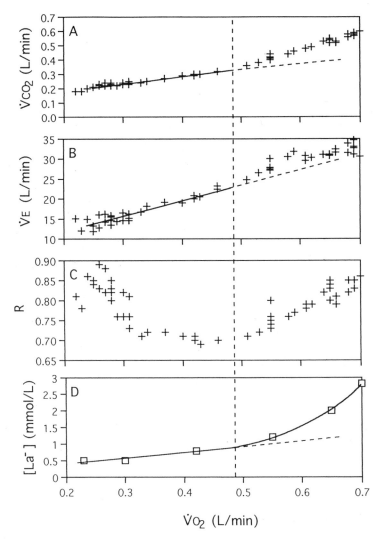

Figure 3. Plots of $\dot{V}CO_2$(A), $\dot{V}E$ (B), R (C) and [La$^-$] (D) against $\dot{V}O_2$ for a subject undergoing the modified one minute incremental exercise test protocol.

was not included in the analysis to obviate error due to non-linearity of gas exchange in the early phase.

Plasma lactate concentration was measured in arterialised-venous blood and was plotted as [La$^-$] against $\dot{V}O_2$ to allow comparison with the other variables.

4. RESULTS

Examples of plots of [La$^-$], $\dot{V}CO_2$, R and $\dot{V}E$ against $\dot{V}O_2$ are shown for a subject whose peak $\dot{V}O_2$ was similar to that shown in Figure 1 (Figure 3). In this subject the modified incremental protocol was used. A clear break point was seen in the plots of [La$^-$],

$\dot{V}CO_2$ and $\dot{V}E$ accompanied by a rising R. Thus, although this occurred at a $\dot{V}O_2$ seen with unloaded pedalling at 60 rpm, the data allowed discrimination of the LT.

In all cases, hyperventilation at the putative LT - determined from the V-slope - was ruled out by showing the mixed expired CO_2 concentration becoming steady or continuing to rise. There was an increase in $\dot{V}E$ /$\dot{V}O_2$ and mixed expired oxygen concentration at LT in half of the subjects. In all cases the value determined for the LT by ventilation and gas exchange criteria was consistent with that determined separately by assay of plasma [La$^-$].

It is of note that in 12 of our subjects, the LT was determined to occur at a $\dot{V}O_2$ at or below that seen during "unloaded" cycling at 60 rpm. Mean LT was 0.65 ($\pm$0.17) L/min; mean peak$\dot{V}O_2$ was 1.1 ($\pm$0.44) L/min; LT 64 ($\pm$12) % peak $\dot{V}O_2$. In this group of patients there was a strong relationship between FEV_1 and peak $\dot{V}O_2$ and a weaker one between FEV_1 and LT. For the linear regression of peak $\dot{V}O_2$ on FEV_1, r = 0.88, p< 0.0001. For the linear regression of LT on FEV_1, r = 0.79, p< 0.001. LT and peak $\dot{V}O_2$ were also related: r = 0.82, p<0.001.

In 13 subjects a single V-slope break point was identified. In two subjects, however, the V-slope plot did not clearly conform to the expected pattern. Two break points were discernible in the relationship, the first from gradients of 0.66 and 0.68 (S_1) and the second from gradients of 0.88 and 0.99 (S_2). In these patients, the [La$^-$] was used to confirm the LT. This occurred at the *first* V-slope break point.

5. DISCUSSION

The unloaded pedalling phase of an exercise testing protocol can, in itself, constitute a significant metabolic challenge in patients such as those with severe COPD. This consequently presents a problem in determining the LT. However, by modifying the protocol we have been able to provide a gas exchange profile at very low work rates which improves resolution of the LT. This is possible without access to an ergometer which provides truly unloaded pedalling and fully variable work loads. In our subjects, the peak $\dot{V}O_2$ and, to a lesser extent, the LT were related to expiratory air flow, suggesting that in patients with the most severe airflow limitation, a modified approach to incremental exercise testing such as we describe may be appropriate. Patessio et al. (4) have shown that in patients with COPD, performing an unloaded cycling period followed by 10 watt increments each minute, there can be a substantial discrepancy between the lactic acidosis threshold (falling [HCO_3^-]) and the LT determined from gas exchange when this occurs within the first 2 minutes of incremental exercise; thereafter, the difference between the estimates is much smaller. The present protocol may therefore be of value in delaying the time of onset of the LT thereby allowing greater precision in its estimation from gas exchange variables.

In a sub-group of subjects, the presence of more than one apparent break point in the V-slope did not conform to the criteria for determining the LT. This made interpretation of the thresholds difficult. We had initially thought that the behaviour around the first break point reflected a "pseudo-threshold" (7) resulting from the filling of CO_2 stores or an over-rapid work rate incrementation. However, the [La$^-$] plot confirmed the earlier break point (i.e. that with the uncharacteristically low S_1) as the LT. Thus, it may be that the criterion of S_1 being close to 1 may not be appropriate to all patients. The cause of the low S_1 seen in these patients remains to be elucidated.

Thus, we have found that the non-invasive discrimination of the LT in an incremental exercise test using standard equipment and suitable adjustments to the study protocol can be adequately performed in patients with severe COPD. However in a minority of patients, gas exchange data alone may not be adequate for this purpose.

6. CONCLUSIONS

In the majority of patients with severe COPD and impaired exercise tolerance, the lactate threshold may be determined from gas exchange variables in spite of these subjects' impaired lung mechanics and abnormal pulmonary circulation. However, the exercise test protocol may need to be adjusted in the most severely impaired patients to enable the lactate threshold to be determined. Finally, care must be exercised when interpreting gas exchange data, especially when there is more than one clear break point in the V-slope: measurement of plasma lactate may be necessary to determine the lactate threshold in these cases.

7. ACKNOWLEDGMENT

This work was supported by a South Thames Regional Research and Development Grant.

8. REFERENCES

1. Beaver W. L., K. Wasserman, and B. J. Whipp. Improved detection of lactate threshold during exercise using a log-log transformation. *J. Appl. Physiol.* **59**: 1936–1940, 1985.
2. Beaver W. L., K. Wasserman, and B. J. Whipp. A new method for detecting anaerobic threshold by gas exchange. *J. Appl. Physiol.* **60**: 2020–2027, 1986.
3. Nery L. E., K. Wasserman, J. D. Andrews, D. J. Huntsman, J. E. Hansen, and B. J. Whipp. Ventilatory and gas exchange kinetics during exercise in chronic airways obstruction. *J. Appl. Physiol.* **53**: 1594–1602, 1982.
4. Patessio A., R. Casaburi, M. Carone, L. Appendini, C. F. Donner, and K. Wasserman. Comparison of gas exchange, lactate, and lactic acidosis thresholds in patients with chronic obstructive pulmonary disease. *Am. Rev. Respir. Dis.* **148**: 622–626, 1993.
5. Sue D. Y., K. Wasserman, R. B. Moricca, and R. Casaburi. Metabolic acidosis during exercise in patients with chronic obstructive pulmonary disease. *Chest.* **94**: 931–938, 1988.
6. Wasserman K., B. J. Whipp, S. Koyal, and W. L. Beaver. Anaerobic threshold and respiratory gas exchange during exercise. *J. Appl. Physiol.* **35**: 236–243, 1973.
7. Whipp B. J., N. Lamarra, and S. A. Ward. Required characteristics of pulmonary gas exchange dynamics for non-invasive determination of the anaerobic threshold. In: *Concepts and Formalizations in the Control of Breathing*, edited by G. Benchetrit, P. Baconnier and J. Demongeot. Manchester: Manchester University Press, 1987, pp. 185–200.

29

CARDIOVASCULAR AND METABOLIC RESPONSES TO EXERCISE IN HEART TRANSPLANTED CHILDREN

C. Marconi,[1] M. Marzorati,[1] R. Fiocchi,[2] F. Mamprin,[2] P. Ferrazzi,[2]
G. Ferretti,[3] and P. Cerretelli[3]

[1] I.T.B.A. Section of Physiology
National Research Council, Milano, Italy
[2] Department of Cardiovascular Surgery
Ospedali Riuniti, Bergamo, Italy
[3] Department of Physiology
C.M.U., University of Geneva, Switzerland

1. INTRODUCTION

Heart transplantation has become an accepted therapy for end-stage congestive heart failure not only in adults but also in children (P-HTR), particularly following improvements in the immunosuppressive regimens. However, little is known about the structural evolution of the allograft or its adaptive response to increased metabolic demand. Normal growth of the allograft is especially important in P-HTR, since the possibility to undergo cardiac development commensurate to the increase in body size is essential for the normal physical performance of these patients (10). In addition, adequate growth may hypothetically be associated with some form of cardiac reinnervation, which, as is well known, does not occur in adults. So far, however, experimental evidence of sinus node reinnervation is lacking even for P-HTR. In fact, a study by Hsu *et al.* (2) has shown that the heart rate (HR) response to graded exercise of P-HTR is similar to that of adult HTR (A-HTR).

The purpose of the present investigation was to detect indirect evidence of functional cardiac reinnervation in children, based on the cardiovascular and metabolic response of P-HTR to exercise. In particular, the focus was to analyse the control mechanisms of heart function in P-HTR and the kinetics of pulmonary gas exchange for rest-to-work transients. This is only a preliminary part of a wider project, which comprises also the assessment of heart beat interval variability by both stochastic and deterministic dynamical analysis, as well as an assessment of catecholamine sensitivity of the myocardium.

2. METHODS

2.1. Subjects

The present investigation was conducted on a limited group of subjects (n = 6) aged 13.7 ± 2.4 yr (mean ± SD). The children's parents were asked to provide informed consent prior to the study, which was approved by the institutional Scientific and Ethical Committee.

Measurements were carried out, on average, 36 months (range: 4–63) after transplantation. All patients were investigated on the occasion of their routinely scheduled clinical laboratory tests. At the time of the study, P-HTR were immunosuppressed only with cyclosporine and azathioprine. None of them showed clinical or histological evidence of rejection. The subjects performed only spontaneous physical activity, and none was enrolled in any formal exercise training program.

2.2. Methods

Heart rate was continuously monitored by ECG (Cardioline ETA 150, Remco Italia). Pulmonary ventilation ($\dot{V}E$) and oxygen uptake ($\dot{V}O_2$) were monitored breath-by-breath by means of a computerised O_2-CO_2 analyzer-flowmeter combination (MMC 4400tc, SensorMedics). Volume and gas analysers calibrations were performed prior to each measurement by means of a 3-liter syringe (Hewlett Packard 14278B), at three different flow rates, and by means of gas mixtures of known composition, respectively.

2.3. Experimental Protocols

The subjects were asked to perform:

   a. a graded incremental test on an electronically braked bicycle ergometer (Medifit 1000S, Remco Italia) in which, after a 5-min rest period, an initial constant load (25–40 W, depending on age and body mass) was imposed for 2–3 min, followed by 10 W increments at 1-min intervals up to voluntary exhaustion;
   b. a 5-min constant-load test, requiring ~60% of peak oxygen uptake ($\dot{V}O_2p$).

3. RESULTS

3.1. Steady-State Measurements

Resting HR, $\dot{V}O_2$, and $\dot{V}E$ of P-HTR were 101±5 beats/min, 5.8 ± 0.9 ml 02/kg.min, and 0.17 ± 0.02 L/kg.min, respectively. Peak values (HRp, $\dot{V}O_2p$, and $\dot{V}Ep$) were 172 ± 22 beats/min, 32.4 ± 7.0 ml O2/kg.min, and 1.36 ± 0.25 L/kg.min, respectively.

3.2. Nonsteady-State Measurements

The adjustment rate of both HR and $\dot{V}O_2$ of a typical P-HTR (a 14 year old girl) is shown in Fig. 1 during graded exercise up to voluntary exhaustion (protocol a). It may be seen that starting from a resting value of ~105 beats/min, HR increases rapidly at the onset

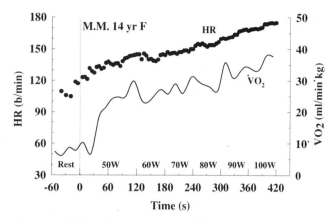

Figure 1. Rate of readjustment of HR and VO$_2$ of a typical P-HTR (a 14 year old girl) during graded exercise to voluntary exhaustion.

of exercise then rising progressively to a peak value of 174 beats/min. The subject could develop up to 100W of mechanical power and the VO$_2$p attained was ~37 ml O$_2$/Kg.min.

The adjustment rate of HR during a 5-min constant-load exercise (~60% VO$_2$p, protocol b) in a P-HTR and, for comparison, in a typical adult HTR taken from the literature (cf. ref. 1) is shown in Fig. 2. Both patients were examined 5 months after transplantation. As may be seen, in a typical P-HTR, HR increases almost immediately at the onset of exercise, reaching a steady-state level in 2–3 min and then stays constant until the end of exercise. At work offset, HR decreases rapidly resuming the pre-exercise value in 3–5 min. All six P-HTR subjects responded similarly. As appears from Fig. 2, the exercise response of P-HTR was different from that of a typical adult HTR (see lower tracing). Indeed, the latter is characterized by: (a) a 30–45 s initial phase during which HR is unchanged or only slightly increased; (b) a retarded linear increase of HR at a rate that is greater the heavier the load; and (c) a slow recovery to pre-exercise levels (5 to 25 min, depending on absolute work rate) (1,4).

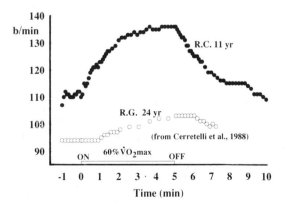

Figure 2. Rate of readjustment of HR during a 5-min constant-load exercise (~60% VO$_2$p) in a typical P-HTR and, for comparison, in a typical adult HTR taken from the literature (c.f. Ref. 3). Both patients were examined 5 months after transplantation.

4. DISCUSSION

As is well known, the innervated heart responds to exercise with an almost instanta-
neous increase in rate, which is sustained initially by the release of the parasympathetic
tone and thereafter by an increase of sympathetic activity (5). In contrast, typical adult
transplanted patients are characterized by a blunted HR response, mainly depending on the
level of circulating catecholamines (3).

The peak HR of P-HTR (~170 beats/min) appears to be definitely higher than that of
adult HTR. Indeed, from a survey of the literature, the average HRp of 432 adult HTR
(average age: 39 yr, range: 29–52 yr), most of them cycling in the upright position, was
~133 (range: 110–148) beats/min. At variance with the above figure, higher values (159
beats/min) were recorded by Savin et al. (7) on a group of 15 A-HTR running on a treadmill up to voluntary exhaustion 3 years after transplantation.

On 31 P-HTR (mean age 13 ± 4 yr) examined 1.3 years after transplantation, Hsu et
al. (2) found HRp of 136 ± 22 beats/min. This average figure is ~30 beats/min lower than
that found in the present P-HTR group and lies in the range prevailing for A-HTR. A sub-
group of 6 P-HTR that was examined by Hsu et al. (2) were characterized by a normal exercise capacity, showed a relatively high HRp (147±19 beats/min); this, however, is
definitely lower than that reported for the present study (~170–175 beats/min).

No information is available from the literature regarding the adjustment rate of HR
of P-HTR at the onset of submaximal constant-load exercise. Thus, the present finding of
a relatively fast HR response at the onset of exercise appears to be new.

The average $\dot{V}O_2p$ of the 6 investigated P-HTR is significantly higher than previously reported by Hsu et al. (2). These authors found an average $\dot{V}O_2p$ of 20±6
ml/kg.min in a group of 31 P-HTR performing a symptom-limited exercise protocol ~15
months after surgery. It is noteworthy that the 6 P-HTR examined by Hsu et al. (2), besides having a higher peak HR, exhibited a higher than average $\dot{V}O_2p$ (i.e. 26.3 ± 5.4
ml/kg.min), which is similar to that found in the present study (32.4 ± 7 ml/kg.min).

5. CONCLUSIONS

The novel finding emerging from the present study is that P-HTR may exhibit cardiovascular and metabolic responses to submaximal and maximal exercise similar to those
of healthy sedentary subjects. This is contrary to most of the previous experimental evidence. This unexpected finding is rather surprising and intriguing. Indeed, it raises a major
issue - the possibility that the allograft, particularly in children, may resume some form of
nervous control. Although sympathetic reinnervation occurs commonly in animal models,
it was believed not to occur in the human allograft, even though, recently, some indirect
evidence of sympathetic reinnervation of transplanted human hearts has been recently provided (7–9).

In any case, it should be pointed out that a faster rate of adjustment of HR in response to exercise, as well as a greater HRp and $\dot{V}O_2p$, may be caused by changes other
than sinus node reinnervation. Indeed, the type of medical treatment may influence the activity of the pacemaker and its sensitivity to catecholamines, particularly with respect to
the use of prednisone and also some individual features of the patient (such as age at time
of transplantation, type and length of disease, and physical fitness). Further investigations
are therefore necessary to gain insight into the mechanisms controlling the function of the
transplanted heart during exercise.

6. ACKNOWLEDGMENT

This work was partially supported by C.N.R. of Italy, Special Target Biotechnology and Bioinstrumentation, and by Grant no. 32–040397.94 of the Fonds National Suisse for the Scientific Research.

7. REFERENCES

1. Cerretelli, P., B. Grassi, A. Colombini, B. Carù, and C. Marconi. Gas exchange and metabolic transients in heart transplant recipients. *Respir. Physiol.* **74**: 355–371, 1988.
2. Hsu, D.T., R.P. Garofano, J.M. Douglas, R.E. Mochler, J.M. Quaegebeur, W.M. Gersony, and L. Addonizio. Exercise performance after pediatric heart transplantation. *Circulation* **88** (part 2): 238–242, 1993.
3. Perini, R., C. Orizio, A. Gamba, and A. Veicsteinas. Kinetics of heart rate and catecholamines during exerice in humans. *Eur. J. Appl. Physiol.* **66**: 500–506, 1993.
4. Pope, S.E., E.B. Stinson, G.T. Daughter, J.S. Schroeder, N.B. Ingels, and E.L. Alderman. Exercise response of the denervated heart in long-term cardiac transplant recipients. *Am. J. Cardiol.* 46: 213–218, 1980.
5. Rowell, L.B., and D.S. O'Leary. Reflex control of the circulation during exercise: chemoreflexes and mechanoreflexes. *J. Appl. Physiol.* **69**: 407–418, 1990.
6. Savin, W.M., W.L. Haskell, J.S. Schroeder, and E.B. Stinson. Cardiorespiratory responses of cardiac transplant patients to graded, symptom-limited exercise. *Circulation* **62**: 55–60, 1980.
7. Schwaiger, M., G.D. Hutchins, V. Kalff, K. Rosenspire, M.S. Hahka, S. Mallette, G.M. Deeb, G.D. Abrams, and D. Wieland. Evidence for regional catecholamine uptake and storage sites in the transplanted human heart by positron emission tomography. *J. Clin. Invest.* **87**: 1681–1690, 1991.
8. Wilson, R.F., B.V. Christensen, M.T. Olivari, A. Simon, C.W. White, and D.D. Laxson. Evidence for structural sympathetic reinnervation after orthotopic cardiac transplantation in humans. *Circulation* **83**: 1210–1220, 1991.
9. Wilson, R.F., D.D. Laxson, B.V. Christensen, A.L. McGinn, and S.H. Kubo. Regional differences in sympathetic reinnervation after human orthotopic cardiac transplantation. *Circulation* **88**: 165–171, 1993.
10. Zales, V.R., K.L. Wright, E. Pahl, C.L. Backer, C. Mavroudis, A.J. Muster, and D.W. Benson Jr. Normal left ventricular muscle mass and mass/volume ratio after pediatric cardiac transplantation. *Circulation* **90** (part 2): II-61-II-65, 1994.

30

DEATH IN SPORTS

Contrast between Team and Individual Sports

M. Parzeller and C. Raschka

Klinikum Fulda
Fulda, Germany

1. INTRODUCTION

For many people, sports represent an important component of their leisure time. The positive effects of sports on physical performance and health are indisputable. Sports are therefore frequently used in the prevention and therapy of different diseases. Nevertheless, there are numerous reports of individuals dying from a range of causes while performing sports (1,3–6,8). Furthermore, considerable public interest is aroused by the often sensational reporting of these events in the media (7). In the scientific literature, the cause of death during the performance of sport is usually subdivided into traumatic or non-traumatic categories (4,6). In the majority of cases, cardiac events are held responsible for non-traumatic deaths. The most usual reasons for cardiac death are hypertrophic cardiomyopathy and also other cardiovascular abnormalities including congenital coronary artery anomalies and valvular disease if the athletes are younger, and coronary artery disease if they are older (i.e. over 35 years) (2,5). In this retrospective study, the causes of death were compared between team sports and individual sports.

2. METHODS

In a large retrospective mortality study lasting 11 years from 1981 to 1991, 1619 deaths in athletes were documented. The data were collected in 6 states of West Germany (Hesse, Baden-Württemberg, Saarland, Hamburg, North-Rhine Westphalia and Schleswig-Holstein having a total population of about 37 million residents). The athletes were all members of a sports club, which had taken out life insurance for the athletes in the event of death.

This study analysed the type of sport (team or individual), the causes of death (traumatic, cardiovascular, traffic accident on the way to sports ground, other reasons), the relation to training or competition, and anthropometric factors such as gender and age.

The Physiology and Pathophysiology of Exercise Tolerance
edited by Steinacker and Ward, Plenum Press, New York, 1996

3. RESULTS

The results are summarized in Tables 1 and 2. Death occurred in 57 different kinds of sports. For team sports, 9 types of sports were involved with a total of 625 cases. For individual sports, 47 sports were included with 994 cases.

The proportion of death between team and individual sports was 38.60 to 61.40%. Most cases for team sports are reported for soccer (501), handball (79), volleyball (19), German fist ball (12), and basketball (7). In individual sports, they were documented for tennis (115), cycling (90), gymnastics (89), table-tennis (70) and bowling (57).

4. DISCUSSION AND CONCLUSIONS

The incidence of sports death is lower for women than for men, both for team sports and individual sports. There are substantial differences in traumatic deaths between these two modes of sport. They largely reflect the highly traumatic risks taken when performing individual sports such as aerial sports (32), cycling (52), and horse-riding (39).

With an incidence of about 61%, cardiovascular complications are the most common cause of death while performing sport. The majority of cases documented for cardiovascular death occur during training in individual sports (357) such as tennis (57) and gymnastics (58). However, in an overview for all kinds of sport, the majority of cardiovascular deaths during training occur in soccer (150). Cardiovascular deaths occurring during competition are mainly seen in team sports such as soccer (190) and handball (33) and during individual sports such as tennis (48) and table tennis (36). Soccer with a 30.95% incidence (relative to all sports deaths) contributes significantly to the epidemiology.

The median age of death for team sports is lower for both male and female athletes, compared to individual sports. The median age of death for incidents occurring on the way

Table 1. Incidence of death for males and females across all sports, and for individual sports and team sports

	All sports	
	male	female
No. cases	1548	71
% of total	95.61	4.39
Median age (yr)	42.05	36.73
	Individual sports	
	male	female
No. cases	935	59
% of total	94.06	5.94
%of all sports	57.75	3.64
Median age (yr)	46.00	39.48
	Team sports	
	male	female
No. cases	613	12
% of total	98.08	1.92
% of all sports	37.86	0.74
Median age (yr)	36.02	23.33

Table 2. Incidence of death, subdivided into eight primary classes

Male and female athletes	Traumatic death: training	Traumatic death: competition	Cardio-vascular death: training	Cardio-vascular death: competition	Traffic accident on the way to sports ground: training	Traffic accident on the way to sports ground: competition	Other death reasons: training	Other death reasons: competition
Team sports	17	10	191	232	104	39	28	4
median age	35.6	34.4	40.6	39.5	23.7	24.1	35.7	25.5
Individ. sports	251	20	353	211	115	26	17	1
median age	39.1	45.5	52.0	51.2	33.0	36.0	41.3	55.0

Male athletes	Traumatic death: training	Traumatic death: competition	Cardio-vascular death: training	Cardio-vascular death:competition	Traffic accident on the way to sports ground: training	Traffic accident on the way to sports ground: competition	Other death reasons: training	Other death reasons: competition
Team sports	16	10	190	232	97	39	25	4
median age	35.0	34.4	40.7	39.5	23.9	24.1	37.6	25.5
Individ. sports	226	18	340	207	105	22	16	1
median age	39.7	46.2	52.0	51.5	33.0	33.5	39.6	55.0

Female athletes	Traumatic death: training	Traumatic death: competition	Cardio-vascular death: training	Cardio-vascular death:competition	Traffic accident on the way to sports ground:training	Traffic accident on the way to sports ground: competition	Other death reasons:training	Other death reasons: competition
Team sports	1	0	1	0	7	0	3	0
median age	45.0	-	27	-	21.0	-	20.3	-
Individ. sports	25	2	13	4	10	4	1	0
median age	33.2	39.0	51.9	36.3	33.1	50.0	68.0	-

to the sports ground is less than 25 yr for team sports. For individual sports, the median age is about 10 years older. This contrasts with a much higher median age of death (about 50 yr) for cardiovascular causes of death during individual sports.

5. REFERENCES

1. Ciampricotti, R. et al. Clinical characteristics and coronary angiographic findings of patients with unstable angina, acute myocardial infarction, and survivors of sudden ischemic death occurring during and after sport. *Am. Heart J.* 123: 1267–1278, 1990.
2. *Der Spiegel* 5 (1.2): 47, 1993.
3. Jokl, E., and L. Meltzer. Acute fatal non-traumatic collapse during work and sport. *Med. Sport* 5: 5–18, 1971.
4. Jung, K., and W. Schäfer-Nolte. Todesfälle im Zusammenhang mit sportlicher Betätigung. *Therapiewoche* 33: 4048–4058, 1983.
5. Maron, B., S. Epstein, and W. Roberts. Causes of sudden death in competitive athletes. *JACC* 7 (1): 204–214, 1986.
6. Müller, M. *Nichttraumatische Sporttodesfälle.* Med. Diss. Universität Erlangen-Nürnberg.
7. Schmid, L., Z. Hornof, and J. Král. *Sportunfälle mit tödlichem Ausgang und Maßnahmen zu ihrer Verhütung.* Berlin: Verlag Volk und Gesundheit, 1962.
8. Virmani, R., M. Rabinowitz M, and H. McAllister. Non traumatic death in joggers. *Am. J. Med.* 72: 874–882, 1982.

PART 5. SPORTS-SPECIFIC LIMITATIONS TO EXERCISE IN HEALTH AND DISEASE

31

PHYSIOLOGICAL LIMITATIONS TO ENDURANCE EXERCISE

R. L. Hughson, H. J. Green, S. M. Phillips, and J. K. Shoemaker

Department of Kinesiology
University of Waterloo
Waterloo, Ontario, N2L 3G1 Canada

1. INTRODUCTION

Endurance performance is dependent on the coordinated responses of the cardiovascular and respiratory systems, muscle metabolism, mechanical efficiency, and thermoregulation. A number of reviews have focused on one, or several, aspects of these responses (1,7,18). Yet, one central tenet of optimizing endurance performance revolves around the efficient aerobic transformation of metabolic substrate into mechanical power output, with delayed depletion of the glycogen reserves (1,10). Thus, it is important to have an efficient oxygen transport system and a metabolic system that supplies appropriate substrates to the mitochondria for oxidative metabolism with minimal concurrent glycolysis, a concept called "tight coupling" of oxidative metabolism (14).

The metabolic conditions in the muscle during constant load submaximal exercise appear to be established early in the non-steady state phase. Linnarson et al. (21) showed that a greater O_2 deficit at a constant submaximal workload during hypoxia was associated with greater concentrations of muscle and blood lactate, and with greater muscle phosphocreatine (PCr) depletion than during normoxic or hyperoxic tests. DiPrampero and Margaria (9) found a progressive depletion of PCr as the absolute O_2 uptake ($\dot{V}O_2$) increased during electrically-induced exercise in dog muscle. Further, they proposed that this was a direct consequence of the increased magnitude of the oxygen deficit incurred at the onset of exercise (9). Thus, it seems likely that individuals who can adapt their oxidative metabolism more rapidly to the demands of the exercise task will benefit greatly later in exercise by establishing an intracellular environment consistent with tighter metabolic control (14).

The magnitude of the oxygen deficit is dependent on the increase in $\dot{V}O_2$ between the baseline work rate and steady state exercise ($\Delta\dot{V}O_2$), and on the rate at which $\dot{V}O_2$ increases at the onset of exercise (i.e. $\dot{V}O_2$ kinetics as given by the time constant, τ). This can be expressed in the relationship: Oxygen deficit = $\Delta\dot{V}O_2 \times \tau$ (29). This will result in an increase in oxygen deficit if the work rate is increased (9,29), or there will be a change in the magnitude of the oxygen deficit if the $\dot{V}O_2$ kinetics are accelerated or slowed. In a range of physiological conditions, we have shown that $\dot{V}O_2$ kinetics can be slowed. For example, with transitions from a mild to a moderate work rate (16), with hypoxia (22), or with beta-blockers (15), the kinetics of $\dot{V}O_2$ are slower compared with their control conditions. With supine exercise, the increase in $\dot{V}O_2$ is slower than in upright exercise, but this $\dot{V}O_2$ kinetic response can be accelerated by application of lower body negative pressure during the supine exercise transition (17). These experiments collectively suggest that $\dot{V}O_2$ kinetics at a constant work rate are determined by the ability of the O_2 transport system to deliver O_2 to the working muscles.

The purpose of this paper is to focus on the O_2 transport responses with exercise and training that might influence the $\dot{V}O_2$ kinetics in the transition from rest to exercise. A group of subjects took part in a training study consisting of 2 h/day of cycling at 60% $\dot{V}O_{2peak}$ that lasted up to 30 days. The data presented show the blood flow and the metabolic responses to a step increase in work rate. As the blood flow kinetics accelerated, so too did the $\dot{V}O_2$ kinetics. The potential interrelationships between O_2 supply at the onset of exercise and metabolism during the steady state of exercise are considered in light of measured changes in muscle oxidative capacity.

2. METHODS

A total of 12 healthy male subjects or subsets of this group, and 11 matched control subjects for the blood flow measurements, took part in this experiment after reading and signing a consent form approved by the Office of Human Research of the university. In the completion of this study, the subjects completed four different types of exercise tests: (a) Incremental exercise tests with work rate increased at 15 W/min for the purpose of measuring $\dot{V}O_{2peak}$. (b) A 90 minute exercise test at a work rate of 60% of the pre-training $\dot{V}O_{2peak}$ for measuring the metabolic response to constant load exercise. (c) Step transitions in work rate from a baseline of 25 W to a work rate equal to 60% pre-training $\dot{V}O_{2peak}$ in which the $\dot{V}O_2$ kinetics responses were observed. (d) Dynamic knee extension exercise raising and lowering a weight equivalent to about 15% maximal voluntary contraction while measurements were made of femoral artery blood flow velocity. The complete experimental details of these studies can be found elsewhere (24,26).

The $\dot{V}O_2$ kinetic responses were evaluated in the transitions between 25 W and the work rate equivalent to 60% pre-training $\dot{V}O_{2peak}$. Breath-by-breath $\dot{V}O_2$ data from four repetitions of the step changes in work rate were ensemble averaged and then fit by a two component exponential model in which the time constant of the second component (τ_2) has been suggested to reflect the muscle level time course of change in oxygen consumption (2), and where the time required to pass through 63% of the difference between baseline and steady state values is expressed by the mean response time (MRT) (20).

Blood flow velocity was obtained by pulsed Doppler measurements using a 4 MHz probe (model 500 V, Multigon Industries Inc., Mt. Vernon, NY) placed over the femoral artery 2–3 cm distal to the inguinal ligament. The time course of change in mean blood velocity (MBV) was evaluated from four repetitions of the step increase in work rate by measuring the time to achieve a 63% increase in flow from the baseline level to the steady state.

3. RESULTS AND DISCUSSION

By the completion of 30 days of training, there were significant increases in two indicators of improved endurance fitness VO_{2peak} and citrate synthase activity (Table 1). However, short-term training did not significantly affect either VO_{2peak} or citrate synthase activity as indicated on tests completed on days 5–8 of training. In spite of no effect on VO_{2peak} or muscle oxidative enzymes with short-term training, there were significant effects on muscle metabolism, VO_2 kinetics and blood flow velocity observed in tests conducted between days 4–10.

3.1. Short-Term Training Responses

The rate of increase in VO_2 at the onset of 60% VO_{2peak} exercise was faster at 4 and 9 days of training than it was prior to training as indicated by both the τ_2 and MRT (Table 1). Consistent with this faster adaptation in oxidative metabolism was a smaller reduction in muscle PCr, and a smaller increase in muscle and blood lactate concentration as measured at the 15 minute point during the 90 minute exercise test on day 5 of training. Also consistent with the faster VO_2 response was the observation of lower blood lactate concentrations noted as early as 2 minutes into the submaximal tests on days 4 and 9. By 6 minutes of constant load exercise, blood lactate had increased by about 2 mmol/l above baseline values in pre-training testing, but by only 1 mmol/l after 4–9 days of training. The data from the current series of experiments are consistent with those previously reported by Green and colleagues from biopsy samples taken after 15 minutes of exercise (3,12) as well as after 3 minutes of exercise (11). The metabolic profiles established in these early biopsy samples were maintained during experiments lasting up to 90 minutes. This supports the hypothesis that the adaptive phase might be critically important in determining the biochemical environment in the later phases of exercise.

Delivery of O_2 to the working muscles is primarily dependent on the adaptation of blood flow. Arterial O_2 content is normally well maintained although small decreases have been observed at the onset of vigorous exercise (23), and highly trained athletes often desaturate at maximal exercise (8). We observed a faster increase in mean blood flow velocity through the femoral artery after 10 days of training (MRT = 8.6±4.4 s) than in the pre-training tests (MRT = 14.2±7.6 s). In both cases, these responses appear to be faster

Table 1. Peak values of oxygen uptake, muscle citrate synthase (CS) activity and parameters for curve fitting describing the increase in VO_2 at the onset of exercise

Training Day	VO_{2peak} (l/min)	CS Activity (ml/kg protein/h)	VO_2-τ_2(s)	VO_2-MRT(s)
Pre-Training	3.52±0.20	3.7±0.3	37.2±4.8	38.1±2.6
4-5	-	3.9±0.3	28.8±5.1*	34.9±2.4*
8-9	3.55±0.20	-	22.9±2.7*	32.5±1.8*
30	3.89±0.18*†	5.7±0.3*†	15.8±1.3*†	28.3±1.0*†

Values are mean±SE (n=7). τ_2 is time constant of the second exponential component, and MRT is mean response time (see Methods).
*significantly different from Pre-training (P<<0.05), †significantly different from training days 4-5 and 8-9 (P<<0.05)

than those of $\dot{V}O_2$, but additional processes including blood flow heterogeneity and redistribution are probably involved in O_2 delivery.

3.2. Long-Term Training Responses

The significant increases in $\dot{V}O_{2peak}$ and muscle citrate synthase activity found after 30 days of 2 h/day endurance training are important in further improving the metabolic responses to exercise at 60% $\dot{V}O_{2peak}$. The rate of increase in $\dot{V}O_2$ was faster at 30 days than it was at days 4 and 9 as well as in pre-training testing (Table 1). Coincident with this was a further reduction in blood lactate so that there was no significant increase in blood lactate above the baseline value during the tests after 30 days of training. Of prime importance to the long-term endurance performance of these individuals were the changes measured in intramuscular metabolism. Muscle lactate concentration at 15 minutes of exercise was 29.8±3.6 mmol/kg dry weight pre-training, 16.8±2.3 after 5 days of training, and 12.3±2.4 after 30 days of training. There was also significantly less PCr depletion after short-term as well as long-term training. PCr concentration at the 15 minute time point during the 60% $\dot{V}O_{2peak}$ exercise was 46.6±2.9 mmol/kg dry weight pre-training, 56.2±3.6 after 5 days of training, and 62.7±2.9 after the 30 days of training. A direct consequence of this smaller depletion of PCr was a lower concentration of inorganic phosphate (Pi) on day 30 (58.4±5.2 mmol/kg dry weight) than in pre-training (82.3±3.3)and a lower calculated free AMP and ADP (5). The lactate/pyruvate ratio was reduced from pre-training (106±28) to 5 days of training (64±12) and 30 days of training (56±7). All of these factors taken together suggest a tighter coupling (3,5,11,14) of metabolic control in the trained state.

3.3. Training, Oxygen Delivery, and Metabolism

Prior to this series of investigations, it had always been noted that faster increases in $\dot{V}O_2$ at the onset of exercise were found in persons with a high $\dot{V}O_{2peak}$, or after at least 5–10 weeks of exercise training (4,13). Presumably in this latter case, there would have been an increase in muscle oxidative enzyme potential as a result of the training program (12). Observation of an increase in mitochondrial enzymes provides a logical mechanism by which the muscle phosphorylation potential can be maintained at a higher level during exercise at a fixed metabolic rate (3,14). Understanding of the metabolic adaptation at the onset of exercise is less complete. Recently, Hochachka and Matheson (14) have hypothesized a rapid recruitment of active enzymes to allow a match of metabolic ATP production rate to metabolic demand. In this scheme, it is possible that, especially at the onset of exercise, the activation of the enzyme catalytic sites could sustain more rapid increases in ATP production if appropriate concentrations of substrate were available. We have hypothesized that an important rate limiting step that sets the rate of increase in $\dot{V}O_2$, even during submaximal exercise such as at 60% $\dot{V}O_{2peak}$, could be the delivery of O_2 as a substrate to the working muscle (16,17), but other substrates could be rate limiting (2,29).

To date, the only available evidence for human muscle metabolism in the transition from rest to exercise comes from studies of relatively small muscle mass using magnetic resonance spectroscopy. The estimated time course of oxidative metabolism closely parallels that of $\dot{V}O_2$ measured at the mouth (2,30). This time course of increase in oxidative metabolism appears to be slower than the rate of increase in O_2 delivery. The quandary lies in attempting to explain the faster $\dot{V}O_2$ kinetics at the onset of exercise after only 4 or

9 days of training in the face of no measurable change in muscle mitochondrial enzyme potential.

The current data are the first to report a faster increase in blood flow following a period of endurance training in human subjects (26). Similar observations have recently been made in animal training studies. In these latter studies, the mechanism suggested to account for the more rapid vasodilation that permits the greater flow has been an endothelium derived relaxation factor, possibly nitric oxide (27). Whether this mechanism accounts for the faster kinetics of the blood flow response in our population of healthy young subjects after 10 days of 2 hours/day of endurance training, or is only involved in a steady state response, is not known.

There are other unknown factors involved in the control of blood flow and the delivery of O_2 at the onset of exercise. Whether blood flow is immediately distributed appropriately to the muscle fibres that are exercising is not known. Nor is the time course known for the process of O_2 extraction from this blood. Although it has recently been suggested that the entire hyperemic response in the first seconds of exercise can be accounted for by the muscle pump (25), this appears unlikely (6,19,28). The muscle pump hypothesis suggests that emptying of the venous vessels will increase the perfusion pressure gradient across the vascular bed (25)]. But, the rapid increase in blood flow even with limbs above the heart where venous pressure is negligible, argues for specific local vasodilation (6,19). Further, the response is graded to match metabolic demand (6,19) suggesting a local vasoactive metabolite produced in proportion to the exercise intensity. It will be important to reconcile this metabolite theory with the current training study in which production of metabolic by-products might be expected to be reduced with improved fitness, yet there was a more rapid increase in blood flow.

4. CONCLUSIONS

Endurance training at 65% $\dot{V}O_{2peak}$ for 2 hours/day for as little as 4–5 days resulted in significant change in a range of metabolic responses that will probably have major impact on the ability to exercise for several hours. The important message from these studies is that these metabolic changes are evident within the first minutes of exercise. This has recently been confirmed in biopsy samples taken after 3 minutes of constant load exercise (11). Thus, the faster $\dot{V}O_2$ kinetics that we observed after 4 days of training was associated with a tighter coupling of metabolic control as shown by less depletion of muscle PCr, and with smaller accumulations of lactate, creatine, and inorganic phosphate (3). Further, there is less depletion of muscle glycogen stores early in exercise (11,12), and this has the consequence of delaying glycogen depletion, a critical factor in fatigue with endurance exercise (1).

The mechanisms responsible for the quick adaptation of metabolism appear to be related to at least two different factors. The acceleration of oxidative metabolism found early in the training program is independent of significant changes in oxidative enzyme potential (12). This does not eliminate the possibility that some factor associated with the activation of enzymes might be involved. But, it suggests that the more rapid increase in muscle blood flow that we observed could be responsible for supplying more O_2 at the onset of exercise to permit a greater aerobic energy yield (16). The continued acceleration of oxidative metabolism noted at 30 days of training is probably related to the increase in mitochondrial enzyme activities.

Endurance exercise is normally classed as being several hours long, but the factors limiting performance appear to have made their mark on energy production within the first minutes of exercise.

5. ACKNOWLEDGMENTS

The original research reported in this paper has been supported by grants from the Natural Sciences and Engineering Research Council of Canada to RLH and HJG.

6. REFERENCES

1. Astrand, P. and K. Rodahl. *Textbook of Work Physiology*. New York: McGraw-Hill Book Co. 1985.
2. Barstow, T. J., S. Buchthal, S. Zanconato, and D. M. Cooper. Muscle energetics and pulmonary oxygen uptake kinetics during moderate exercise. *J. Appl. Physiol.* **77**: 1742–1749, 1994.
3. Cadefau, J., H. J. Green, R. Cussó, M. Ball-Burnett, and G. Jamieson. Coupling of muscle phosphorylation potential to glycolysis during work after short-term training. *J. Appl. Physiol.* **76**: 2586–2593, 1994.
4. Casaburi, R., T. W. Storer, I. Ben-Dov, and K. Wasserman. Effect of endurance training on possible determinants of VO_2 during heavy exercise. *J. Appl. Physiol.* **62**: 199–207, 1987.
5. Connett, R. J. Analysis of metabolic control: new insights using scaled creatine kinase model. *Am. J. Physiol.* **254**: R949-R959, 1988.
6. Corcondilas, A., G. T. Koroxenidis, and J. T. Shepherd. Effect of a brief contraction of forearm muscles on forearm blood flow. *J. Appl. Physiol.* **19**: 142–146, 1964.
7. Coyle, E. F. Integration of the physiological factors determining endurance performance ability. In: *Exercise and Sport Sciences Reviews*, edited by J. O. Holloszy. Baltimore: Williams and Wilkins, 1995, p. 25–63.
8. Dempsey, J. A., P. E. Hanson, and K. S. Henderson. Exercise induced arterial hypoxemia in healthy persons at sea level. *J. Physiol. (Lond)* **355**: 161–175, 1984.
9. di Prampero, P. E. and R. Margaria. Relationship between O_2 consumption, high energy phosphates and the kinetics of the O_2 debt in exercise. *Pflugers Arch.* **304**: 11–19, 1968.
10. Green, H. J. How important is endogenous muscle glycogen to fatigue in prolonged exercise. *Can. J. Physiol. Pharmacol.* **69**: 2971991.
11. Green, H. J., J. Cadefau, R. Cussó, M. Ball-Burnett, and G. Jamieson. Metabolic adaptations to short term training are expressed early in submaximal exercise. *Can. J. Physiol. Pharmacol.* **73**: 474–482, 1995.
12. Green, H. J., S. Jones, M. E. Ball-Burnett, D. Smith, J. Livesey, and B. W. Farrance. Early muscular and metabolic adaptations to prolonged exercise training in humans. *J. Appl. Physiol.* **70**: 2032–2038, 1991.
13. Hickson, R. C., H. A. Bomze, and J. O. Holloszy. Faster adjustment of O_2 uptake to the energy requirement of exercise in the trained state. *J. Appl. Physiol.* **44**: 877–881, 1978.
14. Hochachka, P. W. and G. O. Matheson. Regulating ATP turnover rates over broad dynamic work ranges in skeletal muscles. *J. Appl. Physiol.* **73**: 1697–1703, 1992.
15. Hughson, R. L. Alterations in the oxygen deficit-oxygen debt relationships with beta-adrenergic receptor blockade in man. *J. Physiol. (London)* **349**: 375–387, 1984.
16. Hughson, R. L. and M. A. Morrissey. Delayed kinetics of VO_2 in the transition from prior exercise. Evidence for O_2 transport limitation of VO_2 kinetics. A review. *Int. J. Sports Med.* **11**: 94–105, 1983.
17. Hughson, R. L., H. C. Xing, J. E. Cochrane, and G. C. Butler. Faster increase in oxygen uptake during supine exercise with lower body negative pressure. *J. Appl. Physiol.* **75**: 1962–1967, 1993.
18. Joyner, M. J. Physiological limiting factors and distance running: Influence of gender and age on record performances. In: *Exercise and Sport Sciences Reviews*, edited by J. O. Holloszy. Baltimore: Williams and Wilkins, 1993, p. 103–133.
19. Leyk, D., D. Eβfeld, K. Baum, and J. Stegemann. Early leg blood flow adjustment during dynamic foot plantarflexions in upright and supine body position. *Int. J. Sports Med.* **15**: 447–452, 1994.
20. Linnarsson, D. Dynamics of pulmonary gas exchange and heart rate changes at start and end of exercise. *Acta Physiol. Scand.* Suppl. **415**: 1–68, 1974.
21. Linnarsson, D., J. Karlsson, L. Fagraeus, and B. Saltin. Muscle metabolites and oxygen deficit with exercise in hypoxia and hyperoxia. *J. Appl. Physiol.* **36**: 399–402, 1974.

22. Murphy, P. C., L. A. Cuervo, and R. L. Hughson. Comparison of ramp and step exercise protocols during hypoxic exercise in man. *Cardiovasc. Res.* **23**: 825–832, 1989.

23. Oldenburg, F. A., D. W. McCormack, J. L. C. Morse, and N. L. Jones. A comparison of exercise responses in stairclimbing and cycling. *J. Appl. Physiol.* **46**: 510–516, 1979.

24. Phillips, S. M., H. J. Green, M. J. MacDonald, and R. L. Hughson. Progressive effect of endurance training on $\dot{V}O_2$ kinetics at the onset of submaximal exercise. *J. Appl. Physiol.* **79**: 1914–1920, 1995.

25. Sheriff, D. D., L. B. Rowell, and A. M. Scher. Is rapid rise in vascular conductance at onset of dynamic exercise due to muscle pump. *Am. J. Physiol. Heart Circ. Physiol.* **265**: H1227-H1234, 1993.

26. Shoemaker, J. K., S. M. Phillips, H. J. Green, and R. L. Hughson. Faster femoral artery blood velocity kinetics at the onset of exercise following training. *Cardiovasc. Res.* in press, 1996.

27. Sun, D., A. Huang, A. Koller, and G. Kaley. Short-term daily exercise activity enhances endothelial NO synthesis in skeletal muscle arterioles of rats. *J. Appl. Physiol.* **76**: 2241–2247, 1994.

28. Toska, K. and M. Eriksen. Peripheral vasoconstriction shortly after onset of moderate exercise in humans. *J. Appl. Physiol.* **77**: 1519–1525, 1994.

29. Whipp, B. J. and M. Mahler. Dynamics of pulmonary gas exchange during exercise. In: *Pulmonary Gas Exchange*, edited by J. B. West. New York: Academic, 1980, p. 33–96.

30. Yoshida, T. and H. Watari. ^{31}P-Nuclear magnetic resonance spectroscopy study of the time course of energy metabolism during exercise and recovery. *Eur. J. Appl. Physiol.* **66**: 494–499, 1993.

32

CARDIOPULMONARY AND METABOLIC RESPONSES TO UPPER BODY EXERCISE

Jürgen M. Steinacker

Abteilung Sport- und Leistungsmedizin
Medizinische Klinik und Poliklinik
Universität Ulm, 89070 Ulm, Germany

1. INTRODUCTION

Understanding of physiological responses to upper body exercise is important and meaningful. Muscles of the upper body are not only involved in many daily work routines at home and at work, but also in many kinds of sports (10,14,22,24,34). Upper body exercise can be used also for exercise testing in clinical conditions in which leg exercise is either limited or inconvenient due to amputations, peripheral arterial occlusive disease, arthritis and many other disabilities (2,12,32,36).

Much daily work is mainly of a static type; dynamic work of upper body muscles is more frequently involved in sports (17,22,24,30,36). Since dynamic exercise is preferred for testing cardiopulmonary and metabolic function, the present analysis is focused on this type of exercise.

The more common modes of dynamic upper body exercise are arm-cranking, canoeing and rowing with a fixed seat. While arm-cranking involves mainly arm and shoulder muscles, trunk muscles are additionally involved in rowing with a fixed seat or with canoeing (2,17,23,27,32). Rowing with a sliding seat and swimming are considered as typical whole-body exercises, which stress both upper body and leg muscles (6,16,24).

2. PHYSICAL PERFORMANCE AND METABOLISM

2.1. Maximal Performance

The maximal work rate and maximal oxygen uptake (VO_2max) of upper body exercise have been shown to be lower than that of cycling or of treadmill running (1,4,14,22,25). The reduced maximal performance is clearly related to the smaller mass of upper body muscle compared to leg muscles (22,25,31). Accordingly, it has been demon-

The Physiology and Pathophysiology of Exercise Tolerance
edited by Steinacker and Ward, Plenum Press, New York, 1996

strated that $\dot{V}O_2$ of upper body exercise is clearly related to the muscle mass involved (12).

2.2. Work Efficiency

The net work efficiency of arm-cranking, canoeing and rowing (approximately 18, 15 and 20–22%, respectively) is lower than the work efficiency of cycle ergometer exercise which is 23–25% (5,17,26,31). Therefore, a consistently higher oxygen uptake ($\dot{V}O_2$) has been shown at a given submaximal work rate for these types of upper body exercise compared to treadmill or cycle ergometer exercise (1,4,5,25).

The difference in work efficiency between upper body and leg exercise may be due to different mechanical constraints (22,30,36), but there are also physiological causes. It may be argued, for example, that arm cranking is an unusual work pattern, with arm muscles having different innervation patterns or different recruitment patterns of motor units (22). However, we have demonstrated: (a) a similar work efficiency for rowing in both trained oarsmen and cyclists (26), and (b) no significant difference in work efficiency of arm cranking between trained canoeists and controls (17; unpublished observations).

2.3. The Effects of Muscle Mass on the Recruitment of Muscle Fibres

Work efficiency can be influenced by recruitment of fibre types and by muscle mass. Generally, in normal subjects, upper body muscles have a higher percentage of fast-twitch fibres, less muscle capillarity and less muscle mass compared to leg muscles (12,15,21). At a given work rate (assuming the same work efficiency), an equal number of motor units must be recruited for leg and upper body work. However, the equal recruitment of motor units indicates that a higher percentage of the muscle fibre pool is recruited for the smaller arm muscles, and that more fast-twitch muscle fibres (which fatigue more rapidly than slow-twitch fibres) are recruited. This argument is supported by the observation that while the $\dot{V}O_2$ of arm-cranking was not significantly higher at low work rates compared to cycling, it was greater for work rates at which lactate accumulation started (7).

2.4. Metabolic Consequences

In normal subjects, the recruitment of a higher percentage of fast-twitch muscle fibres in arm exercise will lead to: (a) a higher dependency on carbohydrate metabolism, (b) increased lactic acid formation, and (c) a reduced net work efficiency (21). Accordingly, higher blood lactate concentrations have been found in all types of upper body exercise in non-specifically trained subjects compared to cycling or treadmill running (7,14,17,25,30).

In contrast, in upper-body trained subjects lactate accumulation does not differ much between upper body, leg or combined exercise (17,26,34). For example, maximal lactate concentration was not significantly different in trained canoeists performing cycling, arm-cranking and canoeing (averaging 10.8, 10.8 and 10.2 mmol l^{-1}, respectively) (17). This presumably reflects training adaptations of the upper body muscles (9,21). Accordingly, in highly trained rowers no clear difference was found in the percentages of fast- and slow-twitch fibres or the muscle capillarity between m. vastus lateralis and m. deltoideus (20).

The different muscle mass and the different recruitment of fast- and slow-twitch muscle fibres may be also responsible for the different $\dot{V}O_2$ kinetics in upper body exercise. The $\dot{V}O_2$ kinetics of upper body exercise in response to a step increase of work rate are significantly slower than those of cycle exercise at the same work rate, the same $\dot{V}O_2$ or lactate concentration (7,8). These delayed $\dot{V}O_2$ kinetics will increase dependency on anaerobic metabolism.

2.5. Combined Upper Body and Leg Exercise

The addition of upper body to leg exercise will increase maximal work rate but this will depend on the training status of the subjects (24). Subjects trained in canoeing, rowing or swimming gain more work rate and more $\dot{V}O_2$ from the addition of arm work to leg work (Δpower ~18% and $\Delta\dot{V}O_2$max ~6%) (24). We demonstrated in canoeists a 9.8% increase in $\dot{V}O_2$max during rowing compared to cycling, but $\dot{V}O_2$max was 14.6% lower in arm-cranking and 9.7% lower in canoe ergometer exercise compared to cycling (17). In well trained oarsmen, $\dot{V}O_2$max was reported to be ~3% higher in rowing compared to cycle or treadmill ergometer exercise (6,10,24,26). Again, these effects are linked to the muscle mass which is added to leg work by upper body exercise (4,22,25).

3. CARDIOCIRCULATORY RESPONSES

Since cardiac output is linked to $\dot{V}O_2$ regardless of exercise mode, the characteristics of cardiac output in upper body exercise are similar to those of $\dot{V}O_2$ found in leg exercise, as described above (1,8,19,31), and there was reportedly no difference in right atrial pressure for arm or leg exercise (5). Compared to leg exercise, heart rate is about 20% higher in arm exercise, and stroke volume is somewhat lower (1,5,18,19,33). In trained canoeists and rowers, no significant difference of heart rate was found in arm and leg exercise if the same posture was used (30,34).

The peripheral vascular resistance is higher in arm-cranking compared to leg work (1,5,31) which results in an approximately 20 mmHg higher mean arterial blood pressure in arm-cranking (1,5,19,31). Diastolic arterial pressure increases more than systolic pressure, which remains essentially in the same range for arm and leg work (1,31). If additional muscles are added to upper body work, peripheral vascular resistance decreases. The combination of leg and arm work also has either a similar or a lower vascular resistance and arterial blood pressure than does leg work alone (5,31) (Fig. 1). For example, we have demonstrated a lower systolic arterial pressure and the same diastolic arterial pressure in patients rowing with a fixed seat (i.e. using arm, shoulder and trunk muscles) compared to cycling (27,28). In trained oarsmen, rowing with a sliding seat resulted in lower diastolic and mean arterial pressures compared to cycling in the same position (11). The cardiovascular responses are closely linked to muscle mass, capillarity, the vascular reserve and the $\dot{V}O_2$ of the muscles recruited for the exercise (8), without systematic differences between leg and arm exercise (8,19).

4. VENTILATORY RESPONSES

4.1. Ventilatory Demands

Ventilation ($\dot{V}E$) is tightly coupled to carbon dioxide output ($\dot{V}CO_2$) in arm as well as in leg exercise (1,5,19,20,31). The similarity of the ventilatory response of leg and up-

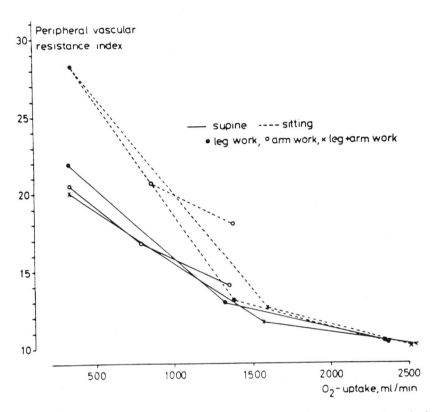

Figure 1. Peripheral vascular resistance index versus oxygen uptake in leg, arm and combined exercise in seated and supine positions. Reproduced from Ref. 5 with permission.

per body exercise was illustrated by Casaburi *et al.* who plotted $\dot{V}CO_2$ and $\dot{V}E$ versus lactate concentration (7). Interestingly, $\dot{V}CO_2$ and VE kinetics are faster than expected in arm exercise (7). It can be speculated that this is related to lower work efficiency and to a smaller protein buffering capacity of the arm.

4.2. Ventilatory Mechanics and Breathing Pattern

Arm exercise may affect breathing patterns at higher work loads, when trunk muscles become more involved in the exercise. During rowing, canoeing and swimming the respiratory musculature is confronted with dual demands: (a) assisting the force generation for the exercise and (b) being the effector for ventilatory control. As a result, the tidal volume achieved may be lower in upper body work. For rowing. tidal volume increased not only to the level at which the compliance curve becomes relatively flat, but increased further into this region implying an increased elastic work (29) (Fig. 2).

In this region of the compliance curve, rising ventilatory demands can be only met by increases in the frequency of breathing. Furthermore, a tight phase-coupling of ventilatory and locomotor rhythms was found during rowing (Fig. 3). All subjects entrained their breathing, so that during the stroke the subjects exhaled and inspired one time between the strokes (29). This entrainment of locomotor rhythm and breathing, together with relatively restrained tidal volumes, will result in two possible strategies to maintain increased venti-

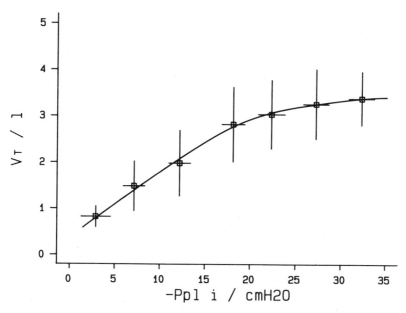

Figure 2. Tidal volume (VT) and inspiratory intrapleural pressure (P_{pl}) in rowing at different work rates in 5 subjects (means and standard deviations). Reproduced from Ref. 29 with permission.

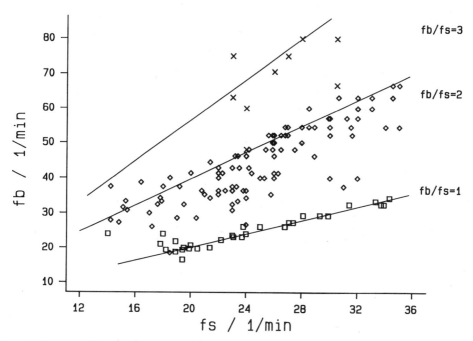

Figure 3. Response of breathing frequency (fb) as a function of stroke frequency (fs). The solid lines represent isopleths of constant fb:fs, fb/fs = 1 (□), 2 (◊), 3 (x). Adapted from Ref. 29.

latory demands: (a) increasing breathing frequency, which is relatively restrained to multiples of frequency of rowing, or (b) increasing the frequency of rowing (reverse entrainment). Both strategies have been demonstrated in ergometer rowing (29). A consequence of a lack of hyperventilation would be a failure to maintain alveolar PO_2 and impairment of CO_2 elimination (29).

5. CLINICAL UPPPER-BODY EXERCISE TESTING

Exercise stress testing is a well-established method for the diagnostic, prognostic and functional assessment of patients. In patients who are unable to perform sufficient leg exercise on a treadmill or cycle ergometer, upper body testing can be a useful alternative. The maximum work load achieved during arm work in patients with coronary heart disease (and without disabilities) was approx. 40% of that during leg work (3,13,23) and also peak $\dot{V}O_2$ was lower (13 versus 18 $ml \cdot min^{-1} kg^{-1}$) (2). Because of the limited work rate in arm-cranking, we have examined rowing with a fixed seat and found maximum work load to be only 10% lower than in cycling. This indicates an advantage of rowing with a fixed seat over arm-cranking for evaluation of overall metabolic capacity as a prognostic factor (27,28). In patients with peripheral arterial occlusive disease, we were able to demonstrate a significantly higher maximum work load (16) and $\dot{V}O_2$ in the rowing versus the cycle test (Fig. 4).

The cardiovascular responses to upper body exercise are not essentially different between patients with coronary heart disease and normals (2,3,13,27,28), although there is a tendency for higher systolic blood pressures in arm cranking (23). In most studies, the maximum rate-pressure product in patients does not differ between upper body and leg exercise (2,27,28); neither do indices of left ventricular wall stress (3).

There is reportedly a higher sensitivity of arm exercise for the detection of ischemia by both ST-depression and angina (23,36), which was achieved only by a higher rate-pressure product (23). Most studies show similar or slightly lower sensitivity and specifity for the detection of coronary heart disease when comparing arm exercise (arm-cranking, row-

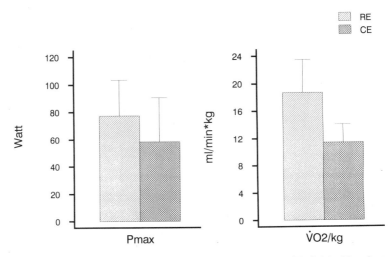

Figure 4. Maximal work load (Pmax) and maximal relative oxygen uptake ($\dot{V}O_2$/kg) in 34 patients with peripheral arterial occlusive disease during cycle (CE) and rowing ergometer (RE) exercise.

ing) with leg exercise (2,13,22,27,28,32). There seems not to be a different occurence of arhythmias (13,22,27,28). Therefore, upper body exercise testing does not confer a particular advantage over leg work for patients in general, but is clinically useful in patients with leg disabilities.

6. CONCLUSIONS

In comparison with leg exercise, upper body exercise is essentially an exercise in which smaller muscle groups are involved. As a result, the cardiopulmonary and metabolic responses to upper body exercise may differ more or less from those to leg exercise because of factors such as muscle mass, fibre type composition and training status. In non-specically trained individuals maximal physical performance, oxygen uptake and work efficiency are lower in upper body exercise compared with leg exercise, while at a given work rate $\dot{V}O_2$, lactate, $\dot{V}E$ and cardiac output are higher. These differences will be diminished in individuals who are, or become, well trained for upper body exercise.

In patients with peripheral arterial occlusive disease and with other disabilities of the legs, upper body testing is useful for clinical diagnostic and prognostic tests.

7. REFERENCES

1. Åstrand P.O., B. Ekblom, R. Messin, B. Saltin, and J. Stenberg. Intra-arterial blood pressure during exercise with different muscle groups. *J. Appl. Physiol.* **20**: 253–256, 1965.
2. Balady G.J., D.A. Weiner, C.H. McCabe, and T.J. Ryan. Value of arm exercise testing in detecting coronary artery disease. *Am. J. Cardiol.* **55**: 37–39, 1985.
3. Balady G.J., E.C. Schick Jr, D.A. Weiner, and T.J. Ryan. Comparison of determinants of myocardial oxygen consumption during arm and leg exercise in normal persons. *Am. J. Cardiol.* **57**: 1385–1387, 1986.
4. Bergh U., I.L. Kanstrup, and B. Ekblom. Maximal oxygen uptake during exercise with various combinations of arm and leg work. *J. Appl. Physiol.* **41**: 191–196, 1976.
5. Bevegård S., U. Freyschuss, and T. Strandell. Circulatory adaptation to arm and leg exercise in supine and sitting position. *J. Appl. Physiol.* **21**: 37–46, 1966.
6. Carey P., M. Stensland, and L.H. Hartley. Comparison of oxygen uptake during maximal work on the treadmill and the rowing ergometer. *Med. Sci. Sports* **6**: 101–103, 1974.
7. Casaburi R., T.J. Barstow, T. Robinson, and K. Wasserman. Dynamic and steady state ventilatory and gas exchange responses to arm exercise. *Med. Sci. Sports Exerc.* **24**: 1365–1374, 1992.
8. Cerretelli P., D. Shindell, D.P. Pendergast, P.E. di Prampero, and D.W. Rennie. Oxygen uptake transients at the onset and offset of arm and leg work. *Resp. Physiol.* **30**: 81.87, 1979.
9. Clausen J.P., K. Klausen, B. Rasmussen, and J. Trap-Jensen. Central and peripheral circulatory changes after training of the arms or legs. *Am. J. Physiol.* **225**: 675–682, 1973.
10. Cunningham D.A., P.B. Goode, and J.B. Critz. Cardiorespiratory response to exercise on a rowing and bicycle ergometer. *Med. Sci. Sports* **7**: 37–43, 1975.
11. Dörfler G., and J.M. Steinacker. Vergleich des Blutdruckverhaltens bei Fahrrad- und Ruderergometrie. Edited by J.M. Steinacker. Berlin: Springer, 1988, pp. 225–229.
12. Enders A.J., M Hopman, and R.A. Binkhorst. The relation between upper arm dimensions and maximal oxygen uptake during arm exercise. *Int. J. Sports Med.* **15**: 279–282, 1994.
13. Hanson P., M. Pease, H. Berkoff, W. Turnipseed, and D. Detmer. Arm exercise testing for coronary artery disease in patients with peripheral vascular disease. *Clin. Cardiol.* **11**: 70–74 1988.
14. Hollmann H., P. Schürch, H. Heck, H. Liesen, A. Mader, R. Rost, and W. Hollmann. Kardiopulmonale Reaktionen und aerob-anaerobe Schwelle bei verschiedenen Belastungsformen. *Dtsch. Z. Sportmed.* **38**: 154–156, 1987.
15. Johnson M.A., J. Polgar, D. Weightman, and D. Appleton. Data on the distribution of fibre types in thirty-six human muscles. An autopsy study. *J. Neurol. Sci.* **18**: 111–129, 1973.

16. Liu Y., J.M. Steinacker, and M. Stauch. Transcutaneous oxygen tension and Doppler ankle pressure during upper and lower body exercise in patients with peripheral arterial occlusive disease. *Angiology* **46**: 689–698, 1995.

17. Mattes T., W. Lormes, M. Grünert-Fuchs, J.M. Steinacker, and M. Stauch. Metabolische und kardiozirkulatorische Maximalwerte bei sportartspezifischer und unspezifischer Ergometerbelastung bei Kanuten. *Sportmedizin*, edited by K. Tittel, K.-H. Arndt, and W. Hollmann. Leipzig, Germany: Joh. Ambr. Barth, 1993, pp. 178–180.

18. Miles D.S., M.N. Sawka, D.E. Hanpeter, J.E.Foster, B.M.Doerr, and M.A.B.Frey. Central hemodynamics during progressive upper- and lower-body exercise and recovery. *J. Appl. Physiol.* **57**: 366–370, 1984.

19. Pendergast D.R.. Cardiovascular, respiratory, and metabolic responses to upper body exercise. *Med. Sci. Sports Exerc.* **21**: 121–125, 1989.

20. Roth W. Ergebnisse sportphysiologischer Studien zur Leistungsentwicklung ausgewählter Sportarten in den Jahren 1964 - 1978 und dem Profil leistungsbestimmender Merkmale sowie der muskelzellulären Grundlagen der spezifischen Leistungsfähigkeit in der Sportart Rudern. Dissertation B, Universität Greifswald, 1979.

21. Saltin B., J. Henrikson, E. Nygaard, and P. Anderson. Fiber types and metabolic potentials of skeletal muscles in sedentary man and endurance runners. *Ann. N.Y. Acad. Sci.* **301**: 3–29, 1977.

22. Sawka M.N. Physiology of upper body exercise. *Exerc. Sports Sci. Rev.* **14**: 175–211, 1986.

23. Schwade J., G. Blomqvist, and W. Shapiro. A comparison of the response to arm and leg work in patients with ischemic heart disease. *Am. Heart J.* **94**: 203–208, 1977.

24. Secher N.H.. The physiology of rowing. *J. Sports Sci.* **1**: 23–53, 1983.

25. Secher N.H., N. Ruberg-Larsen, R.A. Binkhorst, and F. Bonde-Petersen. Maximal oxygen uptake during arm cranking and combined arm plus leg exercise. *J. Appl. Physiol.* **36**: 515–518, 1974.

26. Steinacker J.M., T.R. Marx, U. Marx, and L. Lormes. Oxygen consumption and metabolic strain in rowing ergometer exercise. *Eur. J. Appl. Physiol.* **55**: 240–247, 1986.

27. Steinacker J.M., C. Hübner, and M. Stauch. Hämodynamik und metabolische Beanspruchung bei Ruderergometrie, Fahrrad- und Laufbandergometrie von ambulanten Patienten nach Herzinfarkt. *Herz/Kreisl.* **25**: 239 - 243, 1993.

28. Steinacker J.M., A. Berger, C. Hübner, and M. Stauch. Ruder- vs. Fahrradergometrie. Ein Vergleich in der Risiko-Abklärung vor nicht kardialen Eingriffen. *Münchn. Med. Wchschr.* **135**: 250–254, 1993.

29. Steinacker J.M., M. Both, and B.J. Whipp. Pulmonary mechanics and entrainment of respiration and stroke rate during rowing. *Int. J. Sports Med.* **14**: S15 - S19, 1993.

30. Steinacker J.M. Physiological aspects of training in rowing. *Int. J. Sports Med.* **14**: S3 - S10, 1993.

31. Stenberg, J., P.-O. Åstrand, B. Ekblom, J. Rocye, and B. Saltin. Hemodynamic response to work with different muscle groups, sitting and supine. *J. Appl. Physiol.* **22**: 61–70, 1967.

32. Stratmann H.G., and H.L. Kennedy. Evaluation of coronary artery disease in the patient unable to exercise: Alternatives to exercise stress testing. *Am. Heart J.* **117**: 1344–1365, 1989.

33. Toner M.M., M.N. Sawka, L. Levine, and K.B. Randolf. Cardiorespiratory responses to exercise distributed between the upper and lower body. *J. Appl. Physiol.* **54**: 1403–1407, 1983.

34. Vrijens J., P.Hoekstra, J.Boukaert, and P. van Trank. Effects of training on maximal working capacity and hemodynamic response during leg and arm exercise in a group of paddlers. *Eur. J. Appl. Physiol.* **34**: 113–119, 1975.

35. Wahren J., and S. Bygdeman. Onset of angina pectoris in relation to circulatory adaptation during arm and leg exercise. *Circulation* **44**: 432–441, 1971.

36. Wicks J.R., N.B. Oldridge, B.J. Cameron, and N.L. Jones. Arm-cranking and wheelchair ergometry in elite spinal cord-injured athletes. *Med. Sci. Sports Exerc.* **15**: 224–231, 1983.

33

ISOKINETIC PARAMETERS OF THE TRUNK AS AN INDICATOR OF PHYSICAL CAPACITY

M. Ferrari,[1] F. B. M. Ensink,[2] U. Steinmetz,[2] A. Straub,[2] and A. Krüger[3]

[1] Department of Cardiology and Pulmonology
[2] Department of Anesthesiology
[3] Institute of Sports, Georg-August-University
Goettingen
Germany

1. INTRODUCTION

The isokinetic concept of exercise was described by Hislope and Perrine more than 25 years ago (5). Since the advance of computer technology, isokinetic ergometers are now available for analyzing force continuously during exercise (6), and their reliability and validity has been widely explored (2,3).

Isokinetic trunk ergometry has been used to demonstrate and quantify the benefits of specific training and rehabilitation programs, especially in patients suffering from low back pain (8). However, valid data from individuals with no lumbar pain who participate in standardized physical training programs is not widely available (9). Also, it is not yet known whether nonspecific physical training also improves isokinetic force generation by the trunk musculature. Neither is it clear which parameters are most useful in characterizing the physical capacity of the lumbar spine - which may be defined as the ability to maximally activate a large number of motor units in a given interval of time. In the past, electromyographical analysis has been used to assess this "innervation" capacity. Because of the ability to adapt the work load to the joint angle during motion, isokinetic ergometry allows measurement of the time taken by a muscle group to achieve its maximum (peak) force. The increase of force per unit time during an isokinetic motion can therefore be used as an indicator of the innervation capacity.

Our intention was to test the speed of innervation as a means of distinguishing between trained and physically untrained subjects. We therefore performed a study in a group of experienced sportspersons and compared them to similar subjects who were physically untrained.

The Physiology and Pathophysiology of Exercise Tolerance
edited by Steinacker and Ward, Plenum Press, New York, 1996

2. METHODS

We measured isokinetic parameters for flexion (flex) and extension (ext) of the trunk on a LIDO Back™ Isokinetic System in 27 male recruits before (T1) and after (T2) their basic military training. Eight students served as a control group, and were tested twice over a 3-month period. The positioning of the subject positioning is shown in Figure 1.

The recruits were initially allotted to 2 groups: group 1 (G1, n=14) included those recruits who typically had done less than 2.5 h of sports per week in the past 3 years. Group 2 (G2, n=13) were physically trained recruits (>2.5 h of sports/week).

All 35 volunteers were required to perform 15 repetitions of trunk flexion and extension at a speed of 90°/s per test. The recorded force was digitally corrected for gravity. In the so-called gravity inertia mode, the weight of the upper half of the body was measured before the test. According to the angle of flexion during the exercise test the corresponding force values were added or subtracted (7). The range of motion was set to 80°. Figure 2 shows an original recording of isokinetic trunk ergometry.

All subjects were tested twice (T1, T2), before and after a period of 3 months. The students were asked to keep the amount of physical activity at a constant level for the intervening period. The recruits started their basic military service after test 1. During that time all recruits had to participate in approximately the same amount of physical activity (3). Student's t-test was used to test the differences before and after standardized physical training in each group.

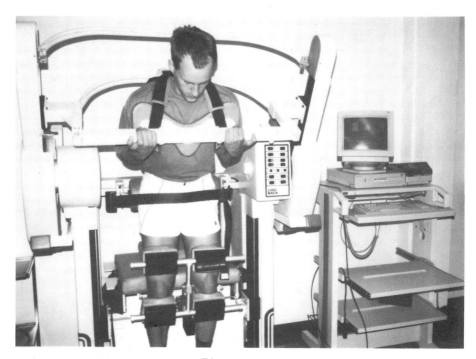

Figure 1. Subject positioning in the LIDO Back™ Isokinetic System. Stabilization was provided by a waist belt and by bolsters on the lower extremities. Data were digitally recorded and analyzed by a personal computer.

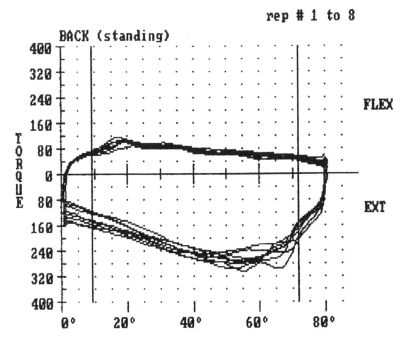

Figure 2. LIDO Back[TM] Isokinetic System. Original recording of isokinetic trunk flexion (Flex) and extension (Ext). Note the good reproducibility of the corresponding curves for all eight consecutive repetitions.

3. RESULTS

The physical data of the subjects are summarized in Table 1. There were no significant changes in body weight during the 3-month study period. All subjects were able to produce a force during flexion and extension over a range of motion of 80° in both tests.

As shown in Table 2, there were no significant changes in the isokinetic parameters among the students. The recruits, however, who were physically untrained (G1) showed an increase in peak torque and in work per repetition. The joint angle of peak torque during flexion of the trunk increased. In group 2, a decrease in peak torque during extension was seen. The joint angle where the initially well trained recruits were able to produce their peak torque did not change significantly, either in extension or in flexion. The work per repetition in this group increased only in flexion.

Table 1. Subject data (mean values ± standard deviation)

	Number	Age (years)	Height (cm)	Weight (kg)
Students	2 female, 6 male	22.4 +/- 2.2	179 +/- 8.8	72 +/- 10.1
Untrained recruits (G 1)	n = 14	20.6 +/- 1.4	180 +/- 6.4	76 +/- 14.5
Trained recruits (G 2)	n = 13	21.5 +/- 1.9	181 +/- 4.7	81.+/- 8.7

Table 2. Results of the isokinetic trunk ergometry (p-values calculated with Student's t-test, mean values ± standard deviation of 15 bends per test per subject)

		peak torque [Nm]			joint angle [dg]			work / repet. [J]		
		Test 1	Test 2	p	Test 1	Test 2	p	Test 1	Test 2	p
Students (n=8)	flex	139.7+/-25.6	142.1+/-23.8	n.s.	17.6+/-11.4	17.9+/-11.8	n.s.	109.4+/-33.0	107.5+/-30.9	n.s.
	ext	337.1+/-76.0	346.8+/-76.9	n.s.	57.9+/-5.8	58.3+/-6.1	n.s.	323.1+/-78.7	318.3+/-73.4	n.s.
G1 (n=14)	flex	119.5+/-37.2	170.9+/-52.6	<0.001	24.1+/-7.1	31.9+/-13.4	<0.05	117.2+/-42.7	149.3+/-43.7	<0.001
	ext	297.8+/-69.3	340.1+/-72.3	<0.01	53.7+/-7.4	53.6+/-5.0	n.s.	307.7+/-67.6	311.0+/-69.5	n.s.
G2 (n=13)	flex	169.8+/-27.9	175.7+/-30.2	n.s.	17.6+/-8.7	21.6+/-8.4	n.s.	146.9+/-30.8	171.2+/-35.0	<0.01
	ext	414.9+/-55.8	382.3+/-38.1	<0.01	55.8+/-4.9	54.9+/-3.9	n.s.	364.0+/-53.0	364.5+/-49.6	n.s.

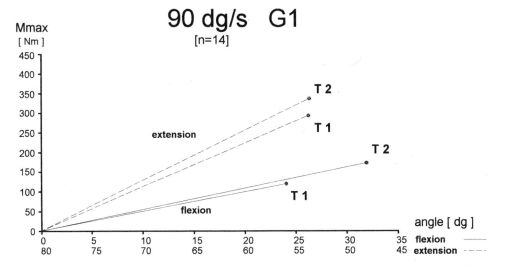

Figure 3. Peak torque (Mmax) to angle at peak torque during flexion and extension in physically untrained (G1) recruits before (T1) and after (T2) their basic military training.

4. DISCUSSION

The values for isokinetic parameters measured in this study did not change significantly in the student (i.e. control) group. As they had kept the amount of physical activity at a constant level throughout the 3-month period, this indicates good reliability for isokinetic trunk ergometry (as performed in this investigation) in healthy subjects.

In untrained recruits taking part in an essentially standardized physical exercise program for 3 months, the isokinetic force values increased whereas the joint angle at peak

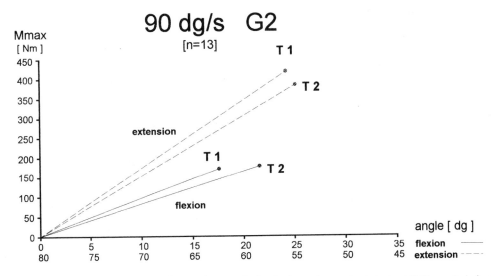

Figure 4. Peak torque (Mmax) to angle at peak torque during flexion and extension in trained (G2) recruits before (T1) and after (T2) their basic military training.

torque remained almost unaltered. The joint angle corresponds to the time when maximal force is produced. The ratio of peak torque (Mmax) to the joint angle of peak torque quantifies the individual speed of innervating a group of muscles. It can therefore be used to describe the innervation capacity. We assume that the innervation capacity did not change during the relatively short period of three months. This ratio did not change in any group.

Although the peak torque of trunk extension decreased in the initially trained recruits during their basic military service, most parameters remained relatively stable in this group. Differences in Mmax for trunk flexion and extension have been reported for athletes depending on their sporting disciplines (1). Among our subjects, Mmax values did not differ significantly. But the Mmax alone clearly cannot quantify the physical capacity (2). We assume that the ratio of Mmax to the angle of peak torque is a more valid indicator for distinguishing physically trained and untrained subjects, as shown in Figures 3 and 4. This ratio remained stable during a training period lasting only 3 months.

The ratio of peak torque divided by the joint angle of peak torque is a new parameter to describe the physical capacity. However, electromyographical validation of this parameter is still lacking. It therefore needs to be investigated further in the context of isokinetic ergometry. Particularly in rehabilitation programs for patients suffering from low back pain, we would expect a further opportunity for isokinetic trunk ergometry to monitor longer-term therapeutical benefits.

5. CONCLUSIONS

We conclude that isokinetic dynamometry of the trunk is a valid and reliable tool to demonstrate and quantify the benefits of physical training in healthy subjects, as well as in patients. The ratio of peak torque to the joint angle at peak torque is proposed to aid in the quantifying of lumber physical capacity. However, further investigations are required to evaluate the robustness of this new isokinetic parameter.

6. REFERENCES

1. Andersson, E, L. Sward, and A. Thorstensson. Trunk muscle strength in athletes. *Med. Sci. Sports Exerc.* 20:587–593, 1988.
2. Delitto, A., S.J. Rose, C.E. Crandell, and M.J. Strube. Reliability of isokinetic measurements of trunk muscle performance. *Spine* 16:800–803, 1991
3. Ferrari, M., F.B.M. Ensink, U. Steinmetz, A. Straub, P. Ahrens, H. Thegeder, and A. Krüger. Adaptation isokinetischer Kraftparameter in der Grundausbildung. *Wehrmed Monatsschr.* 35:42–49, 1994
4. Grabiner, M.D., and J.J. Jeziorowsk. Isokinetic trunk extension discriminates uninjured subjects from subjects with previous low back pain. *Clin. Biomech.* 7:195–200, 1992.
5. Hislop, H.J., and J.J.Perrine. The isokinetic concept of exercise. *Phys. Ther.* 47:114–117, 1967.
6. Jacobs, I., D.G. Bell, and J. Pope. Comparison of isokinetic and isoinertial lifting tests as predictors of maximal lifting capacity. *Eur. J. Appl. Physiol.* 57:146–153, 1988.
7. Malone, T.R. Lido System. *Sport Inj. Manag.* 1:62–76, 1988.
8. Mayer, T.G., R.J Gatchel, N. Kishino, J. Keeley, H. Mayer, P. Capra, and V. Mooney. A prospective short-term study of chronic low back pain patients utilizing novel objective functional measurement. *Pain* 25:53–68, 1986
9. Reid, S., R.G. Hazard, and J.W. Fenwick. Isokinetic trunk-strength deficits in people with and without low-back pain: a comparative study with consideration of effort. *J. Spinal Disord.* 4:68–72, 1991.

34

THE HORSERIDER'S SPINE DURING EXERCISE

Christine Heipertz-Hengst

Seminar für Therapeutisches Reiten
Orthopädische Uni.klinik, D 60528 Frankfurt/Main
Germany

1. INTRODUCTION

There is an extensive need for research into the sports-specific effects of horseback riding, particularly with regard to the rider's spine during exercise. To date, the reactions have been considered only in terms of empirical analysis and by practical experience. On the one hand, horseback riding offers benefits - including its uses in therapy; on the other hand, however, many riders complain of back pain. The purpose of this paper is to explore these issues in greater depth.

2. STRAIN OF HORSEBACKRIDING ON THE SPINE OF THE RIDER

2.1. Methods

2.1.1 Biomechanical Analysis. Electronic measurements of dynamic pressure distribution were made during riding on 80 riders equipped for synchronised high-frequency film analysis and measurement of oscillation. A measuring mat was integrated into the saddle. Working on a capacitative measuring principle using dielectric media (5), the device transmitted the force and impulse of the movement process to a transducer connected to a computer. This system allowed on-line analysis of the measured pressure data, providing coloured pictures, graphs and tables (Fig. 1a). Cinematographical recording of markers placed on both horse and rider was followed by high-speed film biomechanical analysis which allowed the display of vertical and horizontal curves of the movement process in a three-dimensional coordinate system (4) (Fig.1b and c). Oscillation measurements of the horseback (6) were carried out using piezoelectric transducers that indicated acceleration and extension at the C7 and sacral levels (Fig.1d). In brief, therefore: the movements of the horse cause specific impulses on the rider's spine; these are specific for each pace (walk, trot, canter) and speed; they are different in direction, frequency and amplitude; they can be defined as *vibration* and *motion* within three dimensions, in extreme circumstances enhanced to shock, compression, and bending.

The Physiology and Pathophysiology of Exercise Tolerance
edited by Steinacker and Ward, Plenum Press, New York, 1996

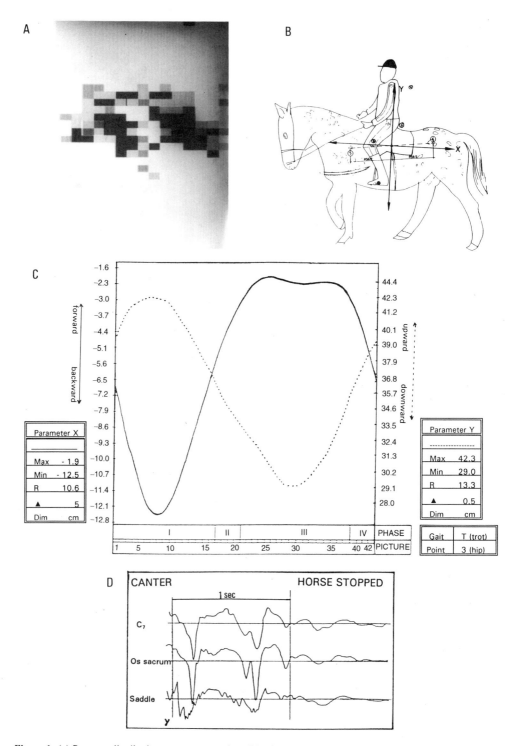

Figure 1. (a) Pressure distribution, output on monitor. (b) High-speed-film movement analysis with markers in a coordinate system. (c) Horizontal and vertical movement of the pelvis during trot. (d) Measurement of oscillation during canter (6).

2.1.2 Sports-Medical Examination. Both orthopaedic and neurological investigations and comparison with the results of therapeutic riding or "hippotherapy" show that the rider's seat on horseback is appropriate for the physiological demands of posture and neuromuscular function, especially with respect to reflexes and muscle tone.

A special study was undertaken to document health and fitness in a representative population (i. e. considering sex, age, skill-level, discipline) of 50 riders. Each completed a comprehensive sports-medical laboratory test that included: cycle ergometry, x-ray analysis if necessary, and twice completing a questionnaire evaluating sports activity. These subjects were drawn from a larger body of routine sports-medical examination of high-level riders (n = 370) between 1974 - 1991 from data of BISp Köln (1). The following results are relevant in this regard: 80% of professional riders suffer from hyperlordosis and statomotoric insufficiencies. It is striking at first glance that almost the same proportion (i.e. 65%) of "just for fun" riders show similar symptoms, whilst middle-level riders have the best spine-status: only 30% manifest vertebral problems. The frequency of back pain during riding is unexpectedly low, but it increases during "lay-off" periods. With regard to the three main disciplines of riding sport (dressage, jumping, eventing), there is an increasing trend of physical power of input (speed and impetus) according to the physical load.

3. TECHNIQUE OF HORSEBACK RIDING

3.1. The Seat - Common Practice

"The rider has to achieve an independent balanced seat, without any means of artificial support, tension, gripping up or stiffness. His aim should be to follow smoothly the horse's movement, keeping his centre of gravity in harmony with that of the horse. ... Only the relaxed rider can sit in a secure balance...There are three principal seat positions: The Dressage Seat, The Light Seat, The Forward (or Jumping) Seat. Incorrect seat positions are caused by tension either throughout or in one part of the body,... Chair seat, ...Split seat; grounds... rider ahead or behind the movement of the horse, rider too high above the horse." (Ref. 6, pp 47 - 56)

3.2. Results and "Scientific" Movement Theory

A seat that satisfies these basic demands is adequate to the needs of sitting on a horse regardless of the style or school of riding technique (e.g. classic english or german, western, south american, icelandic styles). From this starting position (Fig. 2a), the back of the horse moves up and down and the pelvis of the rider swings forward and backward while being tilted back, which represents an extension of the lordosis (Fig. 2b). An incorrect technique is caused by noncoincidence of the rider's actions to the phases of the movements of the horse and/or sitting against the corresponding of the two centres of gravity. The worst influence on the spine results from a backward movement of the rider, while tilting the pelvis forward at the highest point of the movement of the horse; this exaggerates the lordosis (Fig. 2c).

3.3. Consequences

Correctly performed riding in all disciplines provides positive reactions on the rider's spine: muscular stabilisation and strengthening; physiological, axial stimulus of the

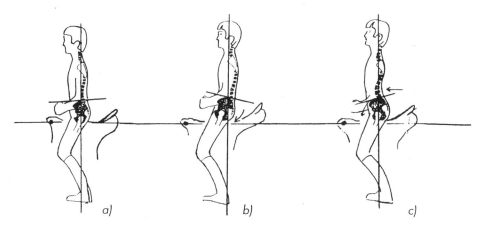

Figure 2. Rider's seat during movements of the horse: 2 a, b Normal technique. 2c Incorrect technique.

vertebral discs (by the rhythmic cycling of pressure and relaxation) and increase of motion-capacity and quality. Inappropriate techniques and an exaggerated quantity (dosage) may cause undue stress and disorders, and even injuries to the horserider's spine: in beginners with poor coordination, this can provoke an enlarged tonus with an inadequate force of muscle contraction; in top-level riders, however, these effects may result from the long-term duration, high density and unbalanced stress. It was surprising that most dressage riders had vertebral problems, and we were able to establish that this phenomenon was most likely caused by their habit of sitting with extremely deeply positioned knees and leaning backward to insure greater influence on the horse instead of "going with its movements" and therefore matching its rhythm.

4. TRAUMATOLOGY

4.1. Sports Damages

Chronic damage and micro-trauma caused by riding exercise are rare. Reflecting on the above mentioned studies, we may assume that the training methods, techniques and intensity will determine whether the effects of riding exercise cause benefits or problems for the horserider's spine.

4.2. Sports Injuries

Risk of injury to the horserider's spine is relatively high due to the physical forces that take place during exercise. For example, cantering at high speed (12 m/s or more) with the rider's seat-position being ca. 165 cm or more leads, in the event of a "crash-landing," to forces of more than 6–10 g - i.e. more than 3–20 times body weight! A fall from or with the horse may cause fractures of the vertebral spine: the literature indicates that this may account for 10.5 to 18.5 % of all riding injuries. The major and most feared complication of this is paraplegia. Two mechanisms during falling can be distinguished: a fall forward over the head of the horse causes injuries of the upper limbs, the head and of the cervical vertebrae; a fall backwards or sidewards causes injuries of the lower limbs and fractures of the vertebral column; and a fall with, or even under, the horse has a similar risk, together with the likelihood of multiple internal trauma and pelvic fracture (2).

4.3. Prophylactics

Appropriate exercise and training methods should incorporate warming up, stretching and gymnastic exercises, supplemented by endurance training to establish better fitness and co-ordination. Professional and top-level riders should interrupt the strain on their spine, and especially the vertebral discs, by twenty-minutes recovery while lying down. In addition, specific sports-medical and sports-physiotherapy assessment and, where necessary, treatment can avoid damage and injuries. There are clear indications and contra-indications for riding exercise, of which the most important are acute and inflammatory stadium and scoliosis of more than 15° (Cobb).

5. THERAPY AND REHABILITATION

The demonstrated positive effects and reactions to the movements of the horse during riding exercise can be used therapeutically in the treatment of orthopaedic and neurological diseases:

5.1. Hippotherapy (Therapeutic Riding)

This is a mode of neurophysiological treatment and also a special kind of physiotherapy.

5.2. Sports Therapy

This incorporates back and trunk training for health and injury prevention and rehabilitation (special programme).

5.3. Principles

These are based on the stimulus of the movements of the horse. This causes an "action-dialogue" between horse and rider, which leads to dynamic reactions and corrections of posture. It is a combination of stabilisation and mobilisation of active and passive movements with the various demands of coordination, perception and balance (Fig. 3).

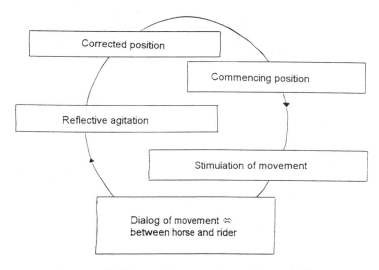

Figure 3. Dialogue of movement between horse and rider.

6. REFERENCES

1. Bundesinstitut für Sportwissenschaften. Data-bank. Köln, 1992.
2. Heipertz, W. Reiten - Sportartspezifische Traumatologie. In: GOTS Manual der Sporttraumatologie. Bern: Huber (In press), 1996.
3. Heipertz-Hengst, C. Das Wechselspiel zwischen Reiter- und Pferderücken in der Bewegung. Zusammen-rücken. Kirchheim: Schürer, 1994, pp 32 - 41.
4. Heipertz-Hengst, C. Analyse der Bewegungen beim therapeutischen Reiten. In: Sportmedizin, edited by K. Tittel, K-A. Arndt and W. Hollmann. Leipzig: Barth, 1993, p 320.
5. Nicol, K. and E. Hennig. A time dependent method for measuring force distribution using a flexible mat as a capacitor. In: Biomechanics V, edited by P. Komi. Baltimore, 1976.
6. Riede, D. Therapeutisches Reiten. München: Pflaum, 1986.
7. German National Equestrian. The Official Handbook. The Principles of Riding. London: Threshold Books, 1987.

35

CHANGES OF HORMONE VALUES DURING AN ULTRA LONG DISTANCE RUN

Christoph Raschka,[1] Regina Schuhmann,[2] Manfred Plath,[2] and Markus Parzeller[1]

[1] Med. Klinik II, Klinikum Fulda
[2] JLU Gießen
Germany

1. INTRODUCTION

The purpose of this study was to examine the hormonal responses of 55 participants (42 males, 13 females) in an ultra long distance run of 1000 km, which consisted of 20 daily runs of 50 km across Germany from Timmendorfer Strand in the north to Mittenwald in the south.

2. METHODS

All of the athletes were experienced long-distance runners (5). Their anthropometric, training and performance data are given in Table 1.

Blood samples were collected and examined before the run and on days 1, 3, 6, 8, 11 and 19 and tested by luminescence-immunoassay (T4, T3, TSH), enzyme-immunoassay (FSH, LH, prolactin) and radio-immunoassay (testosterone, aldosterone, estradiol, glucagon, insulin).

The nutritional intake was determined by means of the computer program diet 2000 and was quantified individually for each participant on every day of the run. The program is based on data of Souci, Fachmann, and Kraut (6). Analysis of variance was used for statistical analysis (5).

3. RESULTS

The testosterone values of both sexes fell after an initial rise, but the changes were not significant (Fig. 1).

Table 1. Anthropometric, training, and performance data

parameter	males		females	
	average	span	average	span
age (in years)				
height (in cm)	174.7	153.7 - 185.5	162.9	154.2 - 169.6
weight (in kg)	69.3	53.2 - 85.4	59.2	49.2 - 72.2
body fat percentage (%)	13.2	7 - 21.8	19.8	15.2 - 26
years of running	9	1 - 25	8	3 - 14
training: km/week	86	30 - 130	76	50 - 100
training: days/week	5	3 - 7	5	4 - 7
ergometry: max. watt	314	200 - 400	206	125 - 250
VO_2max (in l/min)	3.64	2.37 - 4.79	2.6	1.85 - 4
VO_2max (ml/min/kg)	52.9	40.1 - 70.5	41.9	33.6 - 56.6
average daily running time (min/day)	340.3	s = 68.8	390.9	s = 70.7
running time (in km/h)	9.1	s = 1.65	7.9	s = 1.33

(s = standard deviation)

A biphasic undulating decrease was evident for follicule-stimulating hormone (FSH) (Fig. 2): from 39.8 to 37.2 mIU/ml for women, and from 4.9 to 3.9 mIU/ml for men ($p<0.01$). The reaction of luteinizing hormone (LH) was similar (Fig. 3).

Prolactin (Fig. 4) attained a maximum on day 1 (increasing from 5.9 to 13.9 ng/ml) ($p<0.001$), and then subsequently fell although not completely back to control levels). Aldosterone responded similarly (Fig. 5), peaking at 360 pg/ml on day 1 compared to 118 pg/ml prior to the run.

Estradiol rose in the women from 21.1 to 49.8 pg/ml on day 6 ($p<0.001$) and then fell to 9.4 pg/ml by the end of the run (Fig. 6). The men showed a similar response profile: estradiol levels increased from 6.3 to 9.8 pg/ml (peaking, however, on day 3) ($p<0.001$) and then declined to 1.2 pg/ml at the end (Fig. 6).

There were striking increases in thyroid-stimulating hormone (TSH) levels (Fig. 7) from 0.99 mU/l to 1.5 mU/l both on day 3 and day 19 ($p<0.001$). The thyroxine concentra-

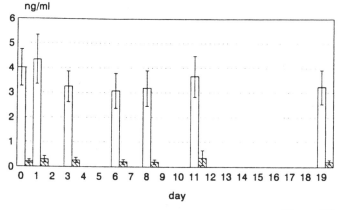

Figure 1. Testosterone responses.

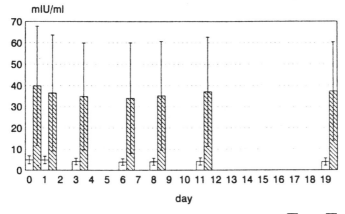

Figure 2. FSH responses.

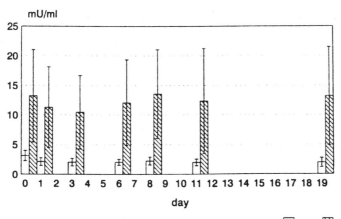

Figure 3. LH responses.

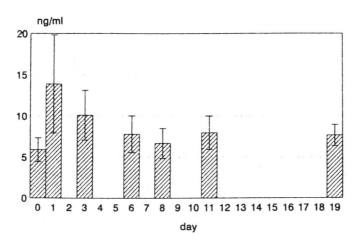

Figure 4. Prolactin responses.

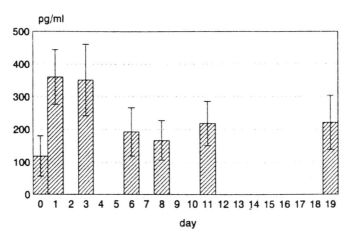

Figure 5. Aldosterone responses.

tions rose above control values (6.0 µg/dl) to 8.9 µg/dl on day 19 (p<0.001) (Fig. 8). Likewise, the fT4 concentrations (Fig. 9) rose from 1.47 ng/dl to 1.75 ng/dl. In contrast , the triiodothyronine levels showed a tendency to decrease from 1.6 ng/ml (day 0) to 1.2 ng/ml (day 19) (p<0.001) (Fig. 10).

The average daily energy intake percentage was 12% protein, 31% fat and 57% carbohydrates. The daily carbohydrate intake increased from 490 g/d in the first third of the run to 640 g/d at the end. The average daily energy intake of the men (women) was 4260 (3033) kcal/d or 17937 (12699) kJ/d, the average nutrient intake being 603 (432) g/d carbohydrates, 323 (248) g/d mono- and disaccharides and 248 (167) g/d polysaccharides. Post-exercise values of blood glucose were higher on the last day, the fructosamine test indicating lower blood glucose values throughout the whole run compared to control values.

The insulin concentrations of the men decreased from 9.6mU/l (day 1) to 5.4 mU/l (day 6) and 7.5 mU/l (day 19) (p<0.001) (Fig. 11).The glucagon levels increased from

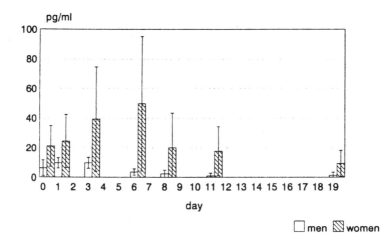

Figure 6. Estradiol responses.

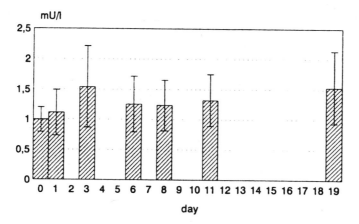

Figure 7. TSH responses.

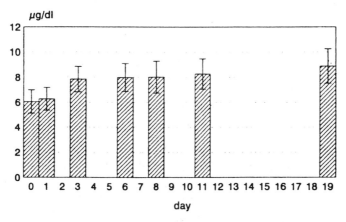

Figure 8. Thyroxine responses.

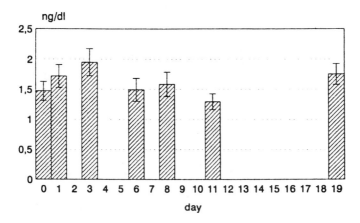

Figure 9. fT4 responses.

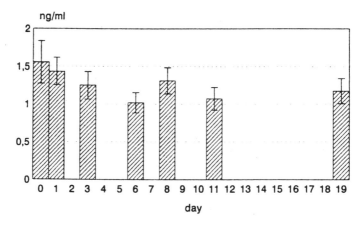

Figure 10. Triiodothyronine responses.

60.7 pg/ml to 160 pg/ml (p<0.001), but fell to 81.5 pg/ml (day 8) and 110 pg/ml (day 19) (Fig. 12).

The concentration of growth hormone (Fig. 13) showed similar developments; these were not significant, however. Cortisol rose from 14.2 µg/dl (day 0) to 37.8 µg/dl (day 19), afterwards consistently falling to 20.8 µg/dl (day 19) (Fig. 14).

4. DISCUSSION

Concerning the descriptors of carbohydrate metabolism, our results indicate slightly higher blood glucose levels in the second stage of the run. This only reflects the tendency of the exercise and post-exercise (1, 2, 4) phase, because the general development (indicated by the fructosamine test) shows falling and, from the point of view of preventive medicine, more favourable values in the second half of the run. This point is supported by

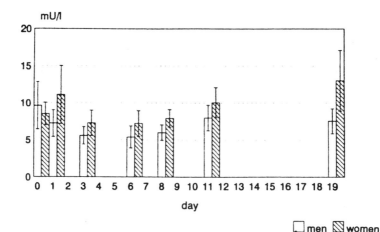

Figure 11. Insulin responses.

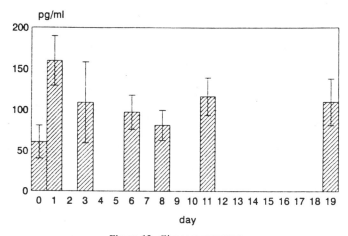

Figure 12. Glucagon responses.

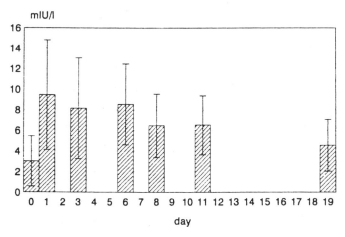

Figure 13. Growth hormone responses.

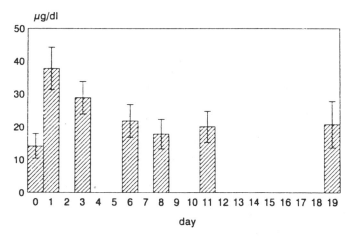

Figure 14. Cortisol responses.

the falling insulin values of the men, because insulin could be considered as an independent cardiovascular risk factor. The rise of insulin antagonists in the early stages of the run (cortisol > glucagon > growth hormone) is inversely proportional to the intake of carbohydrates. Therefore, it might be concluded that already, in the initial period of the run, an intake of carbohydrates in excess of 640 g/d (including ca. 50% mono- and disaccharides and ca. 40% polysaccharides) should be desirable and might be recommended. Ultra long distance running includes also remarkable and appreciable stimulation of thyroid metabolism because of accelerated metabolic functions (3). The observed changes of sexual and adrenal hormone values, however, underlines the conditioning of the endocrine system that occurred during the ultra long distance run as an expression of activation of the adrenocortical system and decrease of sex hormone activity during this long-term exhaustion (1, 2).

5. REFERENCES

1. Dessypris, A., K. Kuoppasalmi, and H. Adlercreutz. Plasma cortisol, testosterone, androstenedione and LH in a non-competitive marathon run. *J. Steroid Biochem.* **7**: 33–37, 1976.
2. Galbo, H. *Hormonal and Metabolic Adaptation to Exercise.* Stuttgart/New York: Thieme Verlag, 1983.
3. Johnannessen, A., C. Hagen, and H. Galbo. Prolactin, growth hormone, thyrotropin, 3.5.3'- triiodothyronine and thyroxine responses to exercise after fat and carbohydrate-enriched diet. *J. Clin. Endocrinol.* **52** (1): 56–61, 1981.
4. Keul, J., B. Kohler, G. von Gluth, U. Luethi, A. Berg, and H. Howald. Biochemical changes in a 100 km run. Carbohydrates, lipids and hormones in serum. *Eur. J. Appl. Physiol.* **47** (2): 181–189, 1981.
5. Raschka, C., and M. Plath. Das Körperfettkompartiment und seine Beziehungen zu Nahrungsaufnahme und klinsch-chemischen Parametern während einer extremen Ausdauerbelastung. *Schweiz. Z. Sportmed.* **40**: 13–25, 1992.
6. Souci, S. W., W. Fachmann, and H. Kraut. Die Zusammensetzung der Lebensmittel. Nährwerttabellen 1986/87. 3. Aufl., WVG, Stuttgart, 1986.

36

HUMAN MUSCLE FIBRE TYPES AND MECHANICAL EFFICIENCY DURING CYCLING

Anthony J. Sargeant [1,2] and Arno C. H. J. Rademaker [2]

[1] Neuromuscular Biology Group
Manchester Metropolitan University
Manchester, England
[2] Institute for Fundamental and Clinical Human Movement Sciences
Vrije University
Amsterdam, The Netherlands

1. INTRODUCTION

In prolonged high intensity exercise it seems advantageous in terms of resisting fatigue to choose relatively fast pedalling rates of ~100 rev/min. This advantage may be due to the greater reserve of power generating capacity at the higher pedalling rates, that is the difference between the power output required for the prolonged exercise and the maximum peak power measured at the same velocity. This is because the maximum power / velocity relationship in cycling exercise has been shown to be parabolic in form with an optimum velocity (Vopt) for maximum power occurring at a pedalling rate of around 120 rev/min (for discussion see ref. 9). The theoretical advantage of choosing fast pedalling rates for sustained high intensity exercise would however be negated if there was a disproportionate increase in the energy cost for the same external power output, that is if the mechanical efficiency decreased. Before considering this issue, however, it is interesting to see what trained cyclists actually do. We therefore asked a group of fit competitive cyclists to cycle around a very large flat indoor arena on their own bicycles at constant speed. Each cyclist performed a series of trials at constant speeds ranging from 20 to 47 km/hr. They were allowed to choose their own gear ratios. The data in Figure 1 shows that as the exercise intensity increased, expressed here as oxygen uptake ($\dot{V}O_2$), so did the freely chosen pedalling rate, an observation that is in agreement with previous reports (3, 5, 7).

It is also worth noting that the mean pedalling rate chosen by the world record holders for the maximum distance achieved in 1 hour is around 105 rev/min (9). So it appears that very fit cyclists performing sustained high intensity exercise do choose very high pedalling rates. This appears in conflict with many exercise physiology texts which state that the optimal pedalling rate for *maximum efficiency* is in the region of 60–70 revs/min. This

The Physiology and Pathophysiology of Exercise Tolerance
edited by Steinacker and Ward, Plenum Press, New York, 1996

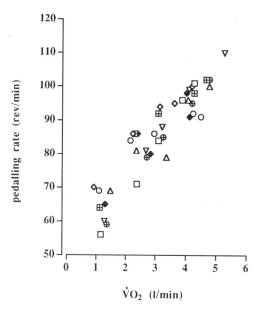

Figure 1. Data for 6 highly trained cyclists ($\dot{V}O_{2max}$ >5 l/min: different symbols for each cyclist) showing the relationship between the pedalling rate that they freely chose and exercise intensity ($\dot{V}O_2$) in a series of trials at constant speed ranging from 20 to 47 km/hr.

would suggest that either (a) the conventional texts regarding optimum pedalling rates for mechanical efficiency do not apply to these very fit cyclists, or (b) the advantage indicated at the beginning of this paper, that is, the relative greater resistance to fatigue consequent upon a greater reserve of power generating capacity outweighs the disadvantage of a reduced mechanical efficiency and the consequent increase in O_2 demand.

2. LABORATORY EXPERIMENTS

In a group of seven fit male subjects ($\dot{V}O_{2max}$ 5.4 ± 0.6 l/min) we studied the effect of five different pedalling rates (40, 60, 80, 100, and 120 rev/min) on the relationship between oxygen uptake and external power output in incremental tests on a cycle ergometer. In the range from 40 to 120 rev/min there was no significant difference in the maximum oxygen uptake attained nor was there a difference in the external power delivered to the ergometer at $\dot{V}O_{2max}$ in the range from 60 to 100 rev/min (Table 1). Thus the 'apparent' mechanical efficiency appears to be unchanged over this latter range, although this does not take account of energy generated from anaerobic metabolism; on the other hand, neither does it take into account the extra energy cost dissipated in simply moving the mass of the legs at faster rates.

In passing, it might be noted that at both the slowest and the fastest pedalling rates studied there was a decrease in the external power. At 120 rev/min we have previously argued that this may be due to the failure to direct leg forces optimally at this very fast pedal rate (1).

2.1. Peak Power in Cycling Compared to Maximum Power

During the multi-stage tests to determine maximum oxygen uptake, mean power (averaged over a complete crank revolution) was measured at the foot pedal interface by

Table 1. Mean ±SD values of external power delivered at $\dot{V}O_{2max}$ (PImax); $\dot{V}O_{2max}$; and PImax expressed as a percentage of the maximum power available at the same pedalling rate, measured on an isokinetic cycle ergometer (2)

pedalling rate (rpm)	40	60	80	100	120
PImax (W)	326±38	362±52	372±52	365±51	306±60
$\dot{V}O_{2max}$ (l/min)	5.08±0.50	5.30±0.73	5.34±0.61	5.40±0.63	5.28±0.78
PImax/max power (%)	55	45	38	35	29

means of strain gauges. In a separate series of tests in which the ergometer was switched to an isokinetic control mode we also measured the *maximum* peak power that subjects could generate at each pedalling rate. These data show how the percentage of maximum power utilised to achieve $\dot{V}O_{2max}$ systematically and significantly decreased as the pedalling rate increased: thus at 40 rev/min, 55% of maximum power is utilised compared to only 29% at 120 rev/min (difference p<.001; Table 1). At first sight this increase in the "reserve" of power generating capability at the faster velocities would seem to indicate that as a strategy to resist fatigue, fast pedalling rates should be chosen. This however takes no account of the relative recruitment and contribution of different fibre type populations to the power output of the whole muscle.

Unfortunately, in relation to human locomotion, little is known (a) about the relative efficiency of different muscle fibre types, or (b) about their power velocity relationships or (c) the degree to which activation frequency, so called "rate-coding", modulates the hierarchical recruitment of motor units in these large locomotory muscles (for a recent review see ref. 10).

3. A MODELLING APPROACH

On the basis of animal muscle experiments it seems reasonable to propose that the mechanical efficiency / contraction velocity relationship for human type I (slow) and type II (fast) muscle fibres would be of the general form shown in Figure 2 (see e.g. 4, 6, 8). Superimposed upon this general form is an indication of the velocities which may approximate to pedalling rates of 60 and 120 rev/min during cycling (see ref. 10). It will be noted that in this model there is a reciprocal change in efficiencies with a cross-over point, at which pedalling rate, in this case 90 rev/min, the efficiencies for the different fibre type populations will be the same. It can be imagined that, as a consequence, the overall efficiency for the leg extensors may change little over a broad range of pedalling rates in exercise intensities at which both fibre type populations are recruited.

On the basis of the earlier observations, we can draw a figure showing how a power output of 350 watts (that is around $\dot{V}O_{2max}$ for our fit subjects) may relate to the maximum power / pedalling rate relationship for the leg extensors in cycling (Figure 3). The possible contribution of the type I fibres to the total power is also shown. By reference back to Figure 2, we can now partition out the proportional contribution of type I and type II fibre populations and weight these according to their efficiencies at each pedalling rate. The outcome from this calculation is that the mechanical efficiency is remarkably constant across all pedalling rates.

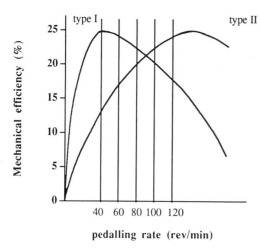

Figure 2. General form of the relationships between mechanical efficiency and velocity for human type I and type II muscle fibre populations. The velocity range equivalent to pedalling rates of 60 and 120 rev/min are derived from ref. 10. Note: (i) The relationships represent a mean since within each population there will be a continuum of properties including that for efficiency. (ii) In the absence of systematic data, no relative differences between maximum efficiencies are given; each population type is normalized to the same maximum.

It should be emphasised that this is a model based on a number of assumptions and generalizations. Nevertheless, it does illustrate why it may not be a disadvantage in terms of mechanical efficiency, and hence oxygen cost, to cycle at fast pedalling rates in prolonged high intensity exercise. There are however some disadvantages to be noted. At *low* exercise intensities the additional oxygen cost of moving the legs at fast pedalling rates will not be compensated by the smaller contribution to power output from the type II fibres which are more efficient at fast contraction velocities. Coordination and the optimal direction of force may also be impaired at very fast pedal rates, as discussed earlier.

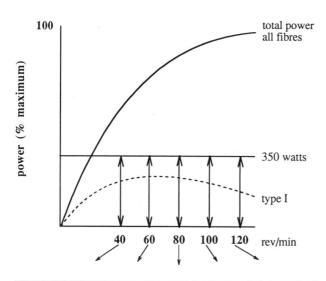

Figure 3. Schematic of the relationship of power to pedalling rate. The total maximum power (that is generated by all fibres) is given by the solid line. The maximum possible contribution from the type I fibre population is given by the dashed curve (assuming a 50% type I fibre type composition) and the level of power close to that to elicit $\dot{V}O_{2max}$ is also shown. In Table 1 the partitioning of the contributions from type I and type II fibre populations is given for 350 watts power output using the pedalling rate dependent efficiency taken from Figure 2.

Type I	247W at 24%	250W at 27%	247W at 23%	223W at 22%	176W at 18%
Type II	103W at 14%	100W at 18%	103W at 21%	127W at 23%	174W at 24%
ME	21%	22%	22%	22%	21%

On balance however it seems that, as one might expect, there are probably good physiological reasons why very fit competitive cyclists choose fast pedalling rates for sustained high intensity exercise.

4. REFERENCES

1. Bauer, J., A.C.H.J. Rademaker, J.A. Zoladz, and A.J. Sargeant. Is reduced mechanical efficiency at high pedalling rate due to less optimally directed leg forces. *Eur. J. Appl. Physiol.* **69 (3)**: S32 1994.
2. Beelen, A., A.J. Sargeant, and F. Wijkhuizen. Measurement of directional force and power during human submaximal and maximal isokinetic exercise. *Eur. J. Appl. Physiol.* **69**: 1–5, 1994.
3. Carnevale, T.J., and G.A. Gaesser. Effects of pedalling speeds on the power-duration relationship for high intensity exercise. *Med. Sci. Sports Exerc.* **24 (2)**: 242–246, 1991.
4. Goldspink, G. Energy turnover during contraction of different types of muscle. In: *Biomechanics VI*, edited by A. Asmussen and P. Jørgensen, Baltimore University Park Press, 1978, pp. 27–39.
5. Hagberg, J.M., J.P. Mullin, M.D. Giese, and E. Spitznagel. Effect of pedalling rate on submaximal exercise responses of competetive cyclists. *J. Appl. Physiol.* **51 (2)**: 447–451, 1981.
6. Lodder, M.A.N., A. de Haan, and A.J. Sargeant. Effect of shortening velocity on work output and energy cost during repeated contractions of the rat EDL muscle. *Eur. J. Appl. Physiol.* **62**: 430–435, 1991.
7. Pugh, L.G.C.E. The relation of oxygen intake and speed in competion cycling and comparative observations on the bicycle ergometer. *J. Physiol. (Lond.)* **241**: 795–808, 1974.
8. Rome, L. The design of the muscular system. In: Neuromuscular Fatigue, edited by A.J. Sargeant and D. Kernell. Amsterdam: Royal Netherlands Academy of Arts and Sciences - Elsevier/North-Holland, 1993, pp. 129–136.
9. Sargeant, A.J. Human power output and muscle fatigue. *Internat. J. Sports Med.* **15**: 116–121, 1994.
10. Sargeant, A.J., and D.A. Jones. The significance of motor unit variability in sustaining mechanical output of muscle. In: *Neural and Neuromuscular aspects of Muscle Fatigue.* Edited by S. Gandevia, R.M. Enoka, A.J. McComas, D.G. Stuart, and C.K. Thomas. New York: Plenum Press, 1995, pp. 323–338.

37

SLEEP APNEA AND CARDIOVASCULAR RISK FACTORS IN BODY BUILDING AND ABUSE OF ANABOLIC STEROIDS*

Nikolaus Netzer, Martin Huonker, Peter Werner, Hugo Prelicz, Josef Keul, and Heinrich Matthys

Section of Pulmonary Medicine and Section of Sports Medicine
Department of Medicine
University Hospital, Freiburg University
Freiburg, Germany

1. INTRODUCTION

From examination of nocturnal breathing pauses in the Sleep Laboratory, sleep apnea syndrome (SAS) can frequently be diagnosed. In at least 80% of such cases, obstructive SAS can be diagnosed, while for the remainder disorders of central neural respiratory control are likely to be the cause.

Normally, obstruction is caused by hypopharyngeal and oropharyngeal narrowing or by lipid deposits in the muscles of the surrounding soft tissues. This causal relationship was demonstrated in various investigations by Horner *et al.* (6) and has also been documented by means of magnetic resonance imaging (MRI) and computerized tomography (CT) (4,10,12)

An obstruction of the hypopharynx by a functional or structural disturbance of the exterior striated muscles (e.g. the sternocleidomastoid muscle) has not yet been described, to our knowledge, as a cause of nocturnal breathing pauses. We speculate, however, that this may well be the etiology of the nocturnal hypoventilation syndrome in the case that we present below.

2. CASE REPORT

The patient was a 32-year-old professional body-building athlete who had consulted a physician with what the patient perceived to be as symptoms of asthma with shortness of

---

* 1"Presented in part at the 1993 ALA/ATS Meeting in San Francisco, California, USA"

breath at night. Mild snoring and frequent apneas during sleep, as well as increasing day-time tiredness, were reported by a friend of the patient. There was no history of internal disease, and the patient was a non-smoker.

As a result of the pulmonologist's initial examination, allergic asthma was excluded as cause of the dyspnea. Screening for sleep apnea syndrome was done in the pulmonologists's office by means of the Mesam Box (this records four variables: arterial oxygen saturation, heart rate, breathing sounds, and body position). This suggested a mild-to-moderate sleep apnea syndrome with snoring, pauses in snoring, and oxygen desaturations to about 85% SaO_2. The apnea index, estimated from the desaturations and pauses in snoring, was 21 per hour. For further diagnostic clarification, the patient was referred to the Sleep Laboratory and to the Section of Sports Medicine.

The sports medical case history revealed the patient's adherence, at the time of the assessment, to a daily power training regimen of approximately four hours which included bench pressing (up to 2,060 kg) and leg pressing (up to 500 kg). In addition, the patient swam for one hour daily. He admitted to having taken considerable quantities of anabolic steroids for five years in connection with his body building. Since that time, he had become increasingly aware of shortness of breath, snoring with apneas and daytime somnolence. He took the following anabolic steroids regularly: methyltestosterone 100 mg per day; metenololacetate 50 mg per day; a depot preparation of testosterone 250 mg every other day; nortesterone 75 mg every fifth day; somatotropine 10 i.u. per week, and occasionally clenbuterol 7.5 mg.

Upon physical examination, we found an athletic man of 117 kg, 178 cm tall, with extremely hypertrophic skeletal muscles: the thighs and upper arms measured twice the normal circumference (i.e. 48 cm). At rest, his blood pressure was 135/85 mm Hg and his pulse rate was 72 per minute. Except for varicoses of both lower legs, the physical examination was unremarkable. Laboratory data confirmed excessive use of anabolic steroids with a testosterone level of 76.7 ng/ml. Although liver enzymes (SGOT, SGPT), lactate dehydrogenase (LDH) and creatine phosphokinase (CPK) were clearly elevated, the MB fraction of the CPK was normal. Total cholesterol was within normal limits, HDL was significantly depressed, and the LDL borderline was elevated.

Electrocardiography showed normal results and no signs of hypertrophy of the left ventricle (Sokolov-Lyon index) were found. A cardiac stress test performed to 250 watts induced no ECG anomalies or symptoms of angina. An appropriate increase in blood pressure with exercise was noted (210/90 mmHg at 250 watts).

The ENT examination disclosed no nasal or pharyngeal obstructions because of enlarged adenoids, and no nasal polyps or hyperplastic mucosa. During a Müller manoeuvre, however, significant narrowing of the hypopharynx was documented.

Pulmonary function testing at rest (body plethysmography), together with spirometry and airway resistance determination, produced normal values: FEV1 = 5.36 l (normal 4.24 l); IVC 5.56 l (5.31 l); FEV1/IVC 94.9% (81.4%); residual volume 2.53 l (1.81 l); specific airways resistance 0.79 kPa.sec (1.01 kPa.sec). Prick's test on the common environmental allergens did not disclose any positive findings. Likewise, the carbachol-challenge test for hyper-reactive airways was also negative.

Nine-channel polysomnography (Sidas 2010) was performed twice and it confirmed the suspected presence of a mild-to-moderate obstructive sleep apnea syndrome with an approximately equal amount of hypopneas and obstructive apneas, together with arterial oxygen saturation going down to almost 80% SaO2. These events occurred primarily in the REM phases of sleep, but also in non-REM sleep. No central apneas were observed. The sleep quality was slightly distorted because of increased arousals. The apnea index was 27 per hour (Fig. 1). During two control polysomnographies with nCPAP (Sullivan CPAP) at a pressure of 7 cm H20, no further obstructive apneas or hypopneas were noticed, with the arterial oxygen saturation remaining stable above 90% and the quality of sleep having returned to normal.

Computer tomography of the neck showed a significant hypertrophy of the neck muscles compared to those of normal males of the same age. Magnetic resonance imaging confirmed the clinically observed hypertrophy of the sternocleidomastoid muscle and the autochthonous muscles of the neck, although the images obtained were of inferior quality due to the difficulty of placing the patient's big neck and his head in the MRI tube (Fig. 2).

Two-dimensional echocardiography showed an absolute heart volume of 1,341 ml (Fig. 3), with an enlarged left ventricular end-diastolic diameter (EDD) of 64 mm (Fig. 4).

The thickness of the septal wall (SRT 13 mm) and that of the posterior wall (PWT 13 mm) were just within normal limits (Fig. 5) and did not support the presence of systolic or diastolic dysfunction. In relation to the patient's body weight, the two-dimensional echo- and cardiographic data (HV/body weight 11.4 ml/kg) were within normal ranges.

3. DISCUSSION

This case report clearly shows two probable, important impacts of body building on the physical health of the adherent. Firstly, there is a probability of inducing an obstructive sleep apnea syndrome because of obstruction of the hypopharynx by the hypertrophied peripharyngeal muscles of the neck. Secondly, massive intake of anabolic steroids may produce a higher risk for damage to the cardiovascular and respiratory systems as much as on the liver (via steroid-induced changes in lipids and liver enzymes). This developed obstructive sleep apnea syndrome or obstructive hypopnea syndrome resulting from massively hypertrophied muscles of the neck was not obvious at first glance, as the ob-

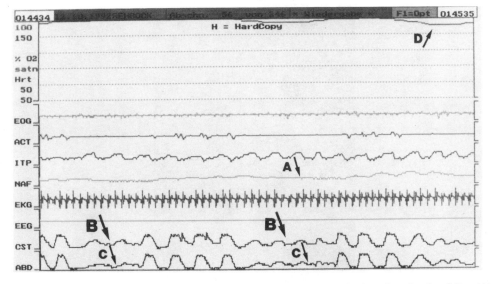

Figure 1. Nine-channel polysomnography shows previous hypopneas, with reduction of nasal and oral flow (A), chest (B) and abdominal (C) effort followed by a decrease of oxygen saturation (D).

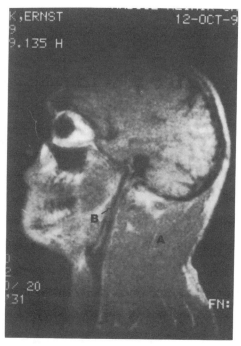

Figure 2. MRI in the sagittal plane shows enlargement of the neck muscles and slight obstruction of the hypopharynx (B).

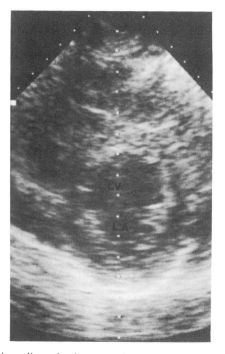

Figure 3. Two-dimensional echocardiography shows an enlarged end-diastolic diameter (EDD). (LV=left ventricle, LA=left atrium).

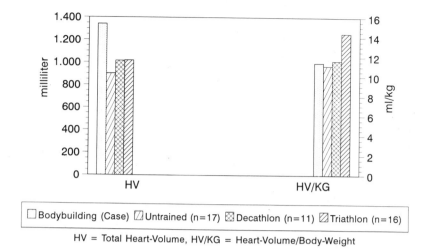

Figure 4. The body builder's heart volume in comparison with heart volumes of decathletes, triathletes, and untrained persons.

struction of the upper airways in this patient was not apparent during daytime. It was only on the account of the significantly decreased muscle tone during the night, particulary during REM sleep, that the aperture of the hypopharynx was narrowed remarkably, leading to the development of hypopneas and also to apneas (7,9,11)

This conclusion is derived from the coherence between the patient's history (matching of the onset of snoring and apneas with the intake of anabolic steroid) and the results of the ENT examination (obstruction of the pharynx, presumably reflecting the hypertrophied muscles of the neck, during a Müller manoeuvre), the CT and MRI images, and the findings of the polysomnography (primarily hypopneas during REM sleep) (5).

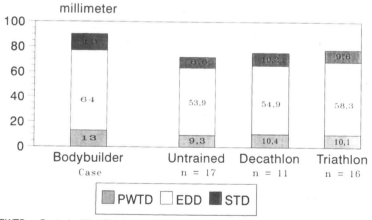

Figure 5. The body builder's left ventricular dimensions in comparison with those of decathletes, triathletes and untrained persons.

Of great interest to us was the similarity of this patient to three other body builders who likewise had a history of sleep apnea syndrome. One was screened positive for that syndrome by means of the Mesam Box, and another had a significantly abnormal polysomnography. However, only one of those three other body builders admitted to taking anabolic steroids; unfortunately, this individual refused to be screened or examined for the presence of sleep apnea syndrome. The significant alterations of the lipid profile, as well as elevation of the liver-enzymes levels, point towards anabolic-induced liver dysfunction and abnormal lipometabolism in this patient. On the account of the observed shift of the cholesterol fraction to a significantly decreased HDL cholesterol, this body builder's risk for arterial sclerosis is substantial. This has been observed by several other investigators who studied body builders (1,14,15).

The echocardiographic data relating to the absolute values of these body builders were significantly elevated compared to normal persons, and also power-trained (decathlon) and endurance-trained athletes (triathlon) (3,13). While their echocardiographic data are within normal ranges (taking body weight into account), even the so-called hypertrophy index (PWT + ST/EDD) for the patient presented here is significantly elevated compared to that of endurance-trained triathletes. The observed echocardiographic anomalies of our patient have not yet resulted in left ventricular dysfunction, but there appears to be a significantly elevated risk for mild cardiac ischemia which is further increased on the account of the existing nocturnal hypoxemias and a raised risk of arterial sclerosis.

4. CONCLUSIONS

We conclude that massive hypertrophy of the muscles of the neck resulting from power training and the simultaneous use of anabolic steroids can lead to the development of an obstructive sleep apnea syndrome. This syndrome, in conjunction with other side-effects of excessive use of anabolic steroids - such as extreme power training, developing cardiomegaly and decline in LDH cholesterol - increases significantly the risk of myocardial infarctions and strokes in body builders (8). As demonstrated recently by Durant *et al.* (2), 4.2% of all American adolescents use anabolic steroids at one point of their lives in order to encourage the development of a greater muscular mass. The resulting negative consequences for health may be of a much higher magnitude than previously appreciated.

5. REFERENCES

1. Ansell, J.E., C. Tiarks, and V.K. Fairchild. Coagulation abnormalities associated with the use of anabolic steroids. *Am. Heart J.* **125**:367–371, 1993.
2. DuRant R.H., V.J. Rickert, C. Ashwoth Seymore, C. Newman, and G. Slavens. Use of multiple drugs among adolescents who use anabolic steroids. *N. Engl. J. Med.* **328**:922–925, 1993.
3. Fellmann, N., M. Sagnol, M. Bedu, G. Falgairette G, E. Van Prgah, G. Gaillard, P. Jouanel, and J. Couderte. Enzymatic and hormonal responses following a 24 h endurance run and a 10 h triathlon race. *Eur. J. Appl. Physiol.* **57**:545–553, 1988.
4. Green, D.E., A.J. Block, N. Abbey, and D. Hellard. MRI measurement of pharyngeal volumes in asymptomatic snorers compared with nonsnoring volunteers. *Amer. Rev. Respir. Dis.* **139**:A373, 1989.
5. Hoffstein, V., R. Chaban, P. Cole, and I. Rubinstein. Snoring and upper airway properties. *Chest* **94**:87–89, 1988.
6. Horner, R.L., R.H. Mohaddin, D.G. Lowell, S.A. Shea, E.D. Burman, B.B. Longmore, and A. Guz. Sites and sizes of fat deposits around the pharynx in obese patients with obstructive sleep apnea and weight matched controls. *Eur. Respir. J.* **2**:613–622, 1989.

7. Issa, F.G., and C.E. Sullivan. Upper airways closing pressures in obstructive apnea. *J. Appl. Physiol.* **57**:520–527, 1984.

8. Kennedy, M.C., and C. Lawrence. Anabolic steroid abuse and cardiac death. *Med. J. Austral.* **158** (5):346–348, 1993.

9. Lopes, J.M., E. Tabachnik, N.C. Muller, H. Levison, and A.C. Bryan. Total airway resistance and respiratory muscle activity during sleep. *J. Appl. Physiol.* **54**:773–777, 1989.

10. Lowe, A.A., and M. Gionhaku, K. Takeuchi, and Y.A. Fleetham. Threedimensional CT reconstruction of tongue and airway in adult subjects with obstructive sleep apnea. *Am. J. Orthod. Dentofacial Orthop.* **90**:364–374, 1986.

11. Remmers, J.E., W.J. Degroot, and E.K. Sauerland. Pathogenesis of upper airwy occlusion during sleep. *J. Appl. Physiol.* **44**:931–938, 1978.

12. Rirkin, J., V. Hoffstein, J. Kalbfleisch, W. McNicholas, N. Zamel, and A.C. Bryan. Upper airway morphology in patients with idiopathic obstructive sleep apnea. *Am. Rev. Resp. Dis.* **129**:353–360, 1984.

13. Urhausen, A., and W. Kindermann. Echokardiographische Befunde bei Bodybuildern im Vergleich zu Hochausdauertrainierten. Krafttrainings- oder Anabolicaeffekt? *D. Zeitschr. Sport-Med.* **41** (1):12–18, 1990.

14. van Rensburg, J.P., A.J. Kielblock, and A. van der Linde. Physiologic and biochemical changes during a triathlon competition. *Int. J. Sports Med.* **7**:30–35, 1986.

15. Zuliani, U., B. Bernardini, A. Catapano, M. Campana, G. Ceriolo, and M. Spattini. Effects of anabolic steroids, testosterone, and HGH on blood lipids and echocardiographic parameters in body builders. *Int. J. Sports Med.* **10**:62–66, 1988.

38

LACTATE CONCENTRATIONS IN DIFFERENT BLOOD COMPARTMENTS AFTER 6-MIN MAXIMAL EXERCISE IN WELL TRAINED ROWERS

W. Lormes, J. M. Steinacker, and M. Stauch

Abt. Sport- u Leistungsmedizin
Universitätsklinik Ulm, D-89075 Ulm
Germany

1. INTRODUCTION

In exercise tests, lactate (La) concentration is usually measured in whole blood (WB), which is hemolyzed or deproteinated. This approach is only valid for steady state conditions. However, it was demonstrated that in incremental tests (1,3,5,7,9) or in repeated 30 s tests (5,9) lactate concentrations rose faster in plasma (Pl) than in erythrocytes (Ery). After 4 min of maximal cycling (4) an increase of La(Ery) was demonstrated up to 15 min after end of work. Therefore it is of interest to study lactate concentrations after a longer lasting test in well trained subjects.

2. METHODS

Twelve males (17.7 (17.4–18.3) years, 87.0 (77.0–95.0) kg, 192.0 (184–201) cm) and 6 females (17.8 (17.1–18.4) years, 69.1 (63.5–80.0) kg, 180.0 (170–183) cm) performed a 6-min maximal test on a rowing ergometer (ConceptII, Morrisville, USA) (8).

Two capillary blood samples for La were taken from the hyperemized earlobe, one of which was deproteinated and La(WB) was measured photometrically (Test Combination Lactat, Boehringer, Mannheim, Germany), the other remained untreated and La(UB) was measured immediatedly after sampling electrochemically (Lactate Analyzer YSI 27, Yellow Springs Instrument, USA) (3,6,9). Samples were taken before exercise, and 0, 2, 5, 10 (n=18) and 20 (n=8) min after exercise (recovery). Hematocrit (Hct) (5), blood gases (ABL 330, Radiometer, Copenhagen, Denmark) and hemoglobin (Hb) (OSM3 Hemoximeter, Radiometer) were analyzed in capillary samples from the other earlobe. Calculated were [La(Pl)] and [La(Ery)] (1,7).

The Physiology and Pathophysiology of Exercise Tolerance
edited by Steinacker and Ward, Plenum Press, New York, 1996

$$[La(Pl)]=[La(UB)]/(1-Hct) \tag{1}$$

$$[La(Ery)]=([La(WB)]-[La(UB)])/Hct \tag{2}$$

Results were given as median (minimum-maximum). Differences were estimated with Wilcoxon's signed rank test and assumed to be significant if $p \le 0.05$.

3. RESULTS

Power output was 449.0 (423.7–493.6) W for males and 313.7 (299.6–344.9) W for females, respectively. La(WB) increased 12.5 fold from 1.2 mmol/l before exercise to 15.0 mmol/l immediately after exercise, La(UB) increased 18 fold, La(Pl) increased 19.5 fold from 1.3 mmol/l to 24.4 mmol/l. La(Ery) increased only 5 fold. Peak La(WB) was found after 5 min of recovery, while peak La(UB) and La(Pl) appeared immediately after exercise. La concentrations before and after exercise are shown in Fig. 1.

La(WB), La(Pl) and La(Ery) did not change significantly in the first 5 minutes of recovery. The decrease of La between 5 min and 10 min of recovery ranged between 8% (Pl) and 10.5% (Ery). There was no correlation between La(Ery) and La(Pl) (r=0.31). The difference between La(Pl) and La(Ery) (Pl-Ery gradient) was -0.1 (-2.0–0.7) mmol/l before exercise and increased to 17.2 (7.5–22.7) mmol/l after exercise, and decreased to 9.7

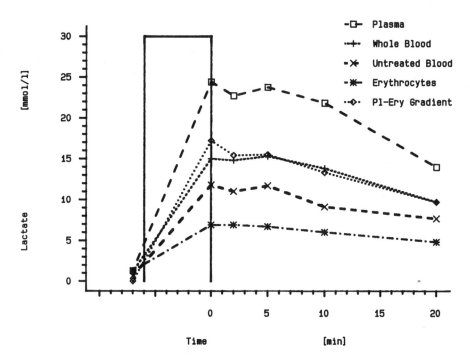

Figure 1. Lactate concentration in plasma, whole blood, untreated blood and erythrocytes and plasma to erythrocytes lactate gradient before exercise and in recovery (medians).

Table 1. Lactate concentrations in whole blood La(WB), in plasma La(Pl), and in erythrocytes La(Ery) before and after 6-min maximal exercise. Median (minimum - maximum). Bolding marks extreme values of every parameter. (*: p<0,05; **: p<0,005)

	La(WB) [mmol/l]	La(Pl) [mmol/l]	La(Ery) [mmol/l]
before Ex.	1.2 (0.5-1.6) **	1.25 (0.7-2.9) **	1.3 (0.2-2.7) **
0' after Ex.	15.0 (11.0-19.0)	**24.4 (16.9-28.6)**	**6.9 (2.4-13.4)**
2'	14.8 (12.2-19.8)	22.7 (18.6-28.5)	**6.9 (3.7-13.5)**
5'	**15.3 (10.3-18.6) **	23.75 (17.3-30.2) **	6.7 (1.6-12.4)
10'	13.6 (9.9-18.2) *	21.8 (13.8-28.1) *	6.0 (2.3-11.6)
20'	9.7 (4.9-14.4)	13.95 (6.7-20.5)	4.8 (2.2-8.7)

(4.0–12.4) mmol/l by 20 minutes of recovery. A correlation was found between Pl-Ery gradient and La(Pl) in recovery: Pl-Ery gradient = 0.81·La(Pl)+0.44, n=78, r=0.82.

The La(Ery)/La(Pl) ratio decreased from 1.1 (0.25–3.86) before exercise to 0.26 (0.10–0.60) after exercise and increased then to 0.4 (0.18–0.44) at 20 min of recovery. There was no correlation between La(Ery)/La(Pl) ratio and La(Pl) (r=-0.21). pH decreased from 7.42 (7.1–7.47) to 7.09 (7.00–7.20) after exercise, with a minimum of 7.06 (6.91–7.15) at 5 min of recovery and increasing to 7.20 (7.04–7.34) at 20 min. The values of Hb, Hct, CO_2-tension and bicarbonate concentration are shown in Table 2.

Hb and Hct did not change in parallel. Peak Hb appeared immediately after exercise, while peak Hct occurred at 5 min of recovery. Both parameters reached their pre-exercise level at 20 min of recovery. Mean corpuscular volume (MCV) did not change during exercise but increased significantly (p=0.02) by 2.6% in the first two minutes of recovery.

4. DISCUSSION

Changes in lactate concentrations in plasma and erythrocytes after an exercise bout of such high intensity as demonstrated appear not to have been reported until now. Bicarbonate buffer capacity was nearly exhausted and pH stayed at very low levels for over 10 min of recovery. La(Ery) was nearly constant during recovery and did not increase after exercise; however, La(Ery) was lower than reported in other studies with shorter exercise duration. In these previous studies, La(Ery) increased further during the initial 5 min of recovery and decreased then significantly (5,6,9). Here, La(Pl) was also very constant dur-

Table 2. Hemoglobin (Hb), hematocrit (Hct), CO_2-tension (pCO_2) and bicarbonate (HCO_3^-) before and after 6-min maximal exercise. Median (minimum - maximum). Bolding marks extreme values of every parameter. (*: p<0,05; **: p<0,005)

	Hb [mg/dl]	Hct [%]	pCO_2 [mmHg]	(HCO_3^-) [mmol/l]
before Ex.	16.3 (14.6-20.0) **	46.4 (42-56.5) **	38.25 (31.8-41.6) **	24.4 (11.3-25.8) **
0' after Ex.	**18.1 (15.7-20.4) **	50.0 (45-59.5)	34.1 (26.2-40.8) **	9.60 (7.9-14.2) **
2'	17.6 (14.7-19.9)	50.1 (44-57)	28.55 (5.6-33.1) *	7.75 (6.5-9.9) **
5'	17.4 (14.6-19.6)	**50.5 (43.8-56) ***	27.1 (24.6-29.7)	**7.5 (5.8-10.0) ***
10'	17.1 (15.2-19.5)	49.3 (43.3-55.5) *	**26.5 (24.0-30.4)**	8.15 (5.5-11.9) *
20'	16.6 (15.4-19.5)	46.7 (41-51.8)	27.45 (23.7-34.5)	10.15 (6.7-18.0)

ing the first 10 min of recovery and decreased slowly afterwards. In studies with shorter exercise duration, the decrease in La(Pl) started earlier and was faster than here (5,6,9,10). In this study, the La(Ery)/La(Pl) ratio was decreased at the end of exercise and further was increasing very slowly. In other studies with shorter exercise duration, at the end of exercise the La(Ery)/La(Pl) ratio was in the range of pre-exercise levels and decreased during the first 15 min of recovery (5,6,9,10). Pl-Ery gradient was in all studies 20 min after exercise much higher than before the test, except one study (1) which reported that the direction of the gradient reversed at 15 min post exercise, however, this finding was not confirmed.

It has been reported that an equilibrium between lactate in erythrocytes and in plasma should be reached after 4 and 7 min (2). Such an equilibrium of La concentrations was not found in this study, although La production and La removal are thought to be constant from 2 and 5 min in a 6-min test (11). However, the relativly low La(Ery) (1,3,5,6,10) may be explained by saturation of active transport mechanisms for La into blood cells by the large amount of lactate in the first 2 min of the test (11). This is supported by a lack of a correlation between La(Ery) and La(Pl), which is to be expected if La exchange mechanisms are saturated. Pl-Ery gradient was decreased approximately 44% in 20 min of recovery, but La removal from erythrocytes was only 30%. This may be due to an effect of low pH-values on La transport from erythrocytes. We found an increasing Pl-Ery gradient at pH lower than 7.15 (Fig. 2).

The slowly increasing La(Ery)/La(Pl) ratio indicates that removal from La(Pl) is not much faster than removal from La(Ery). The low removal of La(Pl) may be also due to de-

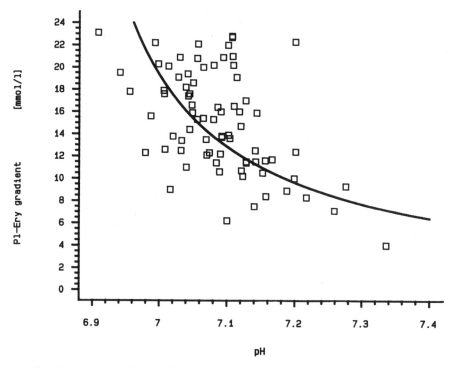

Figure 2. Correlation between Pl-Ery gradient and pH after 6-min maximal exercise on rowing ergometer. Pl-Ery gradient = $1/(-1.77+0.26 \cdot pH)$, n=77, r=0.61.

creased plasma volume, which is indicated by increased Hct, and the high percentage of working muscles in rowing. Therefore, the distribution space for lactate may be small in this experiment.

In this study, lactate in plasma was not measured but calculated. In prior studies (4,6,8) we have shown, that if untreated blood is measured with an electrochemical analyzer without use of hemolyzing agents, the blood cells remain intact and lactate concentration in plasma is measured with account being taken of a hematocrit dependent dilution error: $[La(UB)] = [La(Pl)] \cdot (1-Hct)$. The advantage of this method is the simplicity of measurement and independency from treatment of samples.

In summary, this study supports the contention that after high intense whole body exercise of 6 min duration a significant Pl-Ery gradient persists for more than 20 min of recovery. There is indication of a pH-depending inhibition of La transport, but La clearance may be also inhibited by a small distribution space for lactate.

5. REFERENCES

1. Buono, M. J., and J. E. Yeager. Intraerythrocyte and plasma lactate concentrations during exercise in humans. *Eur. J. Appl. Physiol.* **55**: 326–329, 1986.
2. Daniel, S. S., H. O. Morishima, L. S. James, and K. Adamsons, Jr. Lactate and pyruvate gradients between red blood cells and plasma during acute asphyxia. *J. Appl. Physiol.* **19**(6): 1100–1104, 1964
3. Foxdal, P., B. Sjödin, H. Rudstam, C. Östman, B. Östman, and G. C. Hedenstierna. Lactate concentration differences in plasma, whole blood, capillary fingerblood and erythrocytes during submaximal graded exercise in humans. *Eur. J. Appl. Physiol.* **61**: 218–222, 1990.
4. Harris, T. H., and G. A. Dudley. Exercise alters the distribution of ammonia and lactate in blood. *J. Appl. Physiol.* **66**: 313–317, 1989.
5. Hildebrand, A., W. Lormes, J. Emmert, V. Schneider, J. M. Steinacker, and M. Stauch. Der Einfluß der Blutzellen als Laktatverteilungsraum bei Laufbandbelastung. In: H. Liesen, H., M. Weiß, and M. Baum (Eds.) Regulations- und Repairmechanismen. *Dtsch Ärzte-Verlag, Cologne* 191–194, 1994.
6. Lindinger, M. I., G. J. F. Heigenhauser, R. S. McKelvie, and N. L. Jones. Blood ion regulation during repeated maximal exercise and recovery in humans. *Am. J. Physiol.* **262**: R126-R136, 1992.
7. Lormes, W., J. Emmert, A. Hildebrand, V. Schneider, and J. M. Steinacker. Change of l-lactate concentration in different blood compartments during excercise. *Med. Sci. Sports Exerc.* **25** Suppl: S65, 1993.
8. Lormes, W., R. Buckwitz, H. Rehbein, and J.M. Steinacker. Performance and blood lactate on Gjessing and Concept II rowing ergometer. *Int. J. Sports Med.* **14**: S29-S31, 1993.
9. Lormes, W., A. Hildebrand, J. Emmert, V. Schneider, J.M. Steinacker, and M. Stauch. Laktatkonzentrationen in unterschiedlichen Kompartimenten des Blutes bei Laufbandbelastung. In: Liesen, H., M. Weiß, and M. Baum (Eds.) Regulations- und Repairmechanismen. *Dtsch Ärzte-Verlag, Cologne,* 207–209, 1994.
10. McKelvie, R. S., M. I. Lindinger, G. J. F. Heigenhauser, and N. L. Jones. Contribution of erythrocytes to the control of electrolyte changes of exercise. *Can. J. Physiol. Pharmacol.* **69**: 984–993, 1991
11. Roth, W., E. Hasart, W. Wolf, and B. Pansold. Untersuchungen zur Dynamik der Energiebereitstellung während maximaler Mittelzeitausdauerbelastung. *Med. u. Sport* **23**: 107–114, 1983.

PART 6. ENHANCING EXERCISE TOLERANCE IN HEALTH AND DISEASE

39

CAN AMINO ACIDS INFLUENCE EXERCISE PERFORMANCE IN ATHLETES?

E. A. Newsholme and L. M. Castell

Cellular Nutrition Research Group
Department of Biochemistry
University of Oxford, South Parks Road
Oxford OX1 3QU, United Kingdom

1. INTRODUCTION

Tryptophan is the precursor for the important neurotransmitter, 5-hydroxytryptamine (5-HT). Tryptophan is unique among amino acids in that some of it binds to albumin in the blood: thus, blood contains both bound and free tryptophan. Free tryptophan is thought to compete with some other amino acids, particularly the branched chain amino acids (valine, leucine and isoleucine; BCAA) for entry into the brain across the blood-brain barrier (19). There is evidence that changes in the concentration of plasma free tryptophan may influence the level of 5-HT within the brain (see 14). An increase in the plasma concentration ratio of free tryptophan/BCAA could therefore lead to a marked increase in the rate of entry of tryptophan into the brain, and consequently to an increase in the concentration of 5-HT in some areas of the brain. This has been shown to be the case in the rat (6).

2. NEUROTRANSMITTERS AND RATE OF NEURONAL FIRING

If the concentration of a neurotransmitter in the brain decreases, the rate of neuronal firing in some parts of the brain may be limited, especially if the rate has been high. The consequent inability of one part of the brain to function satisfactorily owing to a decrease in neurotransmitter concentration can result in changes in behaviour (14). For example, depression can be caused by a decrease in the concentration of the monoamine neurotransmitters, noradrenaline, dopamine and/or 5-HT, in some parts of the brain (10).

It is well established that an increase in the level of 5-HT in the brain can result in fatigue and sleep. Thus, an increase in the concentration of 5-HT in certain areas of the brain might ensure a high rate of neuronal firing in a specific part of the brain which could then increase the sensitivity to fatigue and hence influence athletic performance.

One important aspect of this hypothesis for central fatigue is that the free tryptophan concentration can be increased (i.e. the proportion of tryptophan bound to albumin can be decreased) by increasing the plasma concentration of long-chain fatty acids which are also bound to albumin. An increase in the blood catecholamine level increases the rate of adipose tissue lipolysis and this can result in an increase in the concentration of plasma free fatty acids, thus increasing the plasma level of free tryptophan. Since the plasma fatty acid concentration can be increased in endurance exercise, particularly as carbohydrate stores become depleted, this would be expected to increase the free plasma concentration of tryptophan, and this has been observed in both rats (6,11) and human (5,16). Consequently, the plasma concentration ratio of free tryptophan/BCAA could be markedly increased by sustained exercise in humans, and this has been shown to be the case at the end of a marathon (table 1).

The hypothesis that an increase in the concentration of 5-HT in some parts of the brain can cause fatigue predicts, first, that compounds which change the effectiveness of 5-HT should influence fatigue during exercise and, secondly, that ingestion of small quantities of branched chain amino acids during prolonged exercise to maintain the plasma concentration ratio of free tryptophan/BCAA should improve physical performance and decrease mental exertion during exercise (4). Such experiments have been done. Thus, in humans, the administration of paroxetine, a specific 5-HT synaptosomal re-uptake inhibitor reduces the capacity for prolonged exercise (24); a similar effect has been observed in rats using m-chlorophenylpiperazine, a 5-HT agonist (2) and a 5-HT antagonist was shown to delay fatigue in rats (3).

That branched chain amino acids have a central effect is supported by the fact that cognitive activity could either be maintained at the pre-exercise level or improved after exercise when branched chain amino acids were ingested during prolonged exercise (16).

3. POSSIBLE CLINICAL APPLICATIONS FOR BRANCHED CHAIN AMINO ACID SUPPLEMENTATION

3.1. Chronic Fatigue Syndrome

Patients suffering from myalgic encephalomyelitis (ME), now more widely defined as chronic fatigue syndrome (CFS) (23), have normal skeletal muscle function before and after exercise (15). This suggests that central rather than peripheral fatigue may be the cause of the fatigue in this condition (see 13). Could a change in 5-HT levels in the brain or its effectiveness be the cause of the fatigue?

Table 1. Plasma glutamine, branched chain amino acids and free tryptophan concentrations in athletes (n=18) before (Pre) and after (Post) a marathon

	Pre	Immediately Post	1hr Post	16hr Post
Glutamine (uM)	654 ±16	585 ±14[b]	523 ±23[c]	683 ±26
BCAA (uM)	425 ±28	352 ±20[a]	311 ±18[b]	423 ±18
Free tryptophan (uM)	4.9 ±0.26	13.9 ±1.11[c]	9.6 ±0.58[c]	5.1 ±0.33
Plasma concn ratio of free tryp/BCA	1.23 ±0.003	4.16 ±0.36[c]	3.13±0.19[c]	1.21±0.11

[a]p<0.05; [b]p<0.01; [c]p<0.001

3.2. Post-Operative Fatigue after Major Surgery

Plasma albumin levels may be affected by changes in plasma hormone levels in patients undergoing major surgery. Major surgery is very stressful and an important consequence is fatigue, which can last for several weeks after an operation (22). As yet, no physiological mechanism for post-operative fatigue has been identified. Thus, the possibility that a biochemical mechanism might have an important role in this condition was investigated in two groups of patients undergoing major surgery with different surgical procedures (25).

The subjects were either elderly female patients undergoing restorative surgery for neck of femur fracture, or patients undergoing surgery for coronary artery bypass graft which involved cardiopulmonary bypass. Plasma free tryptophan concentrations were markedly increased 48 hr after surgery in the elderly fracture patients, compared with baseline levels pre- and post-surgery. The plasma concentration ratio of free tryptophan/BCAA was also increased after surgery. Relative plasma albumin concentrations were decreased after surgery, as well as the binding constant (Km) of albumin for tryptophan. In the cardiopulmonary bypass patients, plasma free tryptophan levels were markedly increased 24 hr after surgery, compared with baseline levels pre-surgery, and returned to normal five days later. The plasma concentration ratios of free tryptophan/BCAA were increased after cardiopulmonary bypass returning to near baseline levels five days later. Plasma albumin concentrations were decreased significantly after cardiopulmonary bypass returning to baseline five days later. No effects on the binding constant (Km) of albumin for tryptophan were observed after surgery in cardiopulmonary bypass patients, in contrast with elderly patients.

4. THE ROLE OF GLUTAMINE IN THE IMMUNE SYSTEM IN EXHAUSTIVE EXERCISE

For many years, it was considered that both lymphocytes and macrophages obtained most of their energy from the oxidation of glucose and that lymphocytes which had not been subjected to an immune response, i.e. resting lymphocytes, were metabolically quiescent. Recent work has shown that not only do these cells utilise glucose at a high rate, even when quiescent, but that they also use glutamine, and that its rate of utilisation is either similar to, or greater than, that of glucose (1). Surprisingly, very little of the carbon of glucose or that of glutamine is oxidised completely by these cells; glucose is converted almost totally into lactate and glutamine into glutamate, aspartate and alanine. These processes are known as glycolysis and glutaminolysis. High rates of glycolysis and glutaminolysis not only provide energy but also what is termed branched point sensitivity. The latter explains how precision for control of changes in the rate of synthesis of, for example, purine and pyrimidine nucleotides (glutamine and aspartate provide nitrogen and carbon for the bases whereas glucose provides the ribose) may be furnished by the kinetic characteristics at certain branched points in metabolism. A direct consequence of branched point sensitivity is that a decrease in the concentration of glutamine available to these cells would be expected to impair their function and hence decrease the effectiveness of the immune system. This is indeed the case: the response of human lymphocytes to a mitogenic signal is decreased when the level of glutamine in culture medium is decreased, as is phagocytosis and cytokine production by macrophages. Similarly, the presence of glucose is essential for the proliferation of lymphocytes.

The requirement for glutamine will increase dramatically after trauma, major surgery, infection and burns, since there will be increased activity of the immune system, and an increased number of cells involved in proliferation and repair activity. In addition, if the patient is not fed, the intestinal cells will also use endogenously-produced glutamine, so that the plasma level may fall. And there is evidence that glutamine provision corrects or improves an impaired function of some tissues, including the immune system, when that impairment may have been caused by a low plasma level of glutamine. It is speculated that the prime role for net muscle protein breakdown during conditions of trauma may be to maintain a pool of available amino acids (e.g. valine, leucine, isoleucine, glutamine, aspartate) in muscle that can be transaminated (and then metabolised) to provide nitrogen for the synthesis of glutamine. The rapid loss of body protein may therefore be a consequence of the requirement for branched point sensitivity and the need to convert the products of glutamine metabolism by these cells to glucose.

There is evidence that regular exhaustive exercise may be associated with impairment of the immune system. Athletes undergoing intense, prolonged training or participating in endurance races (e.g. the marathon), suffer an increased risk of infection. Prolonged exhaustive exercise can lower the plasma level of glutamine (20), which could be one factor involved in the impaired response of immune cells to opportunistic infections under these conditions (18). The provision of exogenous glutamine after strenuous exercise may, therefore, have a beneficial effect on the immune system; this is currently being investigated in a series of studies.

The studies began with monitoring plasma glutamine concentration over time in normal, sedentary subjects after giving a bolus of glutamine in water. A rapid increase occurred in plasma glutamine concentration, which peaked at 30 min after ingesting the glutamine. In another study the diurnal variation of plasma glutamine levels in normal subjects, both fasting and fed, was monitored (9). Plasma glutamine levels appeared to increase during the day after a protein meal was consumed.

Blood samples were subsequently undertaken before and after exercise from 18 male subjects who participated in a marathon. Some plasma amino acids and acute phase markers were measured, together with numbers of cells of the immune system in the blood.

The plasma glutamine concentration was decreased by 20% one hour after the marathon, and that of BCAA was decreased by 27%, compared with the pre-exercise concentration (table 1).

The plasma free tryptophan level tripled, and there was nearly a 4-fold increase in the plasma concentration ratio of free tryptophan/BCAA immediately after the marathon. An increase in the concentration of plasma free fatty acids correlated with an increase in plasma free tryptophan concentration.

Questionnaires were given to establish the incidence of infection for seven days after exercise in middle-distance, marathon and ultra-marathon runners, and elite rowers, in training and competition. In all these groups blood samples were also taken after exercise. In addition to measurement of plasma glutamine levels, whole-blood counts and lymphocyte proliferation studies *in vitro* were undertaken on some samples.

Infection levels overall were highest in marathon and ultra-marathon runners, and in elite male rowers after 50 min of intensive training. The decrease in plasma glutamine levels observed after marathon running (ca. 20%) was not as marked after training runs or 10 km races (12–15%). A marked increase in white blood cells, in particular in neutrophils, occurred immediately after prolonged, exhaustive exercise; within the next 15 min there was a decrease in total lymphocytes and in the number of T-cells.

Whether the provision of glutamine after exercise will benefit the immune system and result in a decrease incidence of infections is a study that needs to be carried out.

REFERENCES

1. Ardawi, M.S.M. and E.A. Newsholme. Glutamine metabolism in lymphocytes of the rat. *Biochem. J., 212*: 835–842 1983.
2. Bailey, S.P., J.M. Davis and E.N. Ahlborn. Effects of increased brain serotonergic activity on endurance performance in the rat. *Acta Physiol. Scand., 145*: 75 1992.
3. Bailey, S.P., J.M. Davis and E.N. Ahlborn. Neuroendocrine and substrate responses to altered brain 5-HT activity during prolonged exercise to fatigue. *J. Appl. Physiol., 74*: 3006–3012 1992.
4. Blomstrand,E., S. Andersson. P. Hassmen,. B. Ekblom and E.A. Newsholme. Effect of branched-chain amino acid and carbohydrate supplementation on the exercise-induced change in plasma and muscle concentration of amino acids in human subjects. *Acta Physiol. Scand., 153*: 87–96 1995.
5. Blomstrand, E., F. Celsing and E.A. Newsholme. Changes in plasma concentrations of aromatic and branched-chain amino acids during sustained exercise in man and their possible role in fatigue. *Acta Physiol. Scand., 133*: 115 1988.
6. Blomstrand, E., D. Perrett M. Parry-Billings and E.A. Newsholme. Effect of sustained exercise on plasma amino acid concentrations and on 5-hydroxytryptamine metabolism in six different brain regions in the rat. *Acta Physiol. Scand., 136*: 473–481 1989.
7. Blomstrand, E., P. Hassmen B. Ekblom and E.A.Newsholme. Administration of branched-chain amino acids during sustained exercise; effects on preformance and on plasma concentration of some amino acids. *Eur. J. Appl. Physiol., 63*: 83 1991.
8. Blomstrand, E., P. Hassmem and E.A. Newsholme. Effects of branched-chain supplementation on mental performance. *Acta Physiol. Scand., 143*: 225 1991.
9. Castell, L.M., C.T Lui and E.A.Newsholme. Diurnal variation of plasma glutamine and arginine in normal and fasting subjects. *Proc. Nutr. Soc. 54*: 118A 1995.
10. Cowen, P.J., M. Parry-Billings and E.A. Newsholme. Decreased plasma tryptophan levels in major depression. *J. Affective Disorders, 16*: 27–31 1989.
11. Chauloff, F., J.L. Elghozi ,Y. Guenzennec and D. Laude. Effects of conditioned running on plasma, liver and brain tryptophan and on brain 5-hydroxytryptamine metabolism of the rat. *Br. J. Pharmacol., 86*: 33 1985.
12. Davis, J.M., S.P. Bailey, J.A. Woods, F.J. Galiano, M.T. Hamilton and W.P. Bartoli. Effects of carbohydrate feedings on plasma free tryptophan and branched-chain amino acids during prolonged cycling. *Eur. J. Appl. Physiol., 65*: 513–519 1992.
13. Edwards, R.H.T. In: *Biochemical Basis of Fatigue*, edited by H.G Knuttgen. pp. 3–28.Champaign, Il: Human Kinetics, 1983, pp. 3–28.
14. Fernstrom, J.D. and R.J. Wurtman. Control of the brain serotinin levels by diet. *Adv. Biochem. Psychopharm., 11*: 133 1990.
15. Gibson, H., N. Carroll, J.E. Clague and R.H.T. Edwards. Exercise performance and fatiguability in patients with chronic fatigue syndrome. *J. Neurol. Neurosurg.Psych., 56*: 993–998 1993.
16. Hassmen, P., E. Blomstrand, B. Ekblom and E.A. Newsholme. Branched-chain amino acid supplementation during 30-Km competitive run: mood and cognitive performance. *Nutrition, 10*: 1–5 1994.
17. Newsholme, E.A. Application of knowledge of metabolic integration to the problem of metabolic limitations in middle distance and marathon running. *Acta Physiol. Scand. Suppl. 128*: (556) 93 1986.
18. Newsholme, E.A., B. Crabtree and M.S.M. Ardawi. Glutamine metabolism in lymphocytes: its biochemical, physiological and clinical importance. *Q. J. Exp. Physiol., 70*: 473–489, 1985.
19. Pardridge, W. M. Kinetics of competitive inhibition of neutral amino acid transport across the blood-brain barrier. *J. Neurochem., 28*: 103 1977.
20. Parry-Billings, M., R. Budgett, Y. Koutedakis, E. Blomstrand, S. Brooks, C. Williams, P. Calder, S. Pilling, R. Baigrie and E.A. Newsholme. Plasma amino acid concentrations in the overtraining syndrome: possible effects on the immune system. *Med. Sci. Sports Exerc. 24*: 1353–1358 1992.
21. Petruzzello, S. J., D. M. Landers, J. Pie and J. Billie. Effect of branched-chain amino acid supplements on exercise-related mood and performance (Abstract). *Med. Sci. Sports Exerc. 24*: (Suppl), S2 1992.
22. Salmon, P. Nutrition, cognitive performance, and mental fatigue. *Nutrition, 10*: 427–428 1994.

23. Sharpe, M. et al. A report - chronic fatigue syndrome: guidelines for research. *J. Roy. Soc. Med.,* **84**: 118–121 1991.

24. Wilson,W. M. and R. J. Maughan. Evidence for a possible role of 5-hydroxytryptamine in the genesis of fatigue in man: administration of paroxetine, a 5-HT re-uptake inhibitor, reduces the capacity to perform prolonged exercise. *Exp. Physiol.,* **77**: 921 1992.

25. Yamamoto, T., L. M. Castell, J. Botella, A. Young, H. Powell, G.M. Hall and E.A. Newsholme. Alterations in the plasma concentrations of tryptophan and albumin may contribute to post-operative fatigue. *Brain Res Bull.* (In press).

40

ENHANCING EXERCISE TOLERANCE IN PATIENTS WITH LUNG DISEASE

Richard Casaburi

Department of Medicine
Division of Respiratory and Critical Care Physiology and Medicine
Harbor-UCLA Medical Center
1000 W. Carson Street
Torrance, California 90509

1. INTRODUCTION

There is little question that the most disabling symptom of chronic pulmonary disease is exercise intolerance. The patient often becomes homebound, isolated and depressed out of fear of experiencing the dyspnea that exertion brings. The sedentary lifestyle these patients adopt only serves to make the situation worse. The inactive muscles atrophy, which increases the energetic cost of activity and, in fact, may increase dyspnea on exertion.

It is becoming clear that there are two distinct strategies for improving exercise tolerance in patients with lung disease. The scientific basis underlying both of these strategies has evolved and improved in recent years.

The first major strategy to improve exercise tolerance is *psychologic*. Patients become convinced that the shortness of breath they experience is, in itself, harmful. However, in contrast, for example, to the anginal pain experienced by patients with coronary artery disease, dyspnea is not a sign that the body is being damaged. There is growing evidence that patients can be *desensitized* to the symptom of exertional dyspnea.

To date, the most effective way to desensitize to dyspnea on exertion is exercise itself. Haas and his colleagues[11] have recently summarized the hypotheses which have been mustered to explain the desensitizing effects of exercise. The *antidepressant hypothesis* suggests that patients who successfully participate in an exercise program experience positive feedback from mastering something they perceive as difficult. The *social interaction hypothesis* implies that progressive exercise, in a supervised program with others having similar debility, calms unrealistic fears. This would explain why home exercise programs are generally less successful. The *distraction hypothesis* theorizes that, since it is difficult to focus attention on more than one source of information at a time, dyspneic stimuli are perceived as being less intense when the patient's attention is focused on non-dyspneic stimuli. Listening to music or exercising with others distracts the patient from the sensation of dyspnea.

The second major strategy to improve the tolerance of exercise is *physiologic*. The muscles, cardiovascular system and lungs are part of an interlinked system designed to bring oxygen from the atmosphere to the mitochondria and bring CO_2 from the muscle cell out to the atmosphere. This system must gear up to accommodate the 20-fold or so increase in muscle oxygen requirements that exercise normally brings. This transport system is as weak as its weakest link. Any break in the interlinkages will cause exercise intolerance.

2. LIMITATIONS TO EXERCISE TOLERANCE

The patient with lung disease has distinct handicaps in meeting the greatly increased oxygen supply requirements imposed by exercise. First, the work of breathing is increased both by high airways resistance and by hyperinflation. Thus, the increase in ventilation demanded by exercise comes at a high metabolic cost and can lead to fatigue of the respiratory muscles. Making matters worse, the amount of ventilation for a given metabolic rate is substantially higher than in healthy subjects, because the patient with respiratory disease has a lung which is a poor gas exchanger. For these reasons, exercise is often associated with hypoxemia and/or CO_2 retention. Often forgotten is the fact that the cardiovascular system can be impacted by lung disease. The pulmonary vasculature is often destroyed along with the alveolar architecture in both obstructive and restrictive lung disease. As a result, pulmonary vascular resistance cannot fall with exercise, pulmonary artery pressure rises abnormally and in some patients cardiac output increase may be limited - which compromises oxygen delivery to the muscles.

Among patients with chronic lung disease, the responses to exercise in patients with chronic obstructive pulmonary disease have been best studied. Many of these patients are "ventilatory limited" - they are limited in their exercise tolerance by the amount of pulmonary ventilation they can achieve[3]. As the level of exercise rises, so does the ventilation required. When this requirement exceeds the level of ventilation that can be sustained, exercise cannot continue. It is important to realize that asking which of the physiologic abnormalities associated with obstructive lung disease is *the* factor which limits exercise tolerance is essentially a meaningless question. Each physiologic deficit which adds to the ventilatory requirement for exercise or contributes to the ventilatory limitation will contribute to exercise intolerance.

Unfortunately, medical science has identified few therapies which are effective in *either* alleviating the ventilatory limitation or decreasing the ventilatory requirement. Bronchodilator or anti-inflammatory therapies may reduce the work of breathing (potentially forestalling respiratory muscle fatigue) and sometimes lessen ventilation-perfusion imbalance. However, the reversible component of airflow obstruction is usually limited. Supplementary oxygen has several beneficial effects: inhibition of the respiratory chemoreceptors, improvement of oxygen delivery, and hemodynamic improvement through dilation of the pulmonary vasculature. Even in patients who are not severely hypoxemic, supplemental oxygen may increase exercise tolerance[14]. Considerable attention has recently been focused on seemingly contradictory therapies designed to alleviate ventilatory limitation. Respir-atory muscle training seeks to improve the aerobic capacity of the diaphragm and intercostal muscles. Respiratory muscle rest is postulated to reduce the tendency for fatigue. Both modalities show some promise[2] but it is unclear whether either will prove to be practical therapy. Pharmacologic means to depress ventilation have, to date, generally been found only mildly effective (or ineffective) and sometimes produce an unacceptable level of sedation[13]. Lung (or heart-lung) transplantation is likely an effective method to increase exercise tolerance[5], but it seems unlikely to be a realistic option for the vast majority of patients for the foreseeable future. Lung reduction surgery is an interesting new technique which may improve lung mechanics[10], but long term evaluation will be required before this procedure can be recommended with confidence.

3. VENTILATORY REQUIREMENTS FOR EXERCISE

Before discussing exercise training, its important to delineate the factors that contribute to the ventilatory requirement for exercise. We can then consider how exercise training might modify them. Mass balance considerations dictate that the following equation applies:

$$\dot{V}_E = \frac{k \ \dot{V}CO_2}{PaCO_2 \ (1-V_D/V_T)}$$

Therefore the ventilatory requirement is determined by three factors: the rate of CO_2 output ($\dot{V}CO_2$), the level at which arterial PCO_2 ($PaCO_2$) is regulated and the efficiency with which CO_2 is expelled from the lung (quantified as the dead space to tidal volume ratio, V_D/V_T). Each of these three variables undergo characteristic changes with exercise. A key observation is that $PaCO_2$ remains rather precisely regulated over the range of moderate work rates[17]. At higher exercise levels the situation changes. If an incremental exercise test is performed, during heavy exercise the rate of CO_2 output accelerates and arterial PCO_2 falls. This is a direct result of the onset of lactic acidosis. Lactic acid is a strong acid and must be immediately buffered. Bicarbonate is the predominant buffer. This has two distinct consequences. Bicarbonate breaks down to CO_2 and water, thus increasing CO_2 output. The acid stimulus results in a lower $PaCO_2$. So there is an accelerating ventilatory requirement during heavy exercise due both to an accelerating $\dot{V} CO_2$ and a falling $PaCO_2$.

4. EXERCISE TRAINING

Against the background of the determinants of exercise ventilation, the physiologic benefits of exercise training for the patient with lung disease can be considered. Much is known regarding the structural and biochemical changes which occur in the exercising muscles of *healthy* individuals as a result of a program of exercise training[12]. These changes include increases in the number and size of the mitochondria, an increase in the concentration of the enzymes subserving aerobic respiration and, perhaps most importantly, an increase in muscle capillarity. These changes combine to increase the capacity for aerobic work. Therefore, a higher level of exercise can be accomplished before anaerobic metabolism must be called upon to supplement aerobic energy supply. Delaying anaerobic metabolism is valuable because lactic acid is a side product of anaerobic metabolism. This is of importance because, as previously stated, lactic acid stimulates ventilation. Thus, an effective program of exercise training would be expected to decrease the level of lactic acidosis at a given level of exercise and thus decrease the ventilatory requirement. This was shown in a study of healthy subjects who performed constant work rate tests before and after a program of exercise training[9]. For identical levels of heavy exercise, training resulted in substantially decreased end-exercise blood lactate levels. For the heaviest level of exercise studied, end-exercise ventilation was on average almost 40 liters/min lower after training. The reduction in V_E engendered by training was quite highly correlated with the reduction in blood lactate levels ($r = 0.81$).

In considering the applicability of these findings to patients with lung disease, it has been supposed that the obstructive lung disease patient is not a candidate to obtain a true physiologic training response through participation in the exercise programs that are part of pulmonary rehabilitation. First of all, it was thought that most patients cannot tolerate work rates associated with a lactic acidosis. It is true that many patients reach a ventilatory limitation at exercise levels at which healthy subjects would not manifest lactic acidosis. However, it has been discovered that many patients with COPD demonstrate lactic acidosis at surprisingly low work rates[5,15] (some elevate blood lactate levels at a modest walking pace on level ground). In fact, the ability to increase blood lactate correlates very poorly with measures of resting lung function (e.g. FEV_1)[15]. The likely cause is a combination of an extremely sedentary lifestyle (leading to severe deconditioning) plus an uncertain contribution from pulmonary vascular disease which may compromise oxygen delivery to the muscles.

The second reason why pulmonary rehabilitation programs might not induce the changes which forestall the onset of lactic acidosis is that the exercise programs are inadequate. The requirements for an effective exercise training program for healthy subjects are quite well known[1,4]. An adequate program is roughly 5–8 weeks long, exercise sessions must be held 3–5 days per week and the exercise session must last at least 30 minutes. Exercise intensity is an area of controversy that points out how difficult it is to design a training program rationally. The controversy arises, in part, because we are surprisingly ignorant of the actual stimulus that induces the structural and biochemical changes that result from training. There is general consensus that there is a "critical training intensity", i.e. a work rate below which no training effect will be seen no matter how long the program continues. However, there is disagreement as to how best to select the most relevant marker of training intensity[3]. Three variables have been suggested: heart rate, oxygen uptake and blood lactate level. Each has its advantages and shortcomings. Adding to the confusion are recent demonstrations that exercise intensities lower than those previously thought to be below the critical training intensity can induce a physiological training effect[8]. Thus, we have very little in the way of physiologic guidelines to rationally select exercise training intensities.

There is also a substantial literature that reports the results of the exercise programs which are part of pulmonary rehabilitation programs[3]. It is probably not fair to label most of these programs as "training" programs, since most were not designed with a physiologic rationale. A recently composed analysis of the findings of 37 studies[3] is of interest, though few studies drew participants from the complete spectrum of pulmonary patients. Most of the roughly 1000 patients included in these studies were male (81%), elderly (average age = 61 years), severely obstructed (average FEV_1 = 1.1 liters), had moderate hypoxemia (average $PaCO_2$ = 69 torr) and were not CO_2 retainers at rest (average $PaCO_2$ = 42 torr). The clear findings of these studies are that: 1) lung disease does not get better, 2) patients feel that their exercise tolerance has improved and, in fact, they generally are able to sustain a given exercise level for a longer period of time, and 3) during formal exercise testing, improvements in exercise tolerance can be documented sometimes, but not always.

Despite the apparent improvement in exercise tolerance, unequivocal evidence for physiologic improvement in the ability of perform exercise has seldom been obtained. The maximal oxygen uptake ($\dot{V}O_2$ max) during exercise has been shown to increase in about half the studies in which it has been measured[3], but this may have resulted from better motivation and effort after the exercise program (i.e., a psychologic benefit).

We sought to determine if a *physiologic* training effect could be obtained in patients with COPD, specifically whether the ventilatory requirement for exercise of COPD patients could be reduced in the same way as in healthy subjects. We also sought to determine whether it was crucial that they train at high exercise intensity. Nineteen subjects were studied[7] who met the entry criteria that they experienced elevated blood lactate levels during maximal exercise. Exercise testing, including arterial blood gas and lactate analysis, was performed before and after training. The training consisted of five sessions per week for eight weeks. To determine the effect of exercise intensity, one group exercised for 45 minutes per day at high intensity, while the other group exercised at a lower intensity but a longer duration. Participants had predominately moderate disease. The responses to incremental exercise tests were marked by a moderately reduced exercise tolerance and an extraordinarily early onset of lactic acidosis (at an oxygen uptake averaging approximately 0.7 liters/min). Thus these patients had the added ventilatory stimulus of lactic acidosis at very low work rates. Training had a clear effect on this early onset of lactic acidosis. The onset of blood lactate increase was delayed and lactate was lower at any given level of exercise. This had a measurable effect on the ventilatory requirement for exercise. During both incremental and constant work rate exercise, ventilation was lower for a given exercise stress. The averaged results show that, for a given level of exercise, lactate, ventilation, CO_2 output and heart rate were lower after training. This was much more pronounced for the group that trained at the high exercise intensity. As we had seen in healthy subjects, the decreases in ventilation were well correlated with the decreases in blood lactate. We conclude from this study that COPD patients can achieve a training response, and that the reduced lactic acidosis can reduce the ventilatory requirement for exercise as long as the training intensity is high.

These results lead to an important question. Is it necessary to triage patients who present for pulmonary rehabilitation into two groups? One group, who were shown capable of increasing blood lactate during exercise, would be encouraged to participate in a rigorous training program with the aim of reducing the ventilatory requirement for exercise. Patients who were unable to increase blood lactate levels would be assigned to another group, engaging in an exercise program aimed only at acquiring psychologic benefits. It seems likely that many patients with very severe disease who experience intolerable dyspnea when they progressively hyperinflate with exercise would likely be triaged into the latter group. Patients of this type constitute an appreciable portion of those rehabilitated in the United States.

We recently completed a study designed to determine whether reduction of lactic acidosis was the only mechanism by which exercise training could engender a reduced ventilatory requirement for exercise[6]. A total of 51 patients with severe COPD (average $FEV_1 = 0.9$) participated in a rehabilitation program featuring either a "traditional" exercise component (26 patients) or a rigorous training program consisting of 45 minute sessions of cycle ergometer exercise 3 times a week for 8 weeks at work rates targeted to be near maximal targets [16] and advanced as tolerated during the training period. Preliminary results show that the group that underwent the rigorous training program, but not the group who engaged in the "traditional" exercise component, had substantial increases in exercise tolerance as assessed during formal exercise testing (e.g. $\dot{V}O_2$max increased by an average of 16%). Further, in response to identical work rates, the level of ventilation was lower after the rigorous training program. The lower ventilatory requirement for exercise has been tentatively ascribed to a lower V_D/V_T after training - due not to improved ability of the lung to transport gas but to a more efficient slower deeper pattern of breathing.

5. CONCLUSION

These results lead to the tentative recommendation that all patients with lung disease who are able to safely engage in a training program be encouraged to do so. We should seek to perfect physiologically-based principles of exercise prescription which will allow optimization of exercise tolerance of patients disabled by lung disease.

6. REFERENCES

1. American College of Sports Medicine. Position Stand. The recommended quantity and quality of exercise for developing and maintaining cardiorespiratory and muscular fitness in healthy adults. *Med. Sci. Sports Exerc.* **22**: 265–274, 1990.

2. Belman, M.J. Ventilatory muscle training and unloading. In: *Principles and Practice of Pulmonary Rehabilitation* (Casaburi, R. and T.L. Petty, eds.), Philadelphia: Saunders, 1993, pp. 225–240.

3. Casaburi, R. Exercise training in chronic obstructive lung disease. In: *Principles and Practice of Pulmonary Rehabilitation* (Casaburi R., T.L. Petty, eds.), Philadelphia, Saunders, 1993; pp. 204–224.

4. Casaburi, R. Physiologic responses to training. *Clinics in Chest Medicine* **15**: 215–227, 1994.

5. Casaburi, R. Deconditioning. In: *Pulmonary Rehabilitation* (Fishman, A.P. ed.), New York: Marcel Dekker (in press).

6. Casaburi, R., M. Burns, C. Cooper, E. Singer, R. Chang, and J. Porszasz. Physiological benefits of exercise training in severe COPD (Abstract). *Am. J. Respir. Crit. Care Med.* **149**: A598, 1994.

7. Casaburi, R., A. Patessio, F. Ioli, S. Zanaboni, C.F. Donner, and K. Wasserman. Reduction in exercise lactic acidosis and ventilation as a result of exercise training in obstructive lung disease. *Am. Rev. Respir. Dis.* **143**: 9–18, 1991.

8. Casaburi, R., T.W. Storer, C.S. Sullivan, and K. Wasserman. Evaluation of blood lactate elevation as an intensity criterion for exercise training. *Med. Sci. Sports Exerc.* **27**: 852–862, 1995.

9. Casaburi, R., T.W. Storer, and K. Wasserman. Mediation of reduced ventilatory response to exercise after endurance training. *J. Appl. Physiol.* **63**: 1533–1538, 1987.

10. Copper, J.D., E.P. Trulock, A.N. Triantafillou, G.A.Patterson, M.S. Pohl, P.A. Deloney, R.S. Sundaresan and C.L. Roper. Bilateral pneumectomy (volume reduction) for chronic obstructive pulmonary disease. *J. Thoracic Cardiovasc. Surg.* **109**: 106–119, 1995.

11. Haas, F., J. Sasazar-Schicchi, and R. Axen. Desensitization to dyspnea in chronic obstructive pulmonary disease. In: *Principles and Practice of Pulmonary Rehabilitation* (Casaburi, R., TL Petty, eds.), Philadelphia, Saunders, pp. 241–251, 1993.

12. Saltin, B., and P.D. Gollnick. Skeletal muscle adaptability: significance for metabolism and performance. In: *Handbook of Physiology. Skeletal Muscle* (Am. Physiol. Soc.), Washington, DC, pp. 555–631, 1983.

13. Stark, R.D. Dyspnoea: Assessment and pharmacological manipulation. *Eur. Respir. J.* **1**: 290–287, 1988.

14. Stein, D.A., B.L. Bradley, and W.C. Miller. Mechanisms of oxygen effects on exercise in patients with chronic obstructive pulmonary disease. *Chest* **81**: 6–10, 1982.

15. Sue, D.Y., K. Wasserman, R.B. Moricca, and R. Casaburi. Metabolic acidosis during exercise in patients with chronic obstructive pulmonary disease. *Chest* **94**: 931–938, 1988.

16. Reis, A.L., and C.J. Archibald. Endurance exercise training at maximal targets in patients with chronic obstructive pulmonary disease. *J. Cardiopulmon. Rehabil.* **7**: 594–601, 1987.

17. Wasserman, K., B.J. Whipp, and R. Casaburi. Respiratory control during exercise. In: *Handbook of Physiology. Respiration II* (Am. Physiol. Soc.), Washington, DC, pp. 595–619, 1986.

41

EFFECTS OF DRUGS ON EXERCISE TOLERANCE IN PATIENTS WITH CARDIOVASCULAR DISEASE

Martin Stauch

Abteilung Sport- und Leistungsmedizin
University of Ulm, D-89070 Ulm, Germany

1. INTRODUCTION

A consequence of most cardiovascular diseases is exercise limitation. The main causes of exercise intolerance are angina pectoris due to myocardial ischemia, dyspnea in congestive heart failure due to loss of muscle mass, and muscular dysfunction or valvular disease. Besides the heart, peripheral arterial occlusive disease is a major cause for exercise intolerance. Possibilities for drug therapy are limited in this disease.

2. ISCHEMIC HEART DISEASE

2.1. Nitrates

Angina pectoris is diagnostic of coronary arteriosclerosis if it regularly appears during exercise, and can be alleviated promptly after sublingual application or inhalation of nitrates. Application of nitrates, either nitroglycerin or isosorbidedinitrate, increases exercise tolerance. Signs of ischemia in the electrocardiogram appear at higher exercise levels than without the drug. Nitrates have manifold mechanisms of action. For example, there is a reduction of preload due to venous pooling which reduces left ventricular end-diastolic pressure and volume, leading to redistribution of blood flow from epicardial to endocardial vessels. This is enhanced by coronary vasodilation, including collateral vessels at rest and during exercise (2). Afterload reduction is achieved by dilatation of arteriolar resistance vessels. The dominant role in improving exercise tolerance seems to be preload reduction and modification of coronary vascular tone, also in diseased coronary vessels (4, 5). Reduction of blood pressure may lead to an increase in heart rate if short-acting nitrates with immediate efficacy are used. Long-acting nitrates are usually used to improve exercise tolerance. Isosorbidedinitrate has been mostly used for this indication. The efficacy of the oral route for this drug has been questioned as a result of animal experiments, since first-pass metabolites of the dinitrate - 2- and 5-isosorbide-mononitrate - were considered to be without hemodynamic effect (15). The first clinical trials on exercise tolerance and ischemic signs after application of both metabolites showed a significant positive effect (20,21). In fact, these metabolites actually underlie the long-acting action of oral isosorbidedinitrate. 5-isosorbidemononitrate (IS-5-MN) was developed as another long-acting medication for the chronic treatment of myocardial ischemia. Its advantages are a 100% bioavailability with linear dose-response kinetics.

Tolerance to nitrates develops if one tries to maintain a constantly high plasma level (1,23). A "nitrate-free interval" is recommended in chronic therapy (17). Even though nitrate-tolerance seldom is observed by the patient in clinical practice, it is advisable to schedule drug administration for the time of the day when exercise is probable. Periods of rest (e.g. evening and night) may thus be kept free of medication. Drugs in constant-delivery preparations, e.g. transdermal nitroglycerin patches, have been modified to provide peaks and troughs in plasma concentration through interval-dosing regimens: patches can be effective for 12 hours, and then taken off. Alternatively, dosing 20 mg IS-5-MN at 8 am and 3 pm can provide 12 hours of anti-anginal coverage and a 17-hour withdrawal period. The latter has not been associated with significant time-zero or rebound effects (24).

2.2. Molsidomine

Molsidomine is a syndomine derivative with active metabolites SIN-1 and SIN-1A having vasodilator activity, reflecting the release of nitric oxide via a direct increase of cyclic GMP. It is not dependent on sulfydryl groups and therefore avoids tolerance (3, 26). Molsidomine reduces exercise ischemia in the same manner as nitrates do (12), since the mechanism of action also involves release of nitric oxide.

The possibility of combining nitrates with other classes of anti-anginal drugs leaves tolerance as a minor problem in the chronic treatment of patients with exercise ischemia.

2.3. Beta-Blockers

Beta-blockers are the second main pillar of anti-ischemic therapy as well as for all other forms of ischemic heart disease. Beta$_1$-selective agents are ususally preferred. Nonselective beta-blockers tend to decrease the levels of high-density lipoproteins, while increasing triglycerides and low-density lipoproteins. Newer beta-blockers with an additional vasodilator activity, so-called "multiple action beta-blockers", may have a more favorable effect on serum lipoproteins, while maintaining their indications and efficacies for ischemic heart disease. Celiprolol is a long-acting direct vasodilator with partial beta$_2$-agonist activity and low lipid solubility. In comparison to atenolol, it produces less of a reduction in heart rate, but seems to have equal anti-anginal activity (11).

Beta-blockers and nitrates complement each other in a favorable way. The tendency of beta-blockers to increase preload is counteracted by nitrates which act to reduce preload. The tendency of nitrates to increase heart rate, on the other hand, is typically more than counteracted by beta-blockers reducing heart rate, this being the most usual effect. The combination of both drugs increases exercise tolerance more than each drug alone.

2.4. Calcium Channel Blockers

Calcium channel blockers also improve exercise tolerance in ischemic heart disease, but in a less obvious way. They often are co-medication for additional problems such as hypertension that is treated by calcium blockers of the dihydropyridine type. The verapamil type of drug is used more often for patients with ischemia (22), particularly if additional symptoms of atrial arrhythmias are present.

2.5. Loop Diuretics

The principle of preload reduction to improve mycardial exercise ischemia also works with drugs other than nitrates. That is, the loop diuretics frusemide and piretanide, intravenously applied, lower left ventricular filling pressure and reduce electrocardiographic signs of ischemia during cycle ergometer exercise (13,14). Since loop diuretics are frequently used in patients with hypertension, they may have an additional anti-anginal effect if exercise ischemia is present.

2.6. Ace Inhibitors

Angiotensin-converting enzyme (ACE) inhibitors, such as captopril and benazepril, may decrease the number as well as the duration of ischemic episodes in patients with ischemic heart disease, although not affecting exercise training responses (10, 25). They are not standard therapy for anginal patients, unless hypertension and/or congestive heart failure are present.

Angina Pectoris results from ischemia of heart muscle. Consequently, pharmacologic agents that improve blood supply to the heart improve exercise tolerance. The different mechanisms used for this goal are summarized in Table 1.

3. CONGESTIVE HEART FAILURE

To establish a positive effect of drugs on exercise tolerance in congestive heart failure is more difficult. The application of digitalis glycosides yields the most dramatic effect in patients with atrial fibrillation and therefore a greatly increased heart rate. Studies for treatment of congestive heart failure have mostly concentrated on reducing mortality.

Table 1. Drugs improving exercise tolerance by different pathophysiologic mechanisms reducing exercise-induced ischemia

	Nitrates	Beta-blockers	Calcium Antagonists A: Verapamil type	Calcium Antagonists B: Dihydro-pyridine type
Heart rate	increased	reduced	reduced	increased
Contractility	0	reduced	reduced	0
Systolic wall tension a) LV-Volume	decreased			
b) LV-Pressure	decreased	decreased	decreased	decreased
Coronary bloodflow: a) DBP - LVEDP	improved by reduced LVEDP	improved by	improved by	improved by
b) Coronary vascular resistance	reduced, large vessels		reduced	reduced
c) Diastolic duration		prolonged	prolonged	

LV = left ventricle, DBP = diastolic blood pressure, LVEDP = left ventricular end-diastolic pressure

Lately, improvements of exercise tolerance are more frequently determined not only in terms of NYHA classes, but also with measurements of the duration of a walk on the treadmill. Whether the significant increase of walking time in a range of 30 to 80 seconds truly reflects the state of exercise tolerance in daily life may be an open question. Patients must be monitored for other symptoms as well.

For improvement of congestive heart failure consequent to loss of contractile muscle mass (myocardial infarction) or severe dysfunction of cardiac muscle (dilated cardiomyopathy), two principles may be applied. Improving contractile function of the remaining heart muscle is one alternative. Afterload reduction is the other.

3.1. Contractile Stimulation

Digitalis glycosides are the oldest and most prominent agents for the goal of contractile stimulation. Recent studies on ACE inhibitors defined their role in chronic treatment more clearly. During the last 15 years, the application of small doses of beta-blockers has been successful in making heart muscle more sensitive for sympathetic stimuli and thus improving contraction. When titrated carefully from very low doses and used with a firm commitment to long-term treatment, beta-blockers have been shown to prevent further deterioration of heart failure and to improve hemodynamics, exercise tolerance, quality of life and prognosis (18). Among newer beta-blockers, carvedilol (which has vasodilator properties) improves cardiac function but only tended to increase exercise tolerance (16). Due to difficult dosing and the necessity of close monitoring, wide acceptance of this regimen is lacking.

Stimulation of dopaminergic DA_1- and DA_2-receptors with an orally applicable drug, ibopamine, is a new procedure for stimulating myocardial contraction. Low doses appear to exert beneficial neurohormonal, hemodynamic and renal effects, without increased inotropic effects. Improved exercise performance has been documented in mild to moderate heart failure (19), but not always on long-term medication (7). Renal effects with equal pre- and postglomerular filtration appear to be primarily due to its systemic hemodynamic effect (8).

Phosphodiesterase III inhibitors like enoximone and similar substances have been used thus far mostly in very severe congestive failure, when survival but not exercise tolerance is the aim of therapy.

3.2. Afterload Reduction

3.2.1. Ace-Inhibitors. The second principle is directed towards reduction of the work of the heart for a certain workload. ACE inhibitors are established therapy for systolic dysfunction. Nitrates, diuretics, alpha-blockers were previous schemes for this concept of myocardial workload reduction. Comprehensive studies of ACE inhibitors first compared their effects to, or in combination with, digitalis glycoside. This yielded the result lacking for so long that digitalis is beneficial in patients with congestive heart failure and sinus rhythm. Newer studies indicate that ACE inhibitors deliver clinical benefit including improved exercise tolerance also without additional digitalis therapy (6).

3.2.2. Calcium Channel Blockers. This group of drugs should also be suitable agents to reduce afterload. Newer drugs like Amlodipine, Felodipine, Isradipine, Nicardipine and Nisolipine have a higher vascular selectivity, lower negative inotropic effect and less pronounced cardiovascular counter-regulation than first-generation calcium channel blockers.

Also, verapamil has been used for pronounced diastolic dysfunction. Results are mixed. Felodipine, one of the newer substances, has not been shown to be of benefit in patients with mild to moderate heart failure (9). The general use of calcium channel blockers for congestive heart failure might emerge from new studies under way with several of these drugs (27).

4. CONCLUSIONS

Drug therapy to improve exercise tolerance is easy to ascertain and well treatable in exercise-induced myocardial ischemia. In congestive heart failure this is not the case. Walking time on a treadmill is a possibility, but more specific methods such as the measurement of oxygen uptake and the observation of symptoms over longer periods should be employed.

5. REFERENCES

1. Abrams J. Nitrate tolerance and dependence. *Am. Heart J.* **99**: 113–123, 1980.
2. Badger F., G. Brown, C.A. Gallery, E.L.Bolson, and H.T.Dodge. Coronary artery dilatation and hemodynamic responses after isosorbide dinitrate therapy in patients with coronary artery disease. *Am. J. Cardiol.* **56**: 390–395, 1985.
3. Bassenge E., and A. Mulsch. Anti-ischemic actions of molsidomine by venous and large coronary dilatation on combination with antiplatelet effects. *J. Cardiovasc. Pharmacol.* **14**:23–28, 1989.
4. Brown B.G., E.L. Bolson, R.B. Peterson, C.D. Pierce, and H.T. Dodge. The mechanisms of nitroglycerin action: Stenosis vasodilation as a major component of drug response. *Circulation* **65**: 1089–1097, 1981.
5. Brown B.G. Response of normal and diseased epicardial coronary arteries to vasoactive drugs: Quantitative angiographic studies. *Am. J. Cardiol.* **56**: 23E–29E, 1985.
6. Brown E.J., P.H. Chew, A. MacLean, K. Gelperin, J.P. Ilgenfritz, and M. Blumenthal. Effects of fosinopril on exercise tolerance and clinical deterioration in patients with chronic congestive heart failure not taking digitalis. *Amer.J.Cardiol.* **75**: 596–600, 1995.
7. Bussmann W.-D., and K. Wienhöfer. Akut- und Langzeiteffekte von 100 mg N-Methyldopamin (Ibopamin) bei Patienten mit chronischer Herzinsuffizienz Stadium II-III NYHA. *Herz/Kreislauf* **27**:135–138, 1995.
8. Lieverse A.G., D.J. vanVeldhuisen, A.J. Smit, J.G. Zijlstra, S. Maeijer, W.D. Reitsma, K.I. Lie, and A.R.J. Girbes. Renal and systemic hemodynamic effects of ibopamine in patients with mild to moderate congestive heart failure. *J. Cardiovasc.Pharmacol.* **25**:361–367, 1995.
9. Littler W.A., and D.J.Sheridan. Placebo controlled trial of felodipine in patients with mild to moderate heart failure. *Brit.Heart J.* **73**:428–433, 1995.
10. Macgowan G.A., D. O´Callaghan, H. Webb, and J.H. Horgan. The effects of captopril on training in patients with ischemic heart disease. *Clin. Cardiol.* **15**:330–334, 1992.
11. McLenachan J.M., J.T. Wilson, and H.J. Dargie. Importance of acillary properties of beta-blockers in angina: A study of celiprolol and atenolol. *Br. Heart J.* **59**:685–689, 1988.
12. Nechwatal W., M. Stauch, H. Sigel, P. Kress, F. Bitter, H. Geffers, and W.E. Adam. Effects of Molsidomine on global and regional left ventricular function at rest and during exercise in patients with angina pectoris. *Clin. Cardiol.* **4**: 248–253, 1981.
13. Nechwatal W., E. König, J. Isbary, H. Greding, and M. Stauch. Haemodynamic and electrocardiographic effects of frusemide during supine exercise in patients with angina pectoris. *Br. Heart J.* **44**: 67–74, 1980.
14. Nechwatal W., A. Stange, H. Sigel, P. Kress, A. Resch, and M. Stauch. Der Einfluß von Piretanid auf die zentrale Hämodynamik und Belastungstoleranz von Patienten mit Angina pectoris. *Herz/Kreisl.* **14**: 91–96, 1982.
15. Needleman P., S. Lang, and E.M. Johnson. Organic nitrates: Relationship between biotransformation and rational angina pectoris therapy. *J. Pharmacol. Exp. Ther.* **181**: 489–497, 1972.
16. Olsen S.L., E.M. Gilbert, D.G. Renlund, D.O. Taylor, F.D. Yanowitz, and M.R. Bristow. Carvedilol improves left ventricular function and symptoms in chronic heart failure: A double-blind randomized study. *JACC* **25**:1225–31, 1995.

17. Parker J.O., B. Farrell, K.A. Lahey, and J. Moe. Effect of intervals between doses on the development of tolerance to isosorbide dinitrate. *N. Engl. J. Med.* **316**:1440–1444, 1987.

18. Panvilov V., I. Wahlqvist, and G. Osson. Use of beta-adrenoceptor blockers in patients with congestive heart failure. *Cardiovasc. Drugs Therap.* **9**:273–287, 1995.

19. Pouleur, H. Neurohormonal and hemodynamic effects of ibopamine. *Clin. Cardiol.* **18**:I-17-I-21, 1995.

20. Stauch M, Grewe N, and Nissen H. Die Wirkung von 2- und 5-Isosorbidmononitrat auf das Belastungs-EKG von Patienten mit Koronarinsuffizienz. *Verh. Dtsch. Ges. Kreislaufforsch.* **41**, 182–184, 1975

21. Stauch M, and N. Grewe. Die Wirkung von Isosorbiddinitrat, Isosorbid-2- und 5-Monitrat auf das Belastungs-EKG und auf die Hämodynamik während Vorhofstimulation bei Patienten mit Angina pectoris. *Z. Kardiol.* **68**: 687, 1979.

22. Stauch, M., G. Grossmann, A. Schmidt, P. Richter, J. Waitzinger, D. Wanjura, W.E. Adam, and W. König. Effect of gallopamil on left ventricular function in regions with and without ischaemia. *Europ. Heart J.* **8** (Suppl.G) 77–83, 1989.

23. Thadani U., S.F. Hamilton, E. Olson, J.L. Anderson, R. Prasad, W. Voyles, R. Doyle, E. Kirsten, and M. Teague. Duration of effects and tolerance of slow-release isosorbide-5-mononitrate for angina pectoris. *Am. J. Cardiol.* **59**, 756–762, 1987.

24. Thadani U., and P.J. de Vane. Efficacy of isosorbide mononitrate in angina pectoris. *Amer. J. Cardiol.* **70**:67G-71G, 1992.

25. Tzivioni, D., S. Gottlieb, N.S. Khurmi, A. Medina, A. Gavish, and S. Stern. Effect of Benazepril on myocardial ischemia in patients with chronic stable angina pectoris. *Eur. Heart J.* **13**:1129–1134, 1992.

26. Unger P., A. Leone, M.Staroukine, S. Degre, and G. Berkenboom. Hemodynamic response to molsidomine in patients with ischemic cardiomyopathy tolerant to isosorbide dinitrate. *J. Cardiovasc. Pharmacol.* **18**:888–894, 1991.

27. Winter U.J., M.Böhm, and E.Erdmann. Überlegungen zum möglichen Stellenwert der neuen Ca^2-Antagonisten bei der Herzinsuffizienztherapie. *Herz/Kreislauf* **27**:162–167.

42

TRAINING FOR THE ENHANCEMENT OF EXERCISE TOLERANCE IN PATIENTS WITH LEFT VENTRICULAR DYSFUNCTION

Andrew J. S. Coats

National Heart and Lung Institute
Imperial College
London, United Kingdom

1. INTRODUCTION

In normal subjects maximal exercise during standard incremental exercise tests is usually limited at a point where delivery of oxygen to the periphery is maximal. This point is called maximal oxygen uptake, defined as the rate of oxygen uptake where a further increment of workload leads to further anaerobic metabolism, but no further increase in oxygen utilisation. In normals this point is also the point at which cardiac output is maximal.

In patients with chronic heart failure increasing cardiac output by infusion of positive inotropic agents does not increase exercise tolerance (50), suggesting that peripheral factors limit exercise in this syndrome. In heart failure patients already performing maximal leg exercise, the recruitment of extra exercising muscle by the addition of arm exercise leads to an increase in oxygen uptake, whereas in controls the oxygen uptake is already maximal and does not increase with this manoeuvre (24). This indicates that it is a limitation in the peripheral utilisation of oxygen rather than its delivery which is deficient in heart failure.

An extensive literature has described abnormalities in peripheral blood flow control and in endothelial function as well as in skeletal muscle mass, histology, metabolism and function in chronic heart failure (21, 49). These changes acting alone, or in combination, lead to early muscle fatigue and dyspnoea, and may even lead to progressive left ventricular dysfunction via their effects on reflex vasoconstriction and sympatho-excitation (12). The most appropriate therapeutic strategies for treating these abnormalities remain unclear.

It is known that training can enhance exercise performance and in normal subjects this increase owes more to adaptations in the periphery, including peripheral vascular and skeletal muscle function, than it does to any change in cardiac or lung function (6). These observation raise the possibility that exercise training could improve the peripheral factors limiting exercise tolerance in patients with heart failure and left ventricular dysfunction.

The Physiology and Pathophysiology of Exercise Tolerance
edited by Steinacker and Ward, Plenum Press, New York, 1996

2. EXERCISE LIMITATION IN CHRONIC HEART FAILURE

2.1. Left Ventricular Function and Cardiac Output

Chronic heart failure is a multi-system syndrome in which the initiating cause is a reduction in cardiovascular functional reserve, but in which a variety of pathophysiological abnormalities develop in other organ systems (38). Numerous studies have failed to demonstrate any significant correlation between objective measures of left ventricular function and exercise tolerance (22) (Fig. 1). Attempts to improve exercise tolerance by increasing cardiac contractility or by the use of haemodynamically active agents which augment cardiac output have not led to any acute increase in exercise tolerance (19). These results suggest, at least in the well treated non-oedematous patients with chronic heart failure, that limited cardiac output is not acutely limiting exercise tolerance.

2.2. Peripheral Blood Flow

It has been known for many years that patients with heart failure exhibit a sub-normal peripheral blood flow response to both vasodilator medication and to exercise (51). The early studies investigated patients with oedematous heart failure, and were performed before the introduction of vasodilator therapy such as the ACE inhibitors. The putative abnormalities which were thought to produce this limited blood flow included persistent excessive vasoconstrictor drive, a sub-clinical oedema of the resistance blood vessel endothelial layer and a relative paucity of peripheral blood vessels. More recently the endothelial-derived vasodilator and constrictor systems have been shown to be abnormal in heart failure, with a reduction in the nitric oxide vasodilator system and an enhancement of the vasoconstrictor endothelin system (20). Exactly how these more recently described abnormalities fit into the earlier experimental results remains uncertain. In addition, with the more widespread use of ACE inhibitors and with better control of oedema the reduction in blood flow is much less evident.

One other confounding factor is the effect of skeletal muscle wasting, which in this condition can be severe (32). When there is less muscle in a limb, it is inevitable that the

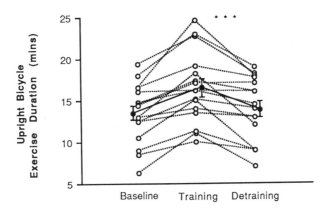

Figure 1. Individual and mean (± SEM) patient upright exercise durations for baseline, training and detraining periods in 17 patients with stable chronic heart failure. *** represents p<0.001 for the comparison between training and detraining periods, n=17. Reproduced from Ref. 11 with permission.

total blood flow to the limb will be reduced, even if the blood flow per unit muscle mass is normal. In studying peripheral blood flow in heart failure there are different techniques used by different groups of workers. Whether the blood flow is measured as absolute flow (such as with thermodilution or Doppler ultrasound methods) or proportionate blood flow (such as with venous occlusion plethysmography or radio-labelled Xenon washout) may lead to different conclusions as to the relative importance of blood flow limitation. When expressed as blood flow per unit active muscle mass and measured in optimally treated patients the reduction in maximal achievable blood flow is not pronounced, but there may be a relative reduction in exercising blood flow due to the abnormalities described above when the competing demands of blood flow are tested by progressive exercise. What can be said, however, is that even when blood flow is taken out of the equation, such as by studying ischaemic exercise tolerance or by augmenting flow by acute administration of vasodilators, then exercise performance and muscle metabolism in heart failure remain profoundly abnormal. Wilson et al. have also recently demonstrated that a proportion of well treated heart failure patients have markedly reduced exercise tolerance, and evidence of early muscular lactate release despite normal skeletal muscle blood flow (49). This shows, that at least in some patients an inherent defect in skeletal muscle metabolism independent of blood flow is operative.

2.3. Skeletal Muscle Function

A variety of skeletal muscle abnormalities have been described in the syndrome of chronic heart failure. These include disordered histology, although without a specific characteristic pattern. Ultrastructurally the mitochondria are abnormal with reduced cristae and depleted oxidative enzymes. In addition there is a shift in fibre type distributions with a predominance of type IIb over IIa fibres (41) similar but not identical to that seen in severe deconditioning.

Metabolic abnormalities have also been described, including early dependence on anaerobic metabolism, excessive early depletion of high energy phosphate pools, and excessive early intra-muscular acidification (33). These changes are thought to be in excess of those attributable to muscle wasting or to deficient blood flow, and may be partially explained by enzymatic changes described in biopsy specimens. These include a reduction in citrate synthase, and beta hydroxy acyl transferase (41). No one as yet has adequately explained the cause of these abnormalities although deconditioning is likely to play a role in some cases. The demonstration of insulin resistance (45), elevated levels of tumour necrosis factor-a, and excessive noradrenaline levels in heart failure may also contribute to skeletal muscle abnormalities which in a severe form are associated with the catabolic state of cardiac cachexia (2).

At a functional level, there is a reduction in the maximal strength of muscle, although perhaps only evident when substantial wasting has occurred, and an early fatiguability. The relative importance of wasting versus qualitative muscle changes remains uncertain. At the macroscopic level, skeletal muscle wasting is seen even in early heart failure, and can be profound. The syndrome of cardiac cachexia has been recognised since ancient times, and in some cases appears to be due to an auto-catabolic process associated with immune cytokine activation and loss of normal anabolic function. Anorexia and intestinal malabsorption may also play a role in some patients. Whatever the mechanism, there does appear to be a reasonable correlation between defects in skeletal muscle and objective exercise limitation in patients with CHF (46). These changes are likely to lead to early muscular fatigue and cessation of exercise because of this symptom.

2.4. Respiratory Muscle Function

Respiratory musculature is abnormal in CHF in a manner similar to skeletal muscle, although it has been far less well studied. Early muscle deoxygenation and the possible development of respiratory muscle fatigue have been described (29, 31). These changes may lead to unpleasant symptoms of dyspnoea in their own right, or to the need for excessive ventilatory effort, to inefficient ventilation with ventilation/perfusion mismatch or to a combination of alterations which lead to dyspnoea. Although the respiratory muscle changes in heart failure are similar to those in skeletal muscle it has been argued that these at least are unlikely to be due to the effect of inactivity alone. Cytokine or neuro-endocrine catabolic effects are more likely to be causative in this setting.

2.5. Symptoms Limiting Exercise

Exercise in CHF is usually limited either by fatigue or dyspnoea and the cause of each has until recently been considered quite separate. Studies of different modes of exercise have shown, however, that the same patient can be limited by different symptoms with subtle changes in the exercise testing procedure, and furthermore that patients limited by each of these two cardinal symptoms do not differ in any their clinical or pathophysiological characteristics (28, 8). These observations suggest that the generation of these symptoms cannot be neatly attributed to different pathophysiological processes, and suggest that we should look in greater detail at the cause of the limiting symptoms.

Regarding the cause of dyspnoea, it has been recognised that elevated pulmonary pressures and pulmonary congestion cannot explain the generation of the symptom in well-diuresed patients. The symptomatic limitation does, however, appear to show a good correlation with the increase in exercise ventilation seen, even when corrected for the carbon dioxide output (4). Thus there is an increased slope in the relationship between ventilation and carbon dioxide output during progressive exercise. The cause of this is unknown although ventilation perfusion mismatch has frequently been proposed (42). This is unlikely to be the sole or predominant cause for there is no obvious abnormality in arterial blood gases during exercise in CHF (7). An alternative explanation is that there is an enhanced sensitivity of ventilatory control mechanisms to progressive exercise in CHF. Two reflexes with enhanced gain in CHF are the arterial chemoreceptors and the muscle "ergoreflex" system (37, 5). Both reflex abnormalities could explain increased ventilation and abnormal sensitivity to dyspnoea during exercise. Ergoreflex hyperactivity, in addition, could link the abnormal skeletal muscle metabolism described above to the ventilatory and respiratory-sensation abnormalities seen in CHF. It could also explain why dyspnoea and fatigue are so difficult to separate in CHF: both may be due to early skeletal muscle metabolic distress during exercise, and both may be sensations carried by the same ergoreflex neural afferent fibres. An overactivity of these fibres and the resultant reflex responses has recently been described in CHF.

Fatigue is much easier to explain. Whether muscle blood flow be limited or muscle metabolism be abnormal the signals from the muscle are perceived as muscular fatigue. It is not known which afferent fibres carry this sensation to the cortex and whether blood-borne chemical messengers also contribute. The best candidate neural afferents are group III and IV fibres, which incidentally are the same fibres implicated in mediating the ergoreflex responses. The chemical triggers for these afferents are unknown but there are suggestions that muscular potassium, adenosine or even stretch receptors responding to flow augmentation may play a role.

3. THERAPEUTIC OPTIONS TO IMPROVE EXERCISE TOLERANCE

3.1. Pharmacological

Diuretics will improve symptoms in grossly oedematous patients but where no pulmonary congestion is present there is no evidence that they improve exercise capacity. Digoxin can, at least in the short term, improve exercise capacity, although whether this is due to positive inotropic effects is uncertain. It has a major a role in the presence of atrial fibrillation as it can prevent an excessive tachycardic response. ACE inhibitors improve prognosis in CHF (13–39). Their ability to improve exercise tolerance, however, is less dramatic, and the effect is slow (19). Some orally acting positive inotropes can increase exercise capacity in the medium term, but usually at the expense of increased mortality and disease progression, so that their use is extremely limited (36, 15). There is increasing interest in the potential role of beta blockers in heart failure, but despite suggestions of beneficial effects on prognosis no convincing improvement in maximal exercise capacity has been demonstrated, although sub-maximal endurance exercise may be improved (47).

3.2. Non-Pharmacological

Cardiac transplantation can improve exercise capacity but only after a considerable delay, and it may require some exercise training of the periphery before the exercise intolerance is overcome. Training alone has repeatedly and convincingly been shown to increase exercise capacity and reduce exercise limiting symptoms in CHF (9).

3.2.1 Exercise Training Prescriptions in Chronic Heart Failure. Patients with left vetricular dysfunction in the absence of the symptoms of were shown to achieve significant improvements in exercise tolerance after participation in a rehabilitation exercise programme (48–16). This was extended to patients with mild to moderate heart failure with demonstrations of improved exercise tolerance, leg blood flow and parameters of ventilatory function after many months of training in uncontrolled retrospective reports (44, 43). These findings were confirmed in patients with stable class II-III heart failure in a controlled comparison of training versus rest (10), even after as little as 8 weeks of training, exercising at 60–80% of peak heart rate for 20–60 minutes 3–5 days per week. These results have been confirmed in many subsequent reports evaluating the mechanisms of the training-induced improvements in exercise capacity.

Training studies to date have been mainly restricted to classes I-III. Class IV patients have been thought to be unable to perform sufficient exercise to obtain worthwhile training effects, at least for whole body dynamic exercise training. The possibility of building up a training effect by the use of single limb training regimes in sequence may have applicability to these more limited patients (26), and this is now an area of considerable research interest. Left ventricular ejection fraction *per se* does not appear to be a reason to prohibit exercise training, as the majority of the benefits are probably mediated via training effects in the periphery, and no trial to date has shown any change in ejection fraction, either beneficial or detrimental. For whole body training the evaluation of maximal oxygen uptake is probably a better guide to the possibilities of exercise training. A maximal oxygen uptake of less than 10 ml/kg/min is probably the lower limit to allow whole body dynamic training to take place.

3.2.2 Aetiology of Heart Failure. It remains possible that some rare forms of cardiomyopathy may show an adverse response to exercise training; these include obstructive valvular disease especially aortic stenosis, obstructive hypertrophic cardiomyopathy or active myocarditities either viral or auto-immune. These conditions should remain contra-indications to exercise therapy.

3.2.3 Choice of Training Regime. It is too early to say how much exercise patients with heart failure should undertake, or which is the best form of exercise; dynamic, isometric or a combination. Patients should be very carefully assessed prior to undertaking exercise training. A full cardiological examination is necessary including tests for the aetiology of heart failure, and the exclusion of active myocarditis or obstructive cardiomyopathy. Any patient exhibiting decompensation of their heart failure should reduce exercise activity and exercise should cease during the episode. Similarly exercise should cease during any inter-current viral or febrile illnesses. Any patient exhibiting exercise-induced serious ventricular arrhythmias should only exercise under supervision and monitoring.

4. EXERCISE TRAINING EFFECTS IN CHRONIC HEART FAILURE

4.1. Left Ventricular Function and Cardiac Output

There has been a suggestion of improved maximal cardiac output in some but not all reports of training in heart failure. No effect on left ventricular ejection fraction has been reported in stable chronic heart failure, although conflicting results have been seen with training in the immediate post-infarct period with severe left ventricular impairment.

4.2. Peripheral Blood Flow

In normal subjects training is known to be able to increase endothelial, large vascular and resistance vessel function (34), and the improved exercise capacity after single limb training undoubtedly depends in part on augmented skeletal muscle blood flow. Although not exhaustively studied in heart failure there is evidence that training can reduce peripheral vascular resistance and increase skeletal muscle blood flow (43). Whether this is predominantly by reducing vasoconstrictor tone, by improving large vessel function or by enhancing endothelial vasodilator function is not known.

4.3. Skeletal Muscle Function

Partial corrections of skeletal muscle abnormalities have been demonstrated after training in heart failure (35–3). These have been seen in single limb and whole body training, in histology, mitochondrial structure (23), oxidative enzymes, magnetic resonance spectroscopy and in both human and animal studies. Although it is likely skeletal muscle wasting can be reversed this has not as yet been documented. The relevance of skeletal muscle improvements to improved exercise tolerance has not been proven despite the likely interaction. The over-activity of the skeletal muscle ergoreflex effects on haemodynamic and ventilatory control during exercise are partially corrected by single limb training (37).

4.4. Respiratory Muscle Function

Specific respiratory muscle training has been reported in chronic heart failure, with specific benefits and being associated with improved exercise tolerance (30). Whether part of a generalised training effect with whole body dynamic exercise or as a specific respiratory muscle training regimen, this appears to be a useful effect.

4.5. Autonomic Nervous System Activity

Exercise training reduces the activiy of the sympathetic and renin angiotensin systems in both normal subjects and in atients with heart failure. Abnormailities in noradrenaline spillover, heart rate variability and exercise heart rate responses are at least patially corrected (11, 25).

4.6. Symptoms Limiting Exercise

Exercise training in heart failure reduces the sensations of both fatigue and dyspnoea, as part of its exercising enhancing effects (11). Although the precise mechanism is unknown it is likely that it is the reversal of the abnormal peripheral pathophysiology which limits exercise in the first place which is the mechanism of this symptomatic improvement. Paramount in this effect is likely to be the profound effects on skeletal muscle function and the muscle ergoreflex. Objective improvements have been noted in both skeletal muscle function and in the control of ventilation during progressive exercise (44, 18).

5. CONCLUSIONS

Exercise training, especially as part of a comprehensive cardiac rehabilitation programme, can increase exercise capacity in compensated stable non-oedematous chronic heart failure. It is associated with improvements in peripheral vascular, muscular and metabolic function. Indices of sympatho-vagal balance may be improved and this suggests a potential protective role in the progression of left ventricular dysfunction and the symptoms of heart failure.

6. REFERENCES

1. Adamopoulos, S., A.J. Coats, F. Brunotte, L. Arnolda, T. Meyer, C.H. Thompson, J.F. Dunn, J. Stratton, G.J. Kemp, G.K. Radda, et al. Physical training improves skeletal muscle metabolism in patients with chronic heart failure. *J. Am. Coll. Cardiol.* **21**:1101–1106, 1993.
2. Anker, S.D., M. Volterrani, J. Swan, T.P. Chua, P.A. Poole-Wilson, and A.J.S. Coats. Hormonal changes in cardiac cachexia, *Circulation* **92**:I-206-I-207, 1995.
3. Brunotte, F., C.H. Thompson, S. Adamopoulos, A. Coats, J. Unitt, D. Lindsay, L. Kaklamanis, G.K. Radda, and B. Rajagopalan. Rat skeletal muscle metabolism in experimental heart failure: Effects of physical training. *Acta Physiol. Scand.* **154**:439–447, 1995.
4. Buller, N.P., and P.A. Poole-Wilson. Mechanism of the increased ventilatory response to exercise in patients with chronic heart failure. *Br. Heart J.* **63**:281–283, 1990.
5. Chua, T.P, A. Amadi, A.L. Clark, D. Harrington, and A.J.S. Coats. Increased chemosensitity to hypoxia at rest and during exercise in chronic heart failure. *Br. Heart J.* **73** (suppl 3):26, 1995.

6. Clausen, J.P. Circulatory adjustments to dynamic exercise and effect of physical training in normal subjects and in patients with coronary artery disease. *Progr. Cardiovasc. Dis.* **18**:459–495, 1976.

7. Clark, A., and A. Coats. The mechanisms underlying the increased ventilatory response to exercise in chronic stable heart failure. *Eur. Heart J.* **13**:1698–1708, 1992.

8. Clark, A.L., J.L. Sparrow, and A.J.S. Coats. Muscle fatigue and dyspnoea in chronic heart failure: two sides of the same coin? *Eur. Heart J.* **16**:49–52, 1995.

9. Coats, A.J.S. Exercise rehabilitation in chronic heart failure. *J. Am. Coll. Cardiol.* **22** (suppl A):172A-177A, 1993.

10. Coats, A.J.S., S. Adamopoulos, T.E. Meyer, J. Conway, and P. Sleight. Effects of physical training in chronic heart failure. *Lancet* **335**:63–66, 1990.

11. Coats, A.J., S. Adamopoulos, A. Radaelli, A. McCance, T.E. Meyer, L. Bernardi, P.L. Solda, P. Davey, O. Ormerod, C. Forfar, et al. Controlled trial of physical training in chronic heart failure. Exercise performance, hemodynamics, ventilation, and autonomic function. *Circulation* **85**:2119–2131, 1992.

12. Coats, A.J.S., A.L. Clark, M. Piepoli, M. Volterrani, and P.A. Poole-Wilson. Symptoms and quality of life in heart failure; the muscle hypothesis. *Br. Heart. J.* **72** (Suppl):S36-S39, 1994.

13. Cohn, J.N., D.G. Archibald, S. Ziesche, J.A. Franciosa, W.E. Harston, F.E. Tristani, W.B. Dunkman, W. Jacobs, G.S. Francis, K.H. Flohr, S. Goldman, F.R. Cobb, P.M. Shah, R. Saunders, R.D. Fletcher, H.S. Loeb, V.C. Hughes, and B. Baker. Effect of vasodilator therapy on mortality in chronic congestive heart failure. *New Engl. J. Med.* 314:1547–1552, 1986.

14. Cohn, J.N., G. Johnson, S. Ziesche, F. Cobb, G. Francis, F. Tristani, R. Smith, W.B. Dunkman, H. Loeb, M. Wong, G. Bhat, S. Goldman, R.D. Fletcher, J. Doherty, C.V. Hughes, P. Carson, G. Cintron, R. Shabetai, and C. Haakenson. A comparison of enalapril with hydralazine-isosorbide dinitrate in the treatment of chronic congestive heart failure. *New Engl. J. Med.* **325**:303–310, 1991.

15. Colucci,W.S., E.H. Sonnenblick, K.F. Adams, M. Berk, S.C. Brozena, A.J. Cowley, J.M. Grabicki, S.A. Kubo, T. LeJemtel, W.A. Littler, et al. Efficacy of phosphodiesterase inhibition with milrinone in combination with converting enzyme inhibitors in patients with heart failure. The Milrinone Multicenter Trials Investigators. *J. Am Coll. Cardiol.* **22**:113A-118A, 1993.

16. Conn, E.H., R.S. Williams, and A.G. Wallace. Exercise responses before and after physical conditioning in patients with severely depressed left ventricular function. *Am. J. Cardiol.* **49**:296–300, 1982.

17. Consensus Trial Study Group. Effects of enalapril on mortality in severe congestive heart failure: results of the Cooperative North Scandinavian Enalapril Survival Study. *New Engl. J. Med.* **316**:1429–1435, 1987.

18. Davey, P., T. Meyer, A. Coats, S. Adamopoulos, B. Casadei, J. Conway, and P. Sleight. Ventilation in chronic heart failure: effects of physical training. *Br. Heart J.* **68**:473–477, 1992.

19. Drexler, H., U. Banhardt, T. Meinertz, H. Wollschläger, M. Lehmann, and H. Just. Contrasting peripheral short-term and long-term effects of converting enzyme inhibition in patients with congestive heart failure. A double-blind, placebo-controlled trial. *Circulation* 79:491–502, 1989.

20. Drexler, H., D. Hayoz, T. Munzel, H. Just, R. Zelis, and H.R. Brunner. Endothelial function in congestive heart failure. *Am. Heart J.* 126:761–764, 1993.

21. Drexler, H., U. Riede, T. Münzel, H. König, E. Funke, and H. Just, Alterations of skeletal muscle in chronic heart failure. *Circulation* **85**:1751–1759, 1992.

22. Franciosa, J.A., M. Park, and T.B. Levine. Lack of correlation between exercise capacity and indexes of resting left ventricular performance in heart failure. *Am. J. Cardiol.* **47**:33–39, 1981.

23. Hambrecht, R., J. Niebauer, E. Fiehn, B. Kalberer, B. Offner, K. Hauer, U. Riede, G. Schlierf, W. Kubler, and G. Schuler. Physical training in patients with stable chronic heart failure: effects on cardiorespiratory fitness and ultrastructural abnormalities of leg muscles. *J. Am Coll. Cardiol.* **25**:1239–1249, 1995.

24. Jondeau, G., S.D. Katz, L. Zohman, M. Goldberger, M. McCarthy, J.P. Bourdarias, and T.H. LeJemtel. Active skeletal muscle mass and cardiopulmonary reserve. Failure to attain peak aerobic capacity during maximal bicycle exercise in patients with severe congestive heart failure. *Circulation* 86:1351–1356, 1992.

25. Kiilavuori, K., L. Toivonen, H. Naveri, and H. Leinonen. Reversal of autonomic derangements by physical training in chronic heart failure assessed by heart rate variability. *Eur. Heart J.* **16**:490–495, 1995.

26. Koch, M., H. Douard, and J.P. Broustet. The benefit of graded physical exercise in chronic heart failure. *Chest* **101**:231S-235S, 1992.

27. Letac, B., A. Cribier, and J.F. Desplanches. A study of left ventricular function in coronary patients before and after physical training. *Circulation* **56**:375–378, 1977.

28. Lipkin, D.P., R. Canepa-Anson, M.R. Stephens, and P.A. Poole-Wilson. Factors determining symptoms in heart failure: comparison of fast and slow exercise tests. *Br. Heart. J.* **55**:439–445, 1986.

29. Mancini, D.M., N. Ferraro, D. Nazzaro, B. Chance, and J.R. Wilson. Respiratory muscle deoxygenation during exercise in patients with heart failure demonstrated with near-infrared spectroscopy. *J. Am. Coll. Cardiol.* **18**:492–498, 1991.

30. Mancini, D.M., D. Henson, J. La Manca, L. Donchez, and S. Levine. Benefit of selective respiratory muscle training on exercise capacity in patients with chronic congestive heart failure [see comments]. *Circulation* **91**:320–329, 1995.

31. Mancini, D.M., D. Henson, J. LaManca, and S. Levine. Evidence of reduced respiratory muscle endurance in patients with heart failure. *J. Am. Coll. CArdiol.* **24**:972–981, 1994.

32. Mancini, D.M., G. Walter, N. Reichek, R. Lenkinksi, K.K. McCully, J.L. Mullen, and J.R. Wilson. Contribution of skeletal muscle atrophy to exercise intolerance and altered muscle metabolism in heart failure. *Circulation* **85**:1364–1373, 1992.

33. Massie, B.M., M. Conway, B. Rajagopalan, R. Yonge, S. Frostick, J. Ledingham, P. Sleight, and G. Radda. Skeletal muscle metabolism during exercise under ischemic conditions in congestive heart failure. Evidence for abnormalities unrelated to blood flow. *Circulation* **78**:320–326, 1988.

34. McAllister, R.M. Endothelial-mediated control of coronary and skeletal muscle blood flow during exercise: Introduction. *Med. Sci. Sports Exerc.* **27**:1122–1124, 1995.

35. Minotti, J.R., E.C. Johnson, T.H. Hudson, G. Zuroske, G. Murata, E. Fukushima, T.G. Cagle, T.W. Chick, B.M. Massie, and M.V. Icenogle. Skeletal muscle response to exercise training in congestive heart failure. *J. Clin Invest.* **86**:751–758, 1990.

36. Packer, M., J.R. Carver, R.J. Rodeheffer, R.J. Ivanhoe, R. DiBianco, S.M. Zeldis, G.H. Hendrix, W.J. Bommer, U. Elkayam, M.L. Kukin, et al. Effect of oral milrinone on mortality in severe chronic heart failure. The PROMISE Study Research Group [see comments]. *N. Engl J. Med.* **325**:1468–1475, 1991.

37. Piepoli, M., A.L. Clark, M. Volterrani, S. Adamopoulos, P. Sleight, and A.J.S. Coats. Contribution of muscle afferents to the hemodynamic, autonomic and ventilatory responses to exercise in patients with chronic heart failure. Effects of physical training. *Circulation* (In press).

38. Poole-Wilson, P.A. Chronic heart failure: causes, pathophysiology, prognosis, clinical manifestations, ivestigations. In: *Diseases of the Heart*, edited by P.A. Poole-Wilson. London: Balliere Tindall, London, 1989, pp. 48–57.

39. SOLVD Investigators. Effect of enalapril on survival in patients with reduced left ventricular ejection fractions and congestive heart failure. *New Engl. J. Med.* **325**:293–302, 1991.

40. Stratton, J.R., J.F. Dunn, S. Adamopoulos, G.J. Kemp, A.J. Coats, and B. Rajagopalan. Training partially reverses skeletal muscle metabolic abnormalities during exercise in heart failure. *J. Appl. Physiol.* **76**:1575–1582, 1994.

41. Sullivan, M.J. H.J. Green, and F.R. Cobb. Skeletal muscle biochemistry and histology in ambulatory patients with long-term heart failure. *Circulation* **81**:518–527, 1990.

42. Sullivan, M.J., M.B. Higginbotham, and F.R. Cobb. Increased exercise ventilation in patients with chronic heart failure: intact ventilatory control despite hemodynamic and pulmonary abnormalities. *Circulation* **77**:552–559, 1988.

43. Sullivan, M.J., M.B. Higginbotham, and F.R. Cobb. Exercise training in patients with severe left ventricular dysfunction: Hemodynamic and metabolic effects. *Circulation* **78**:506–515, 1988.

44. Sullivan, M.J., M.B. Higginbotham, and F.R. Cobb. Exercise training in patients with chronic heart failure delays ventilatory anaerobic threshold and improves submaximal exercise performance. *Circulation* **79**:324–329, 1989.

45. Swan, J.W., C. Walton, I.F. Godsland, A.L. Clark, A.J.S. Coats, and M.F. Oliver. Insulin resistance in chronic heart failure, *Eur. Heart J.* **15**:1528–1532,. 1994.

46. Volterrani, M., A.L. Clark, P.F. Ludman, J.W. Swan, S. Adamopoulos, M. Piepoli, and A.J.S. Coats. Determinants of exercise capacity in chronic heart failure. *Eur. Heart J.* **15**:801–809, 1994.

47. Waagstein, F. Beta blockers in heart failure. *Cardiology* **82** (suppl 3):13–18, 1993.

48. Williams, R.S. Exercise training of patients with ventricular dysfunction and heart failure. *Cardiovasc. Clin.* **15**:218–231, 1985.

49. Wilson, J.R., D.M. Mancini, and W.B. Dunkman. Exertional fatigue due to skeletal muscle dysfunction in patients with heart failure. *Circulation* 87:470–475, 1993.

50. Wilson, J.R., J.L. Martin, and N. Ferraro. Impaired skeletal muscle nutritive flow during exercise in patients with congestive heart failure: role of cardiac pump dysfunction as determined by the effect of dobutamine. *Am. J. Cardiol.* **53**:1308–1315, 1984.

51. Zelis, R., S.H. Nellis, J. Longhurst, G. Lee, and D.T. Mason. Abnormalities in the regional circulations accompanying congestive heart failure. *Progr. Cardiovasc. Dis.* **18**:181–199, 1975.

43

CLINICAL DOSE-RESPONSE EFFECTS OF EXERCISE

E. W. Banister, R. H. Morton, and J. R. Fitz-Clarke

School of Kinesiology
Simon Fraser University
Burnaby, B.C., Canada V5A 1S6

1. INTRODUCTION

Stimuli used to probe the acute physical/physiological response to exercise are well known and characterised so that the pattern and size of their effect may be used diagnostically with some precision (21,22). There is no similar precision of thought on training. Although most training studies show the generally positive benefit of exercise, there is no formal training theory developed for exercise such as the type, quantity or pattern of a training stimulus necessary to produce a prescribed measured effect and the field remains empirical and fallible. Conflicting data on physical, physiological and biochemical measures may be widely observed in studies of the effect of a reduced, maintained, increased or terminated stimulus in a subject undertaking training (4,5,9–12,14). In clinical studies definition of a threshold level of training for reducing such symptoms as sedentary afflictions (8,16), high plasma cholesterol (18) or moderate to medium hypertension (6,20) is still qualitative, expressed variously, in units of distance covered, time spent, or intensity of effort in training and a complex periodisation system specifying the allocation of training time to various levels of activity and relative rest/detraining. Without a theory of training, such empirical investigations remain imprecise and their conclusions speculative and argumentative. A unique feature of the new theory is the proposal of an appropriate quantitative, unit measure of training which must be defined and adopted so that quantitative study of the training response to a stimulus may be unambiguously conducted. The idea will be developed and rationalised further in this paper.

Perhaps the most insightful analysis of the effect of training on physiological biochemistry is contained in a series of papers between 1977 and 1988 authored respectively by Booth (2), Terjung (19), Dudley *et al.* (7) and Mader (13), and illustrated in Figures 1 to 4.

Booth discussed the shape of the response curve of cytochrome *c* concentration in muscle to both a step and ramp change in the duration of daily training. He demonstrated (Fig. 1) the characteristic single exponential rise of an induced biochemical variable to a

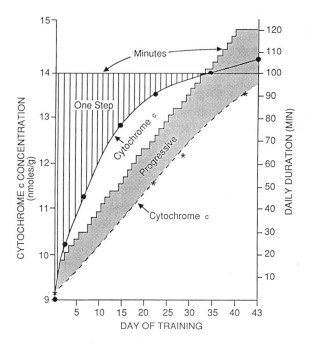

Figure 1. Growth of muscle cytochrome *c* due to a step and ramp increment in daily duration of training. Reproduced from Ref. 2 with permission.

steady-state asymptote for a large step increment in the duration of training with intensity of training presumably held constant, throughout a period of 43 days (d). Analysis of the physiological or biochemical change induced by the ramp dose also shown in Figure 1 is less easy to appreciate and quantify since the dose is always changing and although the underlying kinetics of change are the same (i.e. exponential), the character of the stimulus (e.g step, ramp) dictates the shape of the response curve rise towards the asymptote. The amplitude response of cytochrome *c*, to the same total dose of training in the step and ramp, is best estimated therefore from the pattern of response to the step increase in training stimulus. Thus, in Figure 1, the amplitude response of 14.5 nmol/g cytochrome *c* may be estimated from the step training increment after as little as 28 d of training compared with 43 d for the ramp pattern. Terjung confirmed the shape of the response curve of muscle cytochrome *c* of two different muscle types (a,c) to training, as rising to an asymptote in approximately 28 d when induced by a constant training stimulus (*A* curves: Fig. 2, top and bottom). In this study Terjung also demonstrated the exponential shape of the decay curve of cytochrome *c* to removal of the training stimulus. Interestingly in both of the recovery curves (*C* curves: Fig. 2, top and bottom) recovery begins from a higher concentration of muscle cytochrome *c* than was demonstrated in animals sacrificed at the end of training. This protein excess at the onset of detraining or recovery is consistent with Mader's theoretical modeling described below, of excess protein production at the onset of the "off training" signal.

Dudley et al. (Fig. 3) showed that the asymptotic nature of growth in cytochrome *c* concentration was more sensitive to the metabolic intensity of the training stimulus than simply to its duration. Training carried on for longer than 45–55 min in a session is shown to be ineffective in inducing a greater muscle cytochrome *c* concentration without an ac-

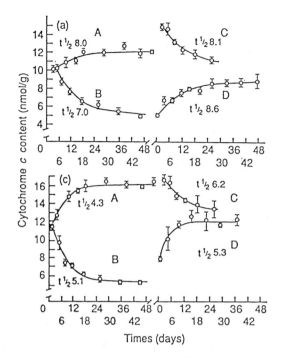

Figure 2. Growth of muscle cytochrome *c* to daily training. Reproduced from Ref. 19 with permission.

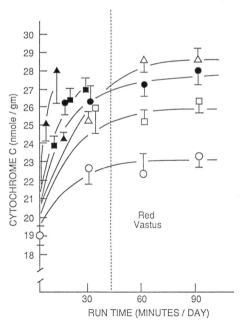

Figure 3. Growth of muscle cytochrome *c* because of increased duration and intensity of training. Reproduced from Ref. 7 with permision.

companying increase in the intensity of training shown by the running speed symbols: slow (o) to fast (Δ).

Mader's elegant theoretical work (Fig. 4) is perhaps of most significance to the clinician and exercise physiologist seeking to develop a dose-response theory of training. Mader proposed that the muscle cellular metabolic "on" response to a step-increase in training, sufficient to stimulate adaptation, is first disruptive to muscle, producing protein specific fragments (PSFs), visible in the circulation (identifiable as a protein "dump") which induces expression of muscle cell mRNA.

Subsequently, a phase-lagged growth of sufficient new protein to enable adaptation to the new metabolic demand on the cell is achieved. The decline and growth of all the metabolites shown in Figure 4 is characterised as exponential by Mader. The "off" metabolic stimulus signal produces a second protein "dump" to the vascular space, despite the immediate cessation of the enhanced metabolic demand, due to a reverse, phase-lagged, exponential decay in cellular metabolic activity to baseline. This mechanism may account for the cytochrome c rebound shown in Terjung's data.

2. A RATIONAL THEORY OF TRAINING

A theory of training must define several elements: (a) the unit of measurement in which training is expressed must be rationalized; (b) the way in which a quantity of training affects responsive elements in a targeted organism must be hypothesised; and (c) theory must be related to reality and the actual pattern of change in an organism defined by the theory, produced by a defined quantity and pattern of a training stimulus, must mirror

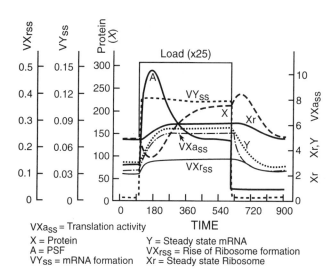

Figure 4. A large on-step increment in training for 25 d produces focal muscle disruption and a high loss of cellular protein (PSFs) which produces maximum activation (A) and expression of mRNA (VYss). A high cellular mRNA content (Y) raises translation (VXass), ribosome formation (VXr) and content (Xr), which leads to cellular protein (enzyme, structural protein) again coming into balance with the enhanced metabolic activity of the cell. The off-stimulus produces a phase-lagged cellular protein overshoot and elimination. At this time, for a limited period, performance will be high as the cell is removed from the disruptive effect of a high sustained training quantity and fatigue decays rapidly. Reproduced from Ref. 13 with permission.

closely the real pattern and amplitude of response of a relevant physical measure by the organism (e.g. a criterion standard distance run).

2.1. Monitoring Training

Conventionally, training is defined qualitatively in terms of its frequency (n), duration (D) and intensity (I) of effort. There is currently no accepted formula for integrating these separate entities into a rational expression of a training stimulus. Usually expression of the volume (V) or quantity (Q) of training and the response to it, is arbitrarily assessed, attributed independently to "n, D or I" and qualitatively discussed relative to each (3,4,9,11,12,14,17).

2.2. Quantification of Training

Here, training is defined as a quantity w(t), expressed in arbitrary units (A.U.) per training session, representing the integrated effect (expressed as a product) of the duration (D) of training in min and the metabolic intensity of the effort (MI). The latter is computed from the product of the exercise delta (Δ) heart rate ratio, (HRex - HRrest)/(HRmax - HRrest), which is sometimes referred to as the fraction of the heart rate reserve, and a metabolic intensity factor (Y) relating the Δ Heart Rate Ratio (ΔHRR) to the exponential rise in a metabolic variable (blood lactate) as the heart rate increases, since the rise in delta heart rate alone does not represent the concomitant degree of metabolic arousal (19). Thus the quantity of training w(t) defined as the Training Impulse is:

$$w(t) = D \times \Delta HRR \times Y \qquad (1)$$

where

$$Y = (\exp) a\Delta HRR \ (a = 1.92 \text{ in males, and } 1.67 \text{ in females}) \qquad (2)$$

Any number of periods within a session or any number of sessions/d or /wk conducted at a different ΔHRR (Fig. 5) may be quantified in this way and integrated to provide a sessional, daily or weekly total Trimp score but usually a daily graphical pattern of training throughout a period (Figs. 9 & 10).

2.3. Physical Assessment of Training Effect: Criterion Performance Measurement

The training response should be continuously evaluated from a subject's performance on a standard physical running task at any distance up to 5 km. This is to assess the fatiguing effect of training and to model the theoretical prediction of physical performance from the daily training impulse against a pattern of real performance in order to validate the theoretical concepts of the training theory (see Figs. 9 & 10 below). Running is a weight-bearing endurance activity considered to be the best type of criterion performance test (CP) to evaluate and provide feedback control of the fatigue developed by effective training in many types of physical activity. For ease of scoring the CP, any run time is easily transformed to a points score by assuming that a run time ordinate Y(x) follows an exponential decay to some limit on a scale, where a world record (WR) scores 1000 points,

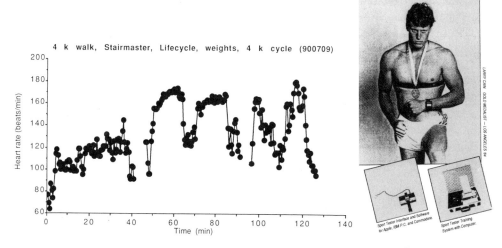

Figure 5. Continuous heart rate recording of a variety of different quantities of training which may be characterized by the product of the ΔHRR and duration of the specific time interval of the activity.

and a time 6 times slower than the WR scores 0 points. L is a defined limit to further improvement by even top class athletes (Fig. 6). Thus X the CP points score is given by:

$$X = 1/b \ln [(a - L)/(Y - L)] \tag{3}$$

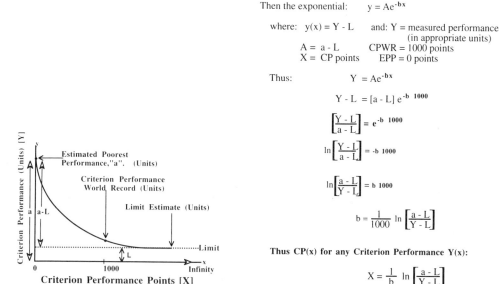

Then the exponential: $y = Ae^{-bx}$

where: $y(x) = Y - L$ and: Y = measured performance
                                        (in appropriate units)
        $A = a - L$ CPWR = 1000 points
        $X = $ CP points   EPP = 0 points

Thus: $Y = Ae^{-bx}$

$$Y - L = [a - L] e^{-b\ 1000}$$

$$\left[\frac{Y - L}{a - L}\right] = e^{-b\ 1000}$$

$$\ln\left[\frac{Y - L}{a - L}\right] = -b\ 1000$$

$$\ln\left[\frac{a - L}{Y - L}\right] = b\ 1000$$

$$b = \frac{1}{1000} \ln\left[\frac{a - L}{Y - L}\right]$$

Thus CP(x) for any Criterion Performance Y(x):

$$X = \frac{1}{b} \ln\left[\frac{a - L}{Y - L}\right]$$

Figure 6. Method of calculating CP from a time trial, an estimated world record worth 1000 points for the distance run and a worst possible time (approximately 6 times slower than the world record) worth 0 points.

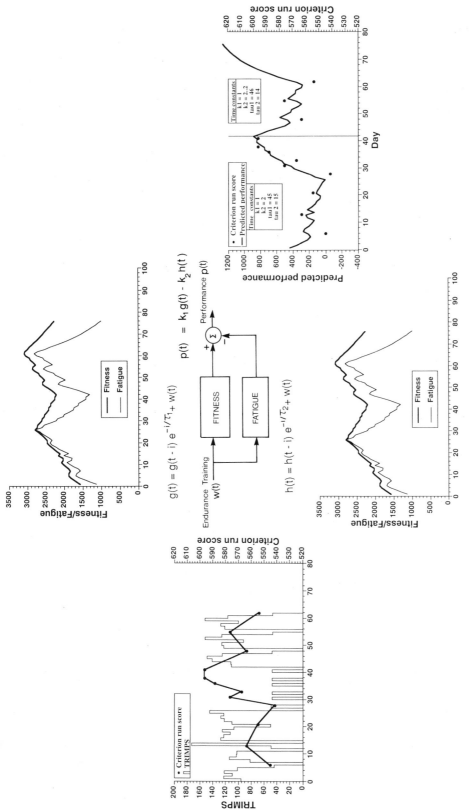

Figure 7. A systems model of the effect on performance of a step on/off quantity of training dose (in arbitrary units). Predicted performance may be modeled to fit the pattern of criterion performance times (CP) and then transformed to CPx scores, by altering the parameter vector k1, k2, τ1, τ2, from the default values (1, 2, 45d, 15d). This defines an individual's parameter vector and enables prediction of his/her future performance from future training. Modeling ideally should be performed bi-weekly to sustain close surveillance of the real performance response to model predicted performance.

2.4. A Theory of Training (Systems Model)

Figure 7 illustrates the main tenets of a training model developed previously (1,3,15). In the model the training impulse, w(t), is transformed by model dimensionless weighting factors, k1 and k2, to represent the training contribution to model fitness (g(t)) and model fatigue (h(t)), respectively. During the interval between training, each model component decays exponentially with a typical decay time constant $\tau 1$ and $\tau 2$, to a residual value. Each transformed training impulse is added at the end of every new day's training session to the previous day's respective residual so that fitness and fatigue grow and decay recurrently throughout a period of training. At any time, the difference between model fitness k1g(t) and fatigue k2h(t) represents predicted performance p(t):

$$p(t) = k1g(t) - k2h(t) \qquad (4)$$

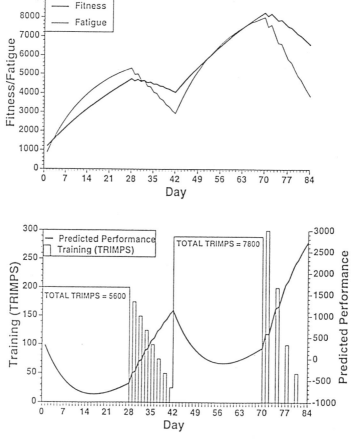

Figure 8. A theoretical plot of the physical response to two incremental doses of training each followed by recovery.

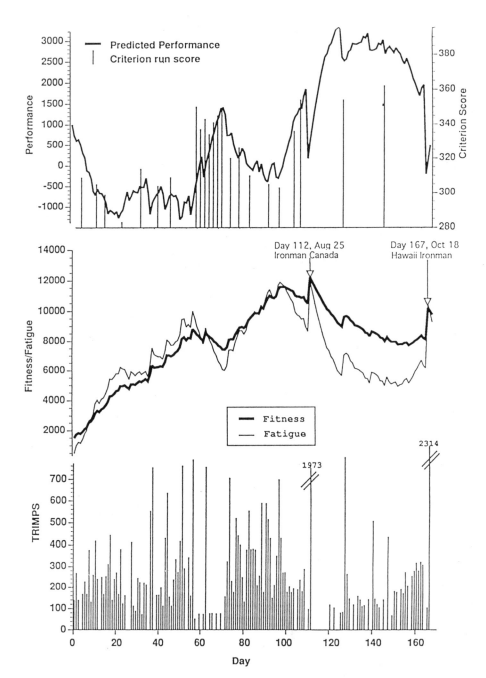

Figure 9. Predicted vs. real performance (top panel) modeled for various on/off blocks of training (bottom panel) in a triathlete completing triathalon competitions.

3. QUANTIFIED EXPRESSION OF THE EFFECT OF TRAINING ON PERFORMANCE

3.1. Format of Training

The step function training stimulus illustrated in Figures 1 to 4 seems both visually and mathematically to be the best training format on which to base a quantified assessment of the physical, physiological and biochemical response to training. The training theory described here proposes that a step perturbation in training stimulus, sufficient to induce a positive exponential growth response in a system's physiological and biochemical variables, produces a decay in physical ability as training fatigue grows. Removal of the stimulus induces a phase-lagged decay of the affected physiological and biochemical variables back to the base-line condition concomitantly with an acute pronounced enhancement of physical performance as the accumulated fatigue of training decays much faster than accumulated fitness because of cellular growth. The on/off block stimulus format of training allows precise study and characterization of the separate "on" and "off" training responses.

3.2. Modeled Effect of Incremental Step Training Separated by Detraining

Figure 8 demonstrates the theoretical, modeled, physical response, using the default parameter vector k1= 1, k2 = 2, $\tau 1$ = 45 d, $\tau 2$ = 15 d, to two incremental 28 d blocks of training each followed by a 14 d exponential decay in training. The initial decay in performance mirrors the disruptive phase of tissue catabolism hypothesized by Mader and the development of fatigue in the performer. Before the end of each training block, it is noticeable that predicted performance begins to improve as it is constrained to do by the mathematical relationship of the model parameters and as is suggested by the Mader model when cellular protein production again begins to come into balance with the enhanced training demand. Once training is markedly reduced, the rebound in physical performance above the baseline condition is as pronounced as is shown in the protein rebound curve of Mader's conceptual model. The transitory rebound in cytochrome c concentration of muscle in heavily trained animals at the onset of the "off-training" signal shown in Figure 2 (19) also probably reflects the increased metabolic potential accrued from training.

3.3. On/Off Effect of Training on Physical Performance in a Triathlete

Figure 9 shows 160 d of a triathlete's training in the stepped on/off format recommended for the quantitative analysis of a training stimulus. The stepped block training (bottom panel) is able to be less well defined in actual practice than in the idealized model of Figure 8, but the on/off features seen in the model of training are easily recognizable. Developing fitness and fatigue are shown in the middle panel. The top panel of the figure shows the alignment of predicted performance from training (solid line) modeled to the criterion performances shown by vertical bars. The athlete was prepared well for a short-course triathlon competition on day 65, which he won. Then followed a period of heavy training prior to a detraining period leading to an Ironman competition on day 112 when again the athlete performed well (finishing in 9.75 hr). Preparation by the athlete on his own for another full course Ironman competition day 167 was poor. A low training density produced a devastating loss of fitness (middle panel: thick line), quite setting aside any value of reduced fatigue (midle panel: thin line) which was un-needed at that point. The athlete then increased the intensity of training, narrowing even further the gap between accrued fitness and fatigue and reducing predicted performance. Performance on day 167 was predictably poorer (finishing time = 11.25 hr).

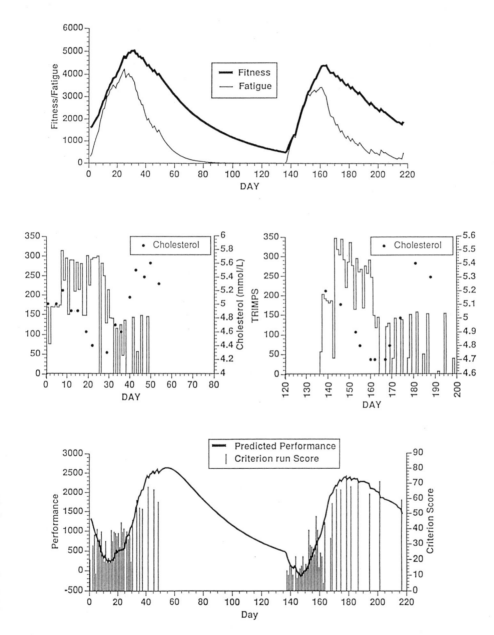

Figure 10. Response of physical and biochemical variables to equal doses of training, each separated by 86 days of non-training.

3.4. On/Off Effect of Training on Physiochemical Variables

Figure 10 illustrates two very similar profiles of 28 d of training and detraining in a subject separated by 86 d of non-training.

The decay of physical performance (bottom panel: vertical bars) and serum total cholesterol (bottom panel: solid circles) during high-quantity training (concentrated in 2 sessions/d) is apparent. The dose/response effect on both the physical and physiochemical measures in each of the two training and detraining periods is remarkably consistent. While the training stimulus and muscle cytochrome c growth have been shown to be in phase, the relationship between both the physical and the physiochemical response to the quantity of training undertaken in Figure 10 are clearly seen to be out of phase. It also seems possible that individual variability in the model parameter vector proposed in the current theory will negate efforts to apply group training methods when seeking the best dose/response effect from exercise for the individual.

4. REFERENCES

1. Banister, E.W., T.W. Calvert, M.V. Savage, and T.M. Bach. A systems model of training for athletic performance. *Aust.J. Sports Med.* 7:57–61, 1975.
2. Booth, F. Effects of endurance exercise on cytochrome c turnover in skeletal muscle. *Ann. NY Acad. Sci.* 301: 431–439, 1977.
3. Calvert, T.W., E.W. Banister, M.V. Savage, and T.M. Bach. A systems model of the effects of training on physical performance. *IEEE Trans. Syst. Man Cybernet.* 6: 94–102, 1976.
4. Costill, D.L., R. Thomas, R.A. Goldbergs, D. Pascoe, C. Lambert, S. Barr, and W.J. Fink. Adaptations to swimming training: influence of training volume. *Med. Sci. Sports Exerc.* 23: 371–377, 1991.
5. Coyle, E.F., W.H. Martin, D.R.Sinacore, M.J. Joyner, J.M. Hagberg, and J.O.Holloszy. Time course of loss of adaptations after stopping prolonged intense endurance training. *J. Appl. Physiol.* 57:1857–1864, 1984.
6. DiCarlo, S.E., H.L. Collins, M.G. Howard, C-Y. Chen, T.J. Scislo, and R.D. Patil. Postexertional hypotension: a brief review. *Sports Med. Train. Rehab.* 5: 17–27, 1994.
7. Dudley, G.A., W.M. Abraham, and R.L. Terjung. Influence of exercise intensity and duration on biochemical adaptations in skeletal muscle. *J. Appl. Physiol.* 53: 844–850, 1982.
8. Ehsani, A.A., D.R. Biello, J. Schult, B.E. Sobel, and J.O. Holloszy. Improvement of left ventricular contractile function by exercise training in patients with coronary artery disease. *Circulation* 7: 350–358, 1986.
9. Hickson, R.C., C. Foster, M.L. Pollock, T.M. Galassi, and S. Rich. Reduced training intensities and loss of aerobic power, endurance and cardiac growth. *J. Appl. Physiol.* 58: 492–499, 1985.
10. Hickson, R.C., and M.A. Rosenkoetter. Reduced training frequencies and maintenance of increased aerobic power. *Med. Sci. Sports Exerc.* 13: 13–16, 1981.
11. Houmard, J.A., T. Hortobagyi, R.A. Johns, N.J. Bruno, C.C. Nute, M.H. Shinebarger, and J.W. Welborn. Effect of short-term training cessation on performance measures in distance runners. *Int. J. Sports Med.* 13: 572–576, 1992.
12. Kirwan, J.P., D.L. Costill, M.G. Flynn, J.B. Mitchell, W.J. Fink, P.D. Neufer, and J. A. Houmard. Physiological responses to successive days of intense training in competitive swimmers. *Med. Sci. Sports Exerc.* 20: 255–259, 1988.
13. Mader, A.J. A transcription-translation activation feedback circuit as a function of protein degradation. *J. Theor. Biol.* 134: 135–157, 1988.
14. McConell, G.K., D.L. Costill, J.J. Widrick, M.S. Hickey, H.Tanaka, and P.B. Gastin. Reduced training volume and intensity maintain aerobic capacity but not performance in distance runners. *Int. J. Sports Med.* 14: 33–37, 1993.
15. Morton, R.H., J.R. Fitz-Clarke, and E.W. Banister. Modeling human performance in running. *J. Appl. Physiol.* 69:1171–1177, 1990.
16. Saltin, B., B. Blomquist, J.H. Mitchell, R.L. Johnson, K. Wildenhall, and C.B. Chapman. Response to submaximal and maximal exercise after bed rest and training. *Circulation* 38: supp 7, 1968.

17. Shephard, R.J. Intensity, duration and frequency of exercise as determinants of the response to a training regime. *Int. zeitsch. agnew Physiol.* **26**: 272–278, 1969.

18. Superko, H.R. Exercise training, serum lipids and lipoprotein particles: is there a change in threshold? *Med. Sci. Sports Exerc.* **23**:667–685, 1991.

19. Terjung, R.L. The turnover of cytochrome c in different skeletal muscle fiber types of the rat. *Biochem. J.* **178**: 569–574, 1979.

20. Tipton, C. In: *Exercise and Spots Science Reviews, vol. 19*, edited by J.O. Holloszy. Baltimore: William and Wilkins, 1991, pp. 447–505.

21. Wasserman, K., J.E. Hanson, D.Y. Sue, and B.J. Whipp. *Principles of Exercise Testing and Interpretation.* Philadelphia: Lea & Febiger, 1987.

22. Whipp, B.J., J.A. Davis, F. Torres, and K. Wasserman. A test to determine parameters of aerobic function during exercise. *J. Appl. Physiol.* **50**: 217–221.

44

CLINICAL IMPROVEMENT OF SKIN MICROCIRCULATION IN PATIENTS WITH CHRONIC VENOUS INCOMPETENCE (CVI) BY PHYSICAL EXERCISE TRAINING

Thomas Klyscz,[1] Irmgard Jünger,[2] Ulrich Jeggle,[1] Martin Hahn,[1] and Michael Jünger[1]

[1] University Hospital of Tübingen
Department of Dermatology
[2] Department of Sports Medicine
Liebermeisterstr. 25, D-72076 Tübingen
Germany

1. INTRODUCTION

In patients with advanced stages of chronic venous incompetence, the disease usually causes a drastic reduction in or, as in the case of arthrogenic obstructive syndrome, an almost total loss of the function of the calf muscle pump. The reduced venous drainage results in impaired cutaneous microcirculation with trophic disturbances of the skin, even to the point of ulceration (1–3,5,6,8). In addition to optimized compression therapy and the surgical elimination of the points of venous incompetence, a key therapeutic role is played by active exercise therapy to improve ankle joint flexibility and strengthen the calf muscle joint pump.

2. PURPOSE OF THE PRESENT STUDY

The goal of our study was the comparative determination of cutaneous microcirculation in patients with CVI in Widmer stages I-III by measuring transcutaneous PO_2 (tcpO$_2$) and laser Doppler flux during a 26-week-long vascular sports programme.

The Physiology and Pathophysiology of Exercise Tolerance
edited by Steinacker and Ward, Plenum Press, New York, 1996

3. EXPERIMENTS

3.1. Patients

Thirty-two patients (12 women and 20 men) with an average age of 57 ± 9 years took part in a half-year study, involving a one-hour exercise therapy session twice a week. Of the patients 17 had post-thrombotic syndrome and 15 a primary varicosis. The control group was composed of 11 individuals in the same age range but with healthy veins.

3.2. Methods

The patients were examined at the beginning of the study and again after 26 weeks. Haemodynamic tests to determine venous drainage performance included the measurement of venous blood pressure and light reflection rheography. Laser Doppler flux measurements were performed using a HeNe laser with a wavelength of 632 nm (Model PF2b Periflux, Perimed, Stockholm, Sweden). The TCM2 oxygen monitor (Radiometer, Copenhagen, tcpO$_2$ at 43°C) was used for the accompanying tcpO$_2$ measurements. The examinations were performed in an air-conditioned room at 22°C on the symptomatic leg of the supine patient. The tcpO$_2$ probe was positioned at the edge of the ulcer. The perimalleolar LDF probe was placed in the florid or healed ulcer area or, in the case of stage I and II patients, in the congested or indurated area proximal to the malleolus medialis.

4. RESULTS

17 of the 32 patients showed significant improvement in venous drainage after 26 weeks compared to the initial examination and were classified as responders. In the 10 CVI patients with florid ulcus cruris, there was a significant reduction of the size of the ulcer during exercise therapy from 817 mm^2 to 121 mm^2 after 26 weeks.

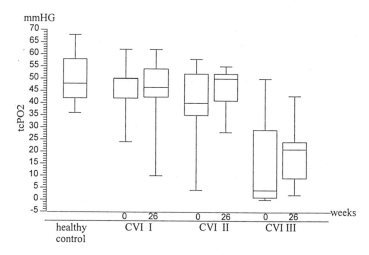

Figure 1. TcPO2-values at rest, measured at inner ankle area.

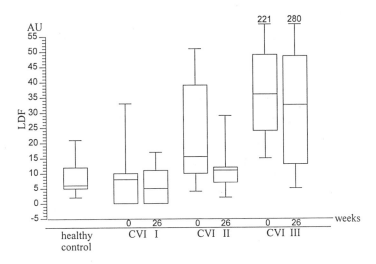

Figure 2. LDF-values at rest, measured at inner ankle area.

The control persons with healthy veins (n=11) had an average resting tcpO$_2$ value of 48 mm Hg. Comparable tcpO$_2$ values were found in stage I CVI patients. Patients in stage II already had a reduced oxygen partial pressure of 40 mm Hg, while stage III patients had pathologically low resting values of 4 mm Hg. These values rose in the course of vascular sports therapy to 50 mm Hg (+10 mm Hg) in CVI stage II patients and 21 mm Hg (+17 mm Hg) in stage III patients. The average resting tcpO$_2$ value of all 32 CVI patients rose during therapy from 36 mm Hg to 41 mm Hg.

The resting LDF value was 6 AU (arbitrary units) in the group of healthy controls. In stage I CVI patients it was 8 AU, in stage II 15.5 AU and in stage II 36 AU. After 26 weeks of vascular sports training the resting LDF signals dropped in CVI stage I from 8 AU to 5 AU, in stage II from 15.5 AU to 11 AU and in stage III from 36 AU to 32.5 AU.

5. DISCUSSION

Cutaneous nutritive perfusion at the lower leg does not differ appreciably between CVI stage I patients and control persons with healthy veins. As the severity of CVI increases, however, the measurable parameters reflecting skin microcirculation change markedly. Franzeck (4) found considerably reduced resting tcpO$_2$ values in patients with venous ulcers in comparison to normal persons. Partsch (10) and Jünger (7) were able to confirm these findings and also found elevated resting LDF values. As a result of these studies we know that the progression of the disease is accompanied by a worsening of cutaneous nutritive perfusion and a less favourable ratio of oxygenation to perfusion. In the present study we were able to confirm the phenomenon that Partsch (10) called "hyperaemic hypoxia", a flux increase with concomitant reduced oxygen partial pressure. We were able to achieve marked improvement of the disturbed perfusion/nutrition ratio in a 26-week-long vascular sports programme. In the most severely affected patients in stage III, the median resting perimalleolar tcpO$_2$ value increased fivefold during therapy from 4 mm Hg to 21 mm Hg. On the average, the 26-week-long vascular sports programme reduced inflammatory hyperperfusion in all 32 patients by half as indicated by the LDF signal, which dropped from 24 AU to 12 AU. The measurable changes were accompanied by an improvement of clinical symptoms and a highly significant healing of venous ulceration.

6. CONCLUSION

All of the CVI patients, especially those in stage III, profited from sports therapy. This was documented in a significant increase in $tcpO_2$ as well as a reduction of inflammatory hyperperfusion. The improved flux/oxygenation ratio, indicating improved cutaneous nutrition, was visible in a highly significant reduction in venous ulceration. There were no cases of clinical deterioration or complications during therapy. A special exercise programme in combination with optimized compression therapy is indicated for CVI in all stages and should become a fixed component of the total therapy concept for patients with chronic venous incompetence.

REFERENCES

1. Campbell, R.R., S.J. Hawkins, P. J. Maddison, and J. P. Reckless. Limited joint mobility in diabetes mellitus. *Ann. Rheum. Dis.* **44** (2): 93–97, 1985.
2. Davies, D.V. Ageing changes in joints. In: *Structural Aspects of Ageing*, edited by G.H. Bourne. New York: Hafner, 1961, p.23.
3. Ehrly, A. M., and H. Partsch. Microcirculatory and hemorheologic abnormalities in venous leg ulcers. In: *Phlebology,* edited by B. Davy and E. Stemmer. UK: John Libby Eurotext, 1989, pp. 142–145.
4. Franzeck, U.K., A. Bollinger, R. Huch, and A. Huch. Transcutaneous oxygen tension and capillary morphologic characteristics and density in patients with chronic venous incompetence. *Circulation* **70**: 806–811, 1984.
5. Gaylarge P.M., H.J. Dodd, and I. Sarkany. Venous leg ulcers and arthropathy. *Br. J. Rheumatol.* **29** (2): 142–144, 1990.
6. Jünger, M., T. Klyscz, M. Hahn, and A. Schiek. Mikroangiopathie der Haut bei chronischer Ischämie. *Phlebologie* **22**: 86–90, 1993.
7. Jünger, M., B. Rahmel, and G. Rassner. Laser-Doppler-Flux in venous ulcers. *Int J Microcirc Clin Exp* **7**: 88, 1988.
8. Klyscz, T., M. Hahn, and M. Jünger. Diagnostische Methoden zur Beurteilung der kutanen Mikrozirkulation bei der chronische Veneninsuffizienz. *Phlebologie* **23**: 141–145, 1994.
9. Klyscz, T., I. Jünger , F. Stracke , O. Schiebel, and M. Jünger. Neuartiges Pedalergometriegerät zur aktiven Bewegunstherapie bei Patienten mit peripheren arteriellen und venösen Zirkulationsstörungen. *Phlebologie* **24**: 1–8, 1995.
10. Partsch, H. Hyperaemic hypoxia in venous ulceration. *Br. J. Dermatol.* **110**: 248–251, 1984.

45

THE EFFECT OF BREATHING PATTERN DURING RESPIRATORY TRAINING ON CYCLING ENDURANCE

Christina M. Spengler, Sonja M. Laube, Marcus Roos, and Urs Boutellier

Exercise Physiology
Swiss Federal Institute of Technology Zurich and University of Zurich
Winterthurerstrasse 190, CH-8057 Zurich
Switzerland

1. INTRODUCTION

Different studies have shown that the respiratory system can limit exercise performance. The maximal duration of exercise at 90–95% of maximal work capacity (W_{max}) was reduced in subjects who had fatigued their respiratory muscles just prior to the exercise test by either breathing against an inspiratory load (8) or by hyperventilating at 66% of their maximal voluntary ventilation (MVV) for 150 min (9). Respiratory endurance training in turn increased exercise duration at 64% of peak oxygen consumption ($\dot{V}O_2$) in sedentary subjects (2) and at 77% $\dot{V}O_{2,peak}$ in trained subjects (1), whereas exercise duration at 90–95% W_{max} was the same as before respiratory training (5, 10). These conflicting results with respect to improvement of endurance by respiratory training may result either from different workloads at which subjects were exercising or from different respiratory training regimes. For example, subjects breathing against resistance can improve respiratory muscle strength whereas subjects performing isocapnic hyperpnea increase respiratory endurance (7). It also seems possible that the specific breathing pattern adopted during the respiratory training could influence the pattern adopted during subsequent exercise. This would be of importance because breathing slowly and deeply is more efficient than rapid shallow breathing during exercise.

Therefore, we investigated whether respiratory training performed at a low respiratory rate (f_R) and high tidal volume (V_T) leads to a more efficient breathing pattern during exercise than respiratory training with a high f_R and a low V_T. Also, we wanted to determine whether training with the slow, deep breathing pattern was more effective than rapid shallow breathing training in prolonging constant work exercise after respiratory training.

The Physiology and Pathophysiology of Exercise Tolerance
edited by Steinacker and Ward, Plenum Press, New York, 1996

Table 1. Spirometric variables, respiratory endurance (RET) times and variables measured during the incremental exercise test both before and after respiratory training (*p<0.05, ** p<00.01)

	group V		group F	
	before	after	before	after
VC (l)	5.8 ± 0.7	6.0 ± 0.7	5.7 ± 0.8	6.0 ± 1.0
peak flow (l·s⁻¹)	11.0 ± 1.5	11.5 ± 1.2	11.0 ± 2.2	10.8 ± 1.1
FEV1 (%)	90.8 ± 7.7	87.9 ± 7.0 *	88.0 ± 9.4	88.3 ± 10.1
MVV (l·min⁻¹)	201 ± 31	237 ± 33 **	189 ± 15	228 ± 23 **
RET time (min)	2.8 ± 0.9	26.8 ± 7.5 **	3.6 ± 1.1	30.0 ± 0.0 **
Wmax (W)	337 ± 25	337 ± 28	349 ± 57	352 ± 51
AT (W)	283 ± 25	284 ± 24	304 ± 49	297 ± 51
lactate at Wmax (mmol·l⁻¹)	10.8 ± 1.6	8.9 ± 2.2 *	9.2 ± 2.4	7.9 ± 1.8 *

2. METHODS

2.1. Subjects

Twenty healthy, physically active, male subjects participated in the study. They were randomly assigned to two groups (group V trained at high V_T and low f_R; group F trained at low V_T and high f_R). There were no differences between the groups in age (26 ± 5 and 26 ± 6 years, respectively), weight (68 ± 7 and 71 ± 5 kg), height (177 ± 6 and 181 ± 5 cm), vital capacity (VC), and MVV (Tab. 1). Apart from the additional respiratory training, subjects were instructed to not change the amount of their habitual weekly physical training.

2.2. Equipment

Respiratory training was performed as isocapnic hyperpnea (achieved by partial rebreathing). The respiratory endurance test (RET) was performed with the training apparatus while VE, V_T, and f_R were monitored with an ergo-spirometric device (OxyconBeta, Mijnhardt BV, Bunnik, Netherlands). Spirometric variables (VC, forced expiratory volume in 1 s [FEV_1], peak flow, and MVV) also were measured with the OxyconBeta. Cycling tests were performed on a bicycle ergometer (Ergo-metrics 800S, Ergoline, Bitz, Germany) while respiratory variables (V̇E, V_T, and f_R) were measured continuously with the OxyconBeta. Blood lactate concentrations were measured using 20 µl blood taken from an earlobe (ESAT6661 analyser, Eppendorf, Hamburg, Germany).

2.3. Study Design

First a W_{max}-test was performed. Subjects started cycling at 100 W, the load was then increased by 30 W every 2 min. At the end of each step, blood lactate concentration was measured. The anaerobic threshold (AT) was determined by averaging the ATs calculated by a modified heart rate deflection method (3), the ventilatory threshold method (11) and the lactate deflection method (6).

At least 3 days later, a cycling endurance test (CET1) was performed. After 3 min of cycling at 120 W, the work load was increased to the subject's AT. Blood was taken every

5 min to analyse lactate concentration. One day later, a respiratory endurance test (RET1) was performed.

Following these tests, the respiratory training phase started. Subjects were training at home for 30 min each day, 5 days a week, for 4 weeks. They came to the laboratory once a week to control their respiratory training with the OxyconBeta system and to determine the weekly increase in target, either V_T for group V or f_R for group F. Subjects of group V were training with a constant breathing frequency of 40 min^{-1}, increasing V_T from an average of 2.9 to 3.8 l. This group started with a mean $\dot{V}E$ of 113 ± 18 l·min^{-1} and increased $\dot{V}E$ up to 150 ± 17 l·min^{-1} during the 4 weeks of training. Group F trained with a constant V_T which was individually selected between 2.2 and 3.4 l. They started their respiratory training with f_R equal to 38 - 42 min^{-1} which was increased up to about 60 min^{-1}. This group started with a mean $\dot{V}E$ of 128 ± 16 l·min^{-1} and increased $\dot{V}E$ up to 171 ± 21 l·min^{-1} during the 4 weeks of respiratory training.

After a break of at least 5 days, a cycling endurance test (CET2) was performed. The following day the subjects performed a respiratory endurance test (RET2). 3–4 days later, spirometric variables were measured, and the incremental cycling test was repeated to determine W_{max}.

2.4. Statistics

Spirometric variables, W_{max}, AT, and lactate concentration from before and after the training period were compared using the paired Wilcoxon signed rank test. The Mann-Whitney U-Test was used to compare data between the two training groups.

For comparison of steady-state values of cycling endurance tests, means of the 10th to 14th min (7th to 11th min of cycling at AT) were calculated for each subject. For blood lactate concentrations, values at the 13th min were compared. To test respiratory data, cycling time, and lactate concentrations for significant changes after respiratory training, Friedman's analysis of variance was used. If significance was found, the paired Wilcoxon's signed rank test was applied. Results are given as means ± s.d.

3. RESULTS

Effect of respiratory training with slow deep (group V) or rapid shallow (group F) breathing on spirometric functions, respiratory endurance and maximal work exercise.

Effect of respiratory training with slow deep (group V) or rapid shallow (group F) breathing on breathing pattern and blood lactate concentration during exercise, and on constant work exercise duration.

4. DISCUSSION

4.1. Spirometry, Respiratory Endurance, and Incremental Exercise Tests (Table 1)

Both groups significantly increased MVV by about 20 % and respiratory endurance by over 700 % during the 4 weeks of respiratory training. This is in accordance with other studies (1, 5, 7, 10). The 2 % reduction in FEV1 in group V is most likely not of physiological relevance in the face of no change in peak flow and an increased MVV. The results of the incremental exercise test demonstrate that subjects did not change the amount of their weekly physical training and that respiratory endurance training does not affect short-term maximal exercise performance. Similarly, in earlier studies, W_{max} did not change as a consequence of the respiratory training (1, 5, 10).

4.2. Cycling Endurance Test (Table 2)

The breathing pattern during exercise did not change after respiratory training in either group but the cycling endurance time increased significantly by about the same amount (group V 36 ± 48 %; group F 28 ± 31 %). One could reason that the breathing pattern during respiratory training was not substantially different between the two groups, nonetheless V_T certainly was at a practicable maximum in group V (since respiratory training needs to be performed at a high intensity, we could not increase V_T above 60% VC in group V). This compares well with ventilation during intensive exercise where V_T tends to plateau at about 65 % VC (4).

Since the breathing pattern during cycling exercise was similar in both groups after respiratory training, our results suggest that prolongation of endurance exercise might be the result of an increased respiratory muscle endurance rather than a change in breathing pattern. Certainly, the decrease in blood lactate concentration could have affected endurance exercise duration as well. However, since the decrease in blood lactate concentration did not affect maximal performance in the incremental exercise test, we favour the possibility that an improved respiratory muscle endurance was more important than reduced blood lactate concentration in prolonging exercise endurance time after respiratory training. Only further studies can elucidate this issue.

Regarding the fact that some investigators have shown an increase in endurance time after respiratory training (1, 2) and others did not (5, 10), the present study suggests that this difference is probably due to different workloads at which exercise endurance was tested rather than the effect of different respiratory training regimes.

Table 2. Respiratory variables and blood lactate concentration during the 10th to 14th min of constant work exercise as well as cycling endurance (CET) time (* p < 0.05)

	CET_1	CET_2
group V		
VE ($l \cdot min^{-1}$)	107.6 ± 17.4	107.6 ± 22.7
VT ($ml \cdot breath^{-1}$)	2941 ± 398	2978 ± 439
f_R ($breaths \cdot min^{-1}$)	36.9 ± 6.4	36.2 ± 7.6
lactate ($mmol \cdot l^{-1}$)	8.7 ± 2.5	7.5 ± 2.4 *
CET time (min)	19.1 ± 4.9	26.0 ± 14.0 *
group F		
VE ($l \cdot min^{-1}$)	109.9 ± 17.6	114.8 ± 14.5
VT ($ml \cdot breath^{-1}$)	3240 ± 625	3267 ± 724
f_R ($breaths \cdot min^{-1}$)	35.3 ± 6.6	36.1 ± 7.0
lactate ($mmol \cdot l^{-1}$)	8.8 ± 2.8	7.5 ± 2.6 *
CET time (min)	16.8 ± 6.0	21.5 ± 9.3 *

4.3. Conclusions

(i) Different breathing patterns during respiratory training did not predict the breathing pattern that was adopted during the subsequent exercise. (ii) Exercise endurance time was prolonged after respiratory training irrespective of the breathing pattern during training.

ACKNOWLEDGMENTS

Financial support was kindly provided by the Swiss National Foundation (grant no. 32–30192.90).

5. REFERENCES

1. Boutellier, U., R. Büchel, A. Kundert, and C. Spengler. The respiratory system as an exercise limiting factor in normal trained subjects. *Eur. J. Appl. Physiol.* **65**: 347–353, 1992.
2. Boutellier, U., and P. Piwko. The respiratory system as an exercise limiting factor in normal sedentary subjects. *Eur. J. Appl. Physiol.* **64**: 145–152, 1992.
3. Conconi, F., M. Ferrari, P. G. Ziglio, P. Droghetti, and L. Codeca. Determination of the anaerobic threshold by a noninvasive field test in runners. *J. Appl. Physiol.* **52**: 869–873, 1982.
4. Dempsey, J. A. Is the lung built for exercise? *Med. Sci. Sports Exerc.* **18**: 143- 155, 1986.
5. Fairbarn, M. S., K. C. Coutts, R. L. Pardy, and D. C. McKenzie. Improved respiratory muscle endurance of highly trained cyclists and the effects on maximal exercise performance. *Int. J. Sports Med.* **12**: 66–70, 1991.
6. Heck, H., G. Hess, and A. Mader. Vergleichende Untersuchung zu verschiedenen Laktat-Schwellenkonzepten. *Dtsch. Z. Sportmed.* **36**: 19–25 and 40–52, 1985.
7. Leith, D. E., and M. Bradley. Ventilatory muscle strength and endurance training. *J. Appl. Physiol.* **41**: 508–516, 1976.
8. Mador, M. J., and F. A. Acevedo. Effect of respiratory muscle fatigue on subsequent exercise performance. *J. Appl. Physiol.* **70**: 2059–2065, 1991.
9. Martin, B., M. Heintzelman, and H.-I. Chen. Exercise performance after ventilatory work. *J. Appl. Physiol.* **52**: 1581–1585, 1982.
10. Morgan, D. W., W. M. Kohrt, B. J. Bates, and J. S. Skinner. Effects of respiratory muscle endurance training on ventilatory and endurance performance of moderately trained cyclists. *Int. J. Sports Med.* **8**: 88–93, 1987.
11. Wasserman, K., and M. B. McIlroy. Detecting the threshold of anaerobic metabolism. *Am. J. Cardiol.* **14**: 844–852, 1964.

46

THE "ENDURANCE PARAMETER RATIO" OF THE POWER-DURATION CURVE AND RACE VARIATION STRATEGY FOR DISTANCE RUNNING

Yoshiyuki Fukuba[1] and Brian J. Whipp[2]

[1] Department of Biometrics
Research Institute for Radiation Biology and Medicine
Hiroshima University, Hiroshima, 734 Japan
[2] Department of Physiology
St.George's Hospital Medical School
Cranmer Terrace, London, SW17 0RE, United Kingdom

1. BACKGROUND AND PURPOSE

The physiological determinants of the running speed, and its variation, that will maximize an endurance athlete's ability to succeed in winning a race are poorly understood. One aspect of the physiological basis of race-pace strategy that has not received appropriate consideration is that of the power-endurance time constraint.

It has been repeatedly demonstrated that the tolerance duration (t: sec) of high intensity cycling is well characterized as a hyperbolic function of power (P: watts) with an asymptote that has been termed "the fatigue threshold" (θ_F: watts), and a curvature constant W' (J).

$$(P - \theta_F) \cdot t = W' \qquad (1a)$$

or

$$P = W' \cdot (1 / t) + \theta_F \qquad (1b)$$

W' appears to represent a constant amount of work which can be performed above θ_F, regardless of its rate of performance. This is notionally equivalent to a constant energy reserve that may be utilized rapidly by exercising at high power outputs, or may be eked out for longer durations by exercising at lower work rates. Recently, this hyperbolic P-t relationship has also been confirmed in running[1] and swimming[2] performance, when speed

The Physiology and Pathophysiology of Exercise Tolerance
edited by Steinacker and Ward, Plenum Press, New York, 1996

(V) is used instead of P. We therefore consider here the consequences of an athlete performing the initial part of the event at a speed which is different from the constant rate that would allow the performance time to be determined by the power-duration relationship. We consider not only the power-duration constraints which limit the athlete's ability to make up the time lost by too slow an early pace, but also the consequences of a more rapid early pace.

2. HYPERBOLIC V-T RELATIONSHIP IN RUNNING

We here use the speed V (m/sec) as an index of power P,

$$(V - V_F) \cdot t = D' \tag{2a}$$

or

$$V = D' \cdot (1 / t) + V_F, \tag{2b}$$

where V_F is now the speed asymptote (m/sec), and D' is expressed in meters (m). As the total running distance (D: m) is:

$$D = V \cdot t, \tag{3}$$

Eq.2 may be rewritten as:

$$D = V_F \cdot t + D'. \tag{4}$$

The relationships of eqs.2a and 4 are graphically represented in Fig.1.

3. MAXIMAL AVERAGE SPEED

Each individual can be assumed to have a specific hyperbolic relationship between V and t. Each is therefore able to run some specific distance X at the maximal average speed ($V_{max(X)}$) which is determined by the crossing point of the particular V-t curve defined by eq.2a, 2b, or 4 (i.e. the "best V-t curve to D'"), and the V-t relation curve to some specific distance X defined by eq.3 (i.e. the "distance curve to X"), as seen in the example of Fig.2a:

$$V_{max(X)} = V_F / \{ 1 - (D' / X) \}, \tag{5}$$

where D' < X.

The running time t ($t_{max(X)}$) to distance X run with a constant speed $V_{max(X)}$ is,

$$t_{max(X)} = (X - D') / V_F. \tag{6}$$

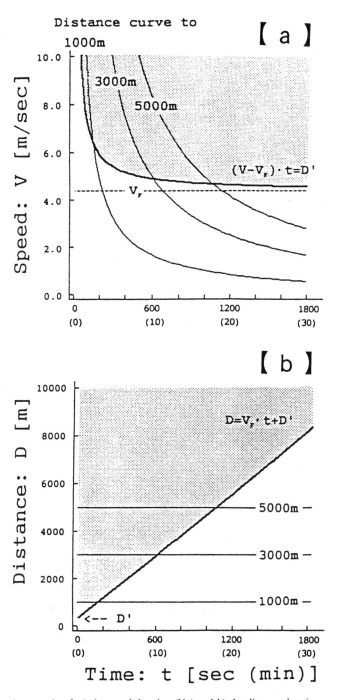

Figure 1. Schematic example of: a) the speed-duration (V-t) and b) the distance-duration curves. The distance curves to X m (X=1000, 3000, 5000 m) are also shown.

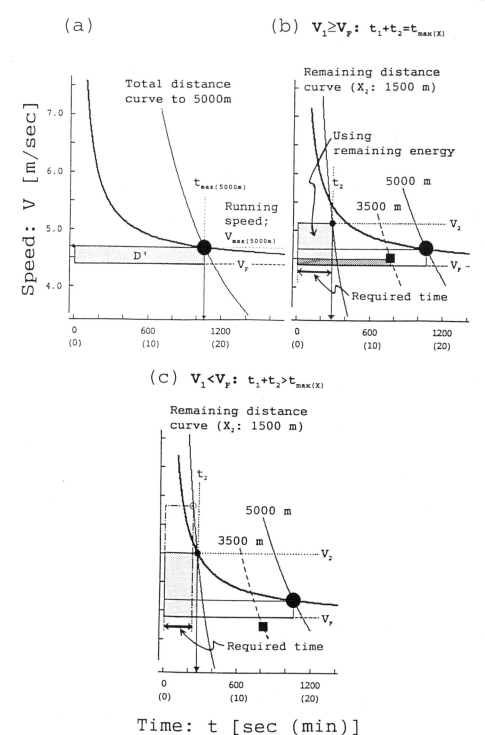

Figure 2. Example of race pace allocation (X1=3500 m, X2=1500 m) in the 5000 m run (X) for a runner whose V_F and D' are 4.4 (m/sec) and 300 (m) respectively, when the runner chooses the speed of a) constant $V_{max(5000m)}$ during the whole X, b) V_1 above V_F during X_1, and c) V_1 below V_F during X_1. See test for the details.

4. ALLOCATION OF RUNNING PACE

In general, runners change the running speed during a race according to their own race-pace strategy, although other runners' tactics can naturally alter this. We shall consider the simple situation in which the runner runs at 2 different speeds (V_1 and V_2) during the parts of 2 divided distances (X_1 and X_2) of the total distance X. We here explain the results by demonstrating the mathematical features of the relationship from both numerical and graphical standpoints. We consider the 5000 m, using a subject whose V_F and D' are 4.4 (m/sec) and 300 (m), respectively. The V_{max} and t_{max} for this race are 4.68 (m/sec) and 1068.2 (sec), respectively (refer the large closed circle in Fig.2).

4.1. $V_F \leq V_1 \leq V_{max(X1)}$

This represents one of a number of plausible allocations of running pace. The initial part X_1 is 3500 m with a speed (V_1) which is relatively slow but still above V_F, eg. 4.50 (m/sec). This speed corresponds to 3.6% below $V_{max(5000)}$. For the initial 3500 m, t_1 and $D'_{(3500)}$ are 777.8 (sec) and 77.8 (m), respectively. These values are determined by the crossing point of the runner's V-t curve to $D'_{(3500)}$ (=77.8; the hatched area in Fig.2b) and the distance curve for 3500 m (see the squared point in Fig.2b). For the second 1500 m, because the runner has access only to the remaining $D'_{(1500)}$ (= 300–77.8 = 222.2; the dotted area in Fig.2b) for the last spurt, he can run only at the speed that is determined by the crossing point of the best V-t curve to $D'_{(1500)}$ (=222.2) and the distance curve to 1500 m (see the small closed circle in Fig.4b). This requires the second 1500 m to be run in a t_2 of 290.4 (sec) and at a velocity V_2 of 5.165 (m/sec). This is +10.3 % greater than $V_{max(5000)}$. As a result, total running time (t_{tot}), ie. the race record, is 1068.2 (sec), and is exactly same to t_{max} which is derived from maximal average speed, $V_{max(5000)}$. That is,

$$t_{tot} = t_1 + t_2 = t_{max(X)}. \qquad (7)$$

4.2. $V_1 < V_F$

Consider, however, if the runner chooses the speed during the initial part which is below V_F, eg. 4.21 (m/sec). This corresponds to 10.0% below $V_{max(5000)}$ for the initial 3500 m of the overall 5000 m distance. Throughout X_1, the runner can reserve the whole D' for the last spurt. Note however that 831.4 (sec) is already taken as t_1 (see the squared point in Fig.4c). In the remaining second 1500 m, the runner can run at the maximal speed ($V_{max(1500)}$) which is determined by the crossing point between the runner's best V-t curve to D' (=300) and the distance curve to 1500 m (see the small closed circle in Fig.4c). Throughout the second part, t_2 is 272.7 (sec) and V_2 is 5.501 (m/sec) which is about +25.0% greater than $V_{max(5000)}$. As a result, total running time (t_{tot}) is 1104.1 (sec); this is some 36 sec longer than t_{max}, because the runner can never run the speed which is required to attain the goal of achieving the time equivalent to t_{max} (see the small open circle in Fig.2c). This is the case for all durations of sub-optimal early race pace which is below V_F. That is,

$$t_{tot} = t_1 + t_2 > t_{max(X)}. \qquad (8)$$

5. GENERAL CONSIDERATION

Here we simulate systematically a two component pace allocation strategy in the 5000 m. We used the real values for a male runner determined from previous study[1] whose V_F and D' are 5.0 (m/sec) and 150 (m) respectively. The runner's V_{max} and t_{tot} are 5.155 (m/sec) and 970 (sec), respectively. We examined the systematic effect of pace allocation, ie. utilizing 2 different speeds (V_1, V_2) in different combinations of X_1-X_2 to the total 5000 m. X_1 and X_2 (m) were chosen to be: 1000 - 4000, 2000 - 3000, 3000 - 2000, and 4000 - 1000. In each X_1-X_2 combination, we calculated the several parameters. V_1 was varied systematically, from -20 % of $V_{max(5000)}$, to the maximum within the limit by D'. As showed in the previous section using a specific numerical and graphical example, it was confirmed that, if V_1 is below V_F, the final race time (ie. T_{tot}) is always longer than $t_{max(5000)}$ (see eq.8).

In Fig.3, the dotted curve shows the expected V_2 to keep same t_{tot} to $t_{max(5000)}$, and solid line shows the calculated V_2.

At the shortest t_1 point (1 in Fig.3), V_2 is same as V_F. This point means that having run at $V_{max(X1)}$ for the initial X_1, the required speed for the remained distance X_2 is V_F. If t_1 took longer, the runner would have to run faster than V_F, at V_1 given by the dotted line - which is the expected V_2 line for reaching the goal with the same running time to $t_{max(5000)}$. However, in the range of t_1 above the deflection point of the solid line, the calculated V_2 (solid line) is dissociated from the expected V_2 (dotted line), because V_2 *cannot* exceed the maximal valve determined by the best V-t curve to the whole D' and distance curve to X_2 (eg. 3 in Fig.3). As a result, this deflection point (2 in Fig.3) means that the athlete must run at V_F for X_1, and $V_{max(X2)}$ for X_2. Therefore, the range between both points (ie. two extreme conditions to keep t_{tot} same as $t_{max(5000)}$) represents the maximal possible range to construct the race-pace as a strategy. In other words, how much speed and/or time can the runner possibly change intentionally between the 2 parts of the race?

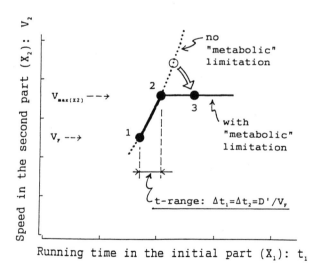

Figure 3. Schematic representation of the relationship of t_1 and V_2 to the "t-range", i.e., the "permitted" the pace allocation for a single change of pace. See text for the details.

The range between two extreme conditions at t_1 and t_2 are described by the following equations,

$$\Delta t_1 = \Delta t_2 = D' / V_F \tag{9}$$

This demonstrates an important property of D' and V_F to the possible time range (t-range) which is permitted for a strategic pace allocation with a single change of pace. That is, this range is not affected by the total distance, X, and any choice of the combination of X_1-X_2 or V_1-V_2, but is purely determined by what may be termed the runner's "endurance parameter ratio", ie. D'/V_F.

Next, to examine the effects of V_F and D' on the t-range, we plotted t_1-V_2 relationship in which V_F or D' was changed systematically. The effect of D' on the t_1-V_2 relationship under the same V_F condition, ie. 5.0 (m/sec) is shown in Fig.4a. The larger D' dramatically extended the t-range and V_2-range for the 5000 m run. However, compared to the effect of D', the effect of V_F was relatively small (Fig.4b). This indicates that D' is relatively important in the decision strategy for the runner's possible range of pace change. In other words, the larger D' the larger is the extent of pace change in the race. However, it should be also noted that V_F is, of course, an essential determinant of $t_{max(X)}$, ie. the actual level of the race time.

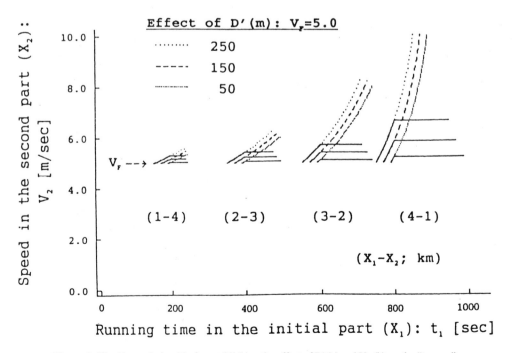

Figure 4. The V_2-t_1 relationship for establishing the effect of D' (a) and V_F (b) on the "t-range".

6. CONCLUDING REMARKS

Our results demonstrate that the speed at the fatigue threshold (V_F) and the curvature constant (D') parameters of the runner's V-t hyperbolic curve each play an important role in the pace allocation strategy of the athlete. That is, 1) when the running speed during any part of the whole running distance is below V_F, the athlete can never attain the goal of achieving the time equivalent to that of running the entire race at constant optimal speed (which is determined by his/her own best V-t curve) even if the runner attempts to make up for the time lost (within the limit of V-t constraint) with a final spurt, and 2) D' is especially important in determining the flexibility of the race pace which the athlete is able to choose intentionally. The ratio of these parameters (ie. D'/V_F) - the "endurance parameter" ratio - may therefore be considered an important determinant of race pace strategy. However, it is important to emphasize that this hyperbolic relationship in unlikely to provide a precise representation of the actual physiological behavior at the very extremes of performance because of distorting factors such as: 1) limitations of muscular (mechanical) force generation for the very highest power requirements, and 2) constraints resulting from substrate provision and thermo-regulatory or body fluid requirements for markedly prolonged exercise[3].

We have also treated the P-t curve as if it is hyperbolic for running: this needs further verification. If it proves not to be, however, an additional term(s) will need to be added to the characterization, but this will not alter the conceptual issue being addressed in this paper. Similarly, we have treated changes of pace as if the initial pace will not change either parameter of the P-t relationship. This also requires experimental verification.

7. ACKNOWLEDGMENT

Y. Fukuba was supported by an overseas research fellowship from Uehara Memorial Foundation in Japan.

8. REFERENCES

1. Hughson, R.L., C.J. Orok, and L.E. Staudt. A high velocity treadmill running test to assess endurance running potential. *Int.J.Sports Med.* **5**: 23–25, 1984.
2. Wakayoshi, K., K. Ikuta, T. Yoshida, M. Udo, T. Moritani, Y. Mutoh, and M. Miyashita. The determination and validity of critical velocity as swimming performance index in the competitive swimmer. *Eur.J.Appl.Physiol.* **64**: 153–157, 1992.
3. Whipp, B.J., and S.A. Ward. Respiratory response of athletes to exercise. In: *Oxford Textbook of Sports Medicine* (M.Harries, L.J.Micheli, W.D.Stanish, and C.Williams, eds.), Oxford University Press, 1994, pp.13–25.

INDEX